W9-AEZ-185

This book comes with access to more content online.

Quiz yourself, track your progress,
and score high on test day!

Register your book or ebook at
www.dummies.com/go/getaccess.

Select your product, and then follow the prompts
to validate your purchase.

You'll receive an email with your PIN and instructions.

Next-Generation
ACCUPLACER®

with Online Practice Tests

by Mark Zegarelli

10/15/19
LN
$29.99

Next-Generation ACCUPLACER® For Dummies® with Online Practice Tests

Published by: **John Wiley & Sons, Inc.**, 111 River Street, Hoboken, NJ 07030-5774, www.wiley.com

Copyright © 2019 by John Wiley & Sons, Inc., Hoboken, New Jersey

Published simultaneously in Canada

No part of this publication may be reproduced, stored in a retrieval system or transmitted in any form or by any means, electronic, mechanical, photocopying, recording, scanning or otherwise, except as permitted under Sections 107 or 108 of the 1976 United States Copyright Act, without the prior written permission of the Publisher. Requests to the Publisher for permission should be addressed to the Permissions Department, John Wiley & Sons, Inc., 111 River Street, Hoboken, NJ 07030, (201) 748-6011, fax (201) 748-6008, or online at http://www.wiley.com/go/permissions.

Trademarks: Wiley, For Dummies, the Dummies Man logo, Dummies.com, Making Everything Easier, and related trade dress are trademarks or registered trademarks of John Wiley & Sons, Inc., and may not be used without written permission. ACCUPLACER is a trademark registered by the College Board, which is not affiliated with, and does not endorse, this product. All other trademarks are the property of their respective owners. John Wiley & Sons, Inc., is not associated with any product or vendor mentioned in this book.

LIMIT OF LIABILITY/DISCLAIMER OF WARRANTY: THE PUBLISHER AND THE AUTHOR MAKE NO REPRESENTATIONS OR WARRANTIES WITH RESPECT TO THE ACCURACY OR COMPLETENESS OF THE CONTENTS OF THIS WORK AND SPECIFICALLY DISCLAIM ALL WARRANTIES, INCLUDING WITHOUT LIMITATION WARRANTIES OF FITNESS FOR A PARTICULAR PURPOSE. NO WARRANTY MAY BE CREATED OR EXTENDED BY SALES OR PROMOTIONAL MATERIALS. THE ADVICE AND STRATEGIES CONTAINED HEREIN MAY NOT BE SUITABLE FOR EVERY SITUATION. THIS WORK IS SOLD WITH THE UNDERSTANDING THAT THE PUBLISHER IS NOT ENGAGED IN RENDERING LEGAL, ACCOUNTING, OR OTHER PROFESSIONAL SERVICES. IF PROFESSIONAL ASSISTANCE IS REQUIRED, THE SERVICES OF A COMPETENT PROFESSIONAL PERSON SHOULD BE SOUGHT. NEITHER THE PUBLISHER NOR THE AUTHOR SHALL BE LIABLE FOR DAMAGES ARISING HEREFROM. THE FACT THAT AN ORGANIZATION OR WEBSITE IS REFERRED TO IN THIS WORK AS A CITATION AND/OR A POTENTIAL SOURCE OF FURTHER INFORMATION DOES NOT MEAN THAT THE AUTHOR OR THE PUBLISHER ENDORSES THE INFORMATION THE ORGANIZATION OR WEBSITE MAY PROVIDE OR RECOMMENDATIONS IT MAY MAKE. FURTHER, READERS SHOULD BE AWARE THAT INTERNET WEBSITES LISTED IN THIS WORK MAY HAVE CHANGED OR DISAPPEARED BETWEEN WHEN THIS WORK WAS WRITTEN AND WHEN IT IS READ.

For general information on our other products and services, please contact our Customer Care Department within the U.S. at 877-762-2974, outside the U.S. at 317-572-3993, or fax 317-572-4002. For technical support, please visit https://hub.wiley.com/community/support/dummies.

Wiley publishes in a variety of print and electronic formats and by print-on-demand. Some material included with standard print versions of this book may not be included in e-books or in print-on-demand. If this book refers to media such as a CD or DVD that is not included in the version you purchased, you may download this material at http://booksupport.wiley.com. For more information about Wiley products, visit www.wiley.com.

Library of Congress Control Number: 2019942822

ISBN 978-1-119-51454-1 (pbk); ISBN 978-1-119-51457-2 (ebk); ISBN 978-1-119-51448-0 (ebk)

Manufactured in the United States of America

V10010950_060719

Contents at a Glance

Table of Contents

Introduction

The American system of community colleges — also called junior colleges — is one of the greatest achievements in American education.

Now, if you've grown up watching *The Big Bang Theory*, you may well side with Sheldon, who never lets up on Penny for being a "community college dropout." (Just remember, he's pretty hard on MIT, Princeton, and Harvard, too!)

But I stand by what I said: Community college is a great achievement for our nation, and an amazing opportunity for you personally.

As the price of education continues to rise, and shows no sign of dropping anytime soon, community college is still the best deal since the nickel hotdog. The average price tag on a semester of community college is something like $1,500, but many students pay far less or even nothing at all due to government assistance through the FAFSA program. And some even *receive* money to help them buy books, a computer, and other important college-related stuff.

You can graduate from community college with an associate's degree (A.A. or A.S). And virtually every community college can help you accumulate 60 credits toward your 120-credit bachelor's degree. Even better, if you maintain a B average at most community colleges, you may find that the doors of your state university just about swing open for you when you graduate.

So, assuming you're with me (and against Sheldon!) that community college is awesome, let's get down to business and talk about the ACCUPLACER.

The ACCUPLACER is more of a speed bump than a hurdle on your road to a college degree. Here are the most important things to know about the ACCUPLACER:

>> **The ACCUPLACER is *NOT* an *entrance* exam.** Your performance on the ACCUPLACER won't keep you out of community college.

>> **The ACCUPLACER *IS* a *placement* exam.** It's an opportunity to jump over non-credit reading, writing, and math courses by showing your community college that you already know this stuff.

In fact, technically speaking, you don't have to take the ACCUPLACER at all. But if you don't, then you'll be giving up an *opportunity* to bypass a bunch of classes you might not have to take. You don't really want to miss out on an opportunity, do you?

About This Book

I wrote this book to help you to do your absolute best on the ACCUPLACER.

If you survived (some prefer the term "graduated") high school, there's probably not much in this book that you don't already know — or at least you *did* know at one time or another. This book offers a quick way to remember what you knew "back then" and to fill in the stuff that you never quite understood or don't remember now.

I've done my best to keep this book limited to the absolute essentials for the ACCUPLACER. If you see it in the book, it'll probably show up on the test. So, the more time and effort you spend working through the information you find here, the better you will do when you take your ACCUPLACER.

Think of it this way: Each section of the ACCUPLACER that you pass allows you to jump over at least one semester (and possibly two or three) of a no-credit community college course. And a college course has about 45 hours of class time, plus whatever time you spend out of class reading, studying, doing homework, and preparing for tests. However you slice it, that looks to me like 75 to 100 hours, or maybe even more, taking a course that doesn't even earn you any credits toward graduation.

And that's just one course.

Or, with the help of this book, you could spend maybe 10 or 20 hours studying on your own, get your skills up to speed, and then pass a section of the ACCUPLACER. (That's right — you don't have to take the entire ACCUPLACER all at once. At most community colleges, you can take one section at a time if that suits you.)

For many of the students I've helped to pass the ACCUPLACER, that seemed like a pretty good deal. So I invite you to come along!

The ACCUPLACER includes five different sections:

>> Reading Test

>> Writing Test

>> Three Math Tests:

- Arithmetic

- Quantitative Reasoning, Algebra, and Statistics (QAS)

- Advanced Algebra and Functions (AAF)

Because the last two math tests have long and confusing names, at times throughout this book I refer to them using the abbreviations "Math QAS" and "Math AAF." Make sense?

Foolish Assumptions

If you've bought this book — or even if you're looking through it while sitting at the café of your favorite bookstore — I'm going to jump to one of three conclusions about you:

>> You need to take the ACCUPLACER to move forward in community college, and want to improve your score.

>> Someone you care about needs to take the ACCUPLACER, and you'd like to help them.

>> You LOVE standardized tests and can't wait to try this one!

That's who I wrote the book for, but if none of this applies to you, you're still more than welcome to buy it!

Icons Used in This Book

Throughout the book, I use a handful of icons to point out various types of information. Here's what they are and what they mean:

REMEMBER

The Remember icon is like a little Post-it note, labeling everything in the book that's key to remember. After reading a chapter, I recommend that you go through it again and test yourself to see how well you remember this information.

EXAMPLE

The Example icon marks where I present an example, followed by a step-by-step solution. Throughout the book, you'll often find an Example icon paired with a Remember icon, indicating that the example goes with the important idea being introduced.

TIP

The Tip icon indicates a quick way to remember important material or perform a task. Use these tips to help you save time and frustration.

WARNING

The Warning icon helps you steer around mistakes that students commonly make. Don't let this happen to you!

TECHNICAL STUFF

The Technical Stuff icon gives information that is often peripheral to the task at hand. Strictly speaking, you don't need this stuff, but in some cases you may find it interesting or useful. Read or skip over as you like.

Beyond the Book

In addition to what you're reading right now, this book comes with a free access-anywhere Cheat Sheet that includes tips to help you prepare for the ACCUPLACER. To get this Cheat Sheet, simply go to www.dummies.com and type **ACCUPLACER For Dummies Cheat Sheet** in the Search box.

You also get access to a 3 full-length online practice tests. To gain access to the online practice, all you have to do is register. Just follow these simple steps:

1. **Register your book or ebook at Dummies.com to get your PIN. Go to** www.dummies.com/go/getaccess.

2. **Select your product from the dropdown list on that page.**

3. **Follow the prompts to validate your product, and then check your email for a confirmation message that includes your PIN and instructions for logging in.**

   If you do not receive this email within two hours, please check your spam folder before contacting us through our Technical Support website at http://support.wiley.com or by phone at 877-762-2974.

Now you're ready to go! You can come back to the practice material as often as you want — simply log on with the username and password you created during your initial login. No need to enter the access code a second time.

Your registration is good for one year from the day you activate your PIN.

Where to Go from Here

You can start anywhere in this book and find something that can help you to do better on the ACCUPLACER.

If you're not quite sure what the ACCUPLACER contains and why you've been asked to take it, Chapters 1, 2, and 3 give you a complete overview of the entire test, and then a look at all five sections.

On the other hand, if you feel like you know all that stuff already and want to get a sense of how well you're likely to do on the test, go to the end of the book, where you'll find two practice tests. If you like, use the first practice as a pre-test, and save the second practice test for after you've worked through the material in the rest of the book.

On the third hand (you've got three hands, right?), if you want to get started improving your score right now, flip to any chapter in Parts 2 through 6. Each of these parts covers one of the five ACCUPLACER sections: reading, writing, and three levels of math. And the last chapter of each part includes practice problems based on the information in that part. Lots o' options!

This last piece of advice should be easy to follow: Don't read what you don't need.

For example, if you're pretty confident you'll pass the ACCUPLACER Writing Test (in my opinion, the easiest of the five sections), then don't bother reading the chapters that cover this section. Additionally, some schools don't require you to take the most difficult of the three math sections, the Advanced Algebra and Functions (AAF) Test. So, if you don't have to take this test, you can skip over Part 6 of this book.

1

Getting Started with the ACCUPLACER

IN THIS PART . . .

Understand the purpose of the ACCUPLACER test and why it's important.

Look at what the ACCUPLACER covers and know how the test is set up.

Prepare for the test and get the best score possible.

Chapter **1**

Getting to Know the ACCUPLACER

I f you've been told that you have to take the ACCUPLACER, you probably have a bunch of basic questions about it, such as the following:

Just what is the ACCUPLACER?

Do I *have* to take it (really)?

When, where, and how do I take it?

What kind of stuff does the ACCUPLACER test?

What happens after I take the test?

In this chapter, I give you answers to the most common questions that students tend to have about the ACCUPLACER.

Knowing What the ACCUPLACER Is and Whether You Really Need to Take It

If you're reading this book, you've probably been advised to take the ACCUPLACER. In this section, I try to anticipate and then answer some of the most basic questions that you may have about the test — including what the heck it is and whether you really need to take it.

What is the ACCUPLACER?

The ACCUPLACER is a placement test for community college (also known as junior college). It's used to assess your current skill level and readiness for the types of schoolwork you'll be required to do in community college — specifically, reading, writing, and math.

Who makes the ACCUPLACER?

The ACCUPLACER is made by the College Board. These are the same folks who created the SAT and the Advanced Placement Program (the AP Tests).

So is the ACCUPLACER like the SAT or ACT?

Yes and no.

The ACCUPLACER tests a lot of the same skills that the SAT and ACT test. However, the SAT and ACT are *entrance tests*. This means that you take these tests *before* you've been accepted to a college. Getting a low score on the SAT or ACT can cause a college to reject your application.

In contrast, the ACCUPLACER is a *placement test*. This means that you take it *after* you've been accepted to community college. A low score on one or more parts of the ACCUPLACER means, at the very worst, that you may need to take one or more non-credit remedial courses.

Do you have to take the ACCUPLACER?

The short answer is no. But if you don't take it, your community college may place you in a set of remedial courses that are too easy for you. Passing the ACCUPLACER — or any of the five sections of the ACCUPLACER — allows you to place out of these non-credit courses, so you can begin earning college credits immediately.

So, you can think of the ACCUPLACER not as a required test that you *must* pass, but rather as an *opportunity* to jump over a bunch of lower-level courses that you may not need to take.

What is the Next-Generation ACCUPLACER?

The current version of the ACCUPLACER is called the *Next-Generation ACCUPLACER*. This name distinguishes it from the older and now defunct version.

REMEMBER

The Next-Generation ACCUPLACER is the *only* version of the test being administered in 2019 and for the foreseeable future. So, heads up: If you've bought any other ACCUPLACER books along with this one, check to make sure that they're *explicitly* for the Next-Generation ACCUPLACER. If not, don't use them! (Or, perhaps, use them to prop open a door or, in a pinch, give them as birthday presents to people you plan never to see again.)

Throughout this book, whenever you read *ACCUPLACER*, you can assume that I'm talking about the *Next-Generation ACCUPLACER*.

Discovering When, Where, and How to Take the ACCUPLACER

If you've read (or even skimmed) this far, I'm going to assume that you've decided to take the ACCUPLACER. Now, you may want to know some specifics about where and how to register for and take the test. That's what this section is about.

How do you register for and take the ACCUPLACER?

The ACCUPLACER is administered by your community college. If you've recently enrolled in a community college, an administrator probably mentioned the ACCUPLACER among a short list of important things to take care of as a new student.

The ACCUPLACER is most often done either by appointment or by just walking into the right office and asking the nice person behind the desk to take it. But the procedure can vary a bit among institutions. If you have any doubts as to how to get started, call your community college to get specifics.

How long does the ACCUPLACER take?

The ACCUPLACER is an *untimed* test, which means that you can take as long as you like. This is in contrast to most other tests, including those that you're used to taking from high school. In Chapter 3, I give you some strategies to make the most of this aspect of the ACCUPLACER.

Do you need to take all sections of the ACCUPLACER on the same day?

No! You can opt to take all five sections at once, or break them up in any way you like. In Chapter 3, I discuss a few strategies to maximize your advantage when taking the ACCUPLACER.

How is the ACCUPLACER administered?

Most often, the ACCUPLACER is administered via the Internet, at a computer located in the administrative office or testing center of a community college. Some schools, however, administer the ACCUPLACER on paper, in its COMPANION format. If you're not sure which format your school uses, the office that administers the test can tell you.

In Chapter 3, I discuss how the online form of the ACCUPLACER is a *computer adaptive test* (CAT), and what this means to you as a test-taker.

What do you need to bring to the ACCUPLACER?

When taking the ACCUPLACER, be sure to bring your student ID. If your school administers the ACCUPLACER using its computer format (this is most common), scrap paper will be provided. You won't need or be allowed to use a calculator for the math portions of the test — though for some questions, a calculator will appear on your computer screen.

If your school offers the ACCUPLACER on paper, in its COMPANION format, then you'll be allowed to use a simple four-function calculator on some portions of the test, which they should provide for you. (Your fancier scientific or graphing calculator, however, is out.)

What accommodations does the ACCUPLACER make for disabled students?

If you have a documented disability that requires special accommodations for taking the ACCUPLACER, contact your community college to let them know, and they'll get you set up.

TIP

If you have an Individualized Education Program (IEP) or a medical condition that allows you to receive extra time on other standardized tests, no worries: The ACCUPLACER is an untimed test, so you (and everybody else!) can take all the time you need to answer the questions. However, just to be practical, I recommend that you arrive early enough in the day that everybody in the administrative office isn't about to leave for the day! (See Chapter 3 for more on this topic.)

Understanding What Happens After You Take the ACCUPLACER

If you've read through the earlier sections of this chapter, you may be wondering what happens after you take the ACCUPLACER. In this section, I discuss the ins and outs and ups and downs of what life after your ACCUPLACER may look like.

How is the ACCUPLACER scored?

Each of the five sections of the ACCUPLACER is scored separately on a scale from 200 to 300. Generally speaking, any score of 280 or above is passing. You don't get a letter grade such as an A or a B, but what you *do* get is a free pass out of taking the no-credit college course associated with that test. Yay!

However, different community colleges draw the pass-fail line in different places. So a score of 278 on one section may be a passing score at your school. Or, a 278 may be a passing score on the Arithmetic Test but not on the Writing Test.

Additionally, some schools allow students who are only a few points below the passing score to take more accelerated no-credit courses or even partial-credit courses. (Who says "close" only counts in horseshoes?)

The person who administers your test is probably the best person to ask for information about what the passing ACCUPLACER scores are at your community college.

If you do well on the ACCUPLACER, what happens?

Every section of the ACCUPLACER that you do well on enables you to skip the no-credit remedial course work in that subject. This means that you *don't* have to spend a semester (or more!) taking a course that adds no credit toward your college degree. Even better, because passing the ACCUPLACER demonstrates your competence in a subject area, you can move on to upper-level courses with the confidence that you're ready to do the work!

If you DON'T do well on the ACCUPLACER, what happens?

REMEMBER

This is key: If you don't do well on the ACCUPLACER, you still won't be kicked out of community college. It's just not that kind of test.

This feature makes the ACCUPLACER different from college entrance tests like the SAT and ACT. Most colleges and universities set a minimum SAT and ACT score. And while there may be some wiggle room in special cases, scoring on the low side definitely lowers your chances of being accepted. And, naturally, more competitive schools tend to require higher scores.

But, as I mention earlier in this chapter, the ACCUPLACER isn't an entrance test. In fact, if you're scheduled to take the ACCUPLACER, this means that you've already been accepted into community college (congratulations!).

Failure to pass any portion of the ACCUPLACER simply means that your community college is going to require you to take at least one remedial no-credit course before they allow you to enroll in a credit course in that subject area. Essentially, they want to set you up for success, to make sure that you have the skills necessary to pass your courses when the time comes.

What if you pass some sections of the ACCUPLACER but not others?

When you pass any section of the ACCUPLACER, you're done with that section forever! You never have to take it again, and you're exempted from taking remedial courses in that area of study.

How many times can you take the ACCUPLACER?

Usually, your community college will give you two chances to pass each section of the ACCUPLACER. In some cases — especially for a good student with good grades — they may stretch this to three times.

The good news is that when you pass any section of the ACCUPLACER, you're officially done with that section, and you don't have to take it again.

Chapter **2**

Taking a Closer Look at What's on the ACCUPLACER

E very question on the ACCUPLACER is a multiple-choice question with four possible answers, A through D.

The ACCUPLACER has a total of five sections:

» Reading Test

» Writing Test

» Three Math Tests:

- Arithmetic

- Quantitative Reasoning, Algebra, and Statistics (QAS)

- Advanced Algebra and Functions (AAF)

In this chapter, I get into a few specifics about what you'll find on each section of the test.

Reading Test

The ACCUPLACER Reading Test is similar to many other reading tests you've probably taken throughout your life. Most of the test requires you to read a passage and then answer one or more comprehension questions. Additionally, a few questions give you a sentence with a word or short phrase missing, and you're asked to fill in the blank.

One passage is a work of fiction, and the rest are nonfiction. Each question presents you with four answers that test you on the following reading skills.

>> **Information and ideas:**

- Reading closely for factual details in the passage

- Determining the central idea (or main idea) — the most important point that the writer is making

- Summarizing — restating the information in the passage in a different way that conveys its meaning effectively

- Understanding relationships among elements within the passage, and especially making inferences about what is not specifically stated but can be reasonably understood

>> **Rhetoric:**

- Word choice — why the writer chose to use a specific word or phrase

- Text structure — how the writer organizes the information that they are presenting

- Point of view — what the writer believes or feels about what they are writing about

- Purpose — why the writer chose to write this passage

- Arguments — how the writer frames their case to persuade the reader

>> **Synthesis:** Given a pair of texts that discuss a common theme from two different perspectives, comparing and contrasting information or rhetoric in the passages

>> **Vocabulary:** Demonstrating comprehension of a specific word or phrase

Writing Test

The ACCUPLACER Writing Test presents you with essays that are in need of editing. Your job is to answer questions, each of which presents you with a choice of four possible ways to express an idea. The questions test your understanding of the following information.

>> **Standard English conventions:**

- Sentence structure — avoiding sentence fragments and run-on sentences, understanding coordination and subordination of clauses, making sentences readable with parallel structure, and avoiding inappropriate verb shifts

- Usage — subject-verb agreement, pronoun-antecedent agreement, clarifying frequently confused words, and following the conventions of English expression

- Punctuation — how and when to use the punctuation that students most often misuse, such as commas, semicolons, colons, and dashes

>> **Expression of ideas:**

- Development — expressing an idea clearly, adding supporting information, and maintaining focus

- Organization — introducing ideas in a logical sequence, and helping the reader to understand your point through the use of introductions, conclusions, and transitions

- Effective language use — choosing the precise word, keeping your language concise (avoiding extra words and redundancy), maintaining a consistent style and tone, and using standard English syntax

Three Math Tests

The three ACCUPLACER math tests break down as follows:

>> Arithmetic

>> Quantitative Reasoning, Algebra, and Statistics (QAS)

>> Advanced Algebra and Functions (AAF)

In this section, I discuss the specific types of math you can expect to find in each math section.

Arithmetic

The Arithmetic Test is the most basic of the three ACCUPLACER math sections. It tests your knowledge and ability in five areas.

>> **Whole number operations:**
- The four basic operations (addition, subtraction, multiplication, and division) applied to whole numbers
- Estimation and rounding numbers
- Applying the order of operations (PEMDAS)
- Word problems that provide real-world context

>> **Fraction operations:**
- The four basic operations (addition, subtraction, multiplication, and division) applied to both fractions and mixed numbers
- Word problems that provide real-world context

>> **Decimal operations:**
- The four basic operations (addition, subtraction, multiplication, and division) applied to decimals
- Estimation and rounding numbers
- Word problems that provide real-world context

>> **Percent:**

- Calculating the percent of a number

- Setting up and solving equations that involve percents

- Percent increase and decrease

- Word problems that provide real-world context

>> **Number comparisons and equivalents:**

- Comparing values on the number line

- Using inequality symbols ($<$, $>$, $\leq$, and $\geq$) to compare values

- Comparing values expressed as fractions, decimals, or percents

Quantitative Reasoning, Algebra, and Statistics

The Quantitative Reasoning, Algebra, and Statistics Test (QAS for short) is the second of the three ACCUPLACER math tests in terms of difficulty. Here's what the QAS covers.

>> **Rational numbers:**

- Calculating with rational numbers (fractions, decimals, and percents)

- Word problems that provide real-world context

>> **Ratio and proportional relationships:**

- Calculating rates, ratios, and proportions

- Word problems that provide real-world context (finding rates and using ratios to set up and solve proportional equations)

>> **Exponents:**

- Calculating exponents and radicals (roots)

- Working with negative and fractional exponents

- Working with scientific notation

>> **Algebraic expressions:**

- Evaluating algebraic expressions given values of the variable or variables

- Simplifying algebraic expressions by combining like terms and distributing

- Word problems that provide real-world context (rewriting a story as an algebraic expression in terms of a variable)

>> **Linear equations:**

- Creating and solving linear equations in one variable

- Creating and solving systems of linear equations in two variables

- Understanding and simplifying linear inequalities

>> **Linear applications and graphs ($y = mx + b$):**

- Understanding and identifying slope and y-intercept

- Graphing basic linear equations

- Working with parallel and perpendicular lines on the graph

- Working with systems of equations on the graph

- Understanding linear inequalities

- Word problems that provide real-world context (rewriting a story as a linear equation)

>> **Probability and sets:**

- Defining a sample space and events within it

- Understanding and calculating simple, compound, and conditional probability

- Understanding basic set notation, including union and intersection

>> **Descriptive statistics:**

- Describing a sample set using visual tools such as boxplots

- Calculating mean and median as measures of center of a sample set

- Finding the shape (skew) and spread (range) of a sample set

>> **Geometry concepts for Pre-Algebra:**

- Calculating the area and perimeter of squares and rectangles

- Calculating the area and circumference of a circle

- Finding the volume of a solid using a formula

>> **Geometry concepts for Algebra 1:**

- Expressing area, perimeter, and volume as algebraic expressions

- Using the Pythagorean theorem ($a^2 + b^2 = c^2$)

- Using the distance formula to calculate length on the xy-graph

- Working with basic geometric transformations (translations, reflections, and rotations)

Advanced Algebra and Functions

The Advanced Algebra and Functions Test (AAF for short) is the most difficult of the three ACC-UPLACER math tests. Not every community college requires you to take this test, so be sure to check with your school before drilling down on the following topics.

>> **Factoring:**

- GCF (greatest common denominator) factoring

- Quadratic factoring (reverse FOILing)

- Factoring using the difference of squares and the sum and difference of cubes

- Factoring cubic equations

>> **Quadratics:**

- Identifying and creating quadratic equations

- Solving quadratics using factoring

- Using the quadratic formula ($x = \dfrac{-b \pm \sqrt{b^2 - 4ac}}{2a}$) to solve quadratics

- Working with quadratic inequalities

- Solving systems of equations that involve a quadratic equation

>> **Functions:**

- Understanding and working with function notation ($f(x)$)
- Evaluating linear functions ($f(x) = mx + b$) and quadratic functions ($f(x) = ax^2 + bx + c$)
- Graphing the most common parent functions
- Understanding basic function transformations
- Word problems that provide real-world context

>> **Radical and rational equations:**

- Understanding radical and rational equations
- Graphing radical and rational equations
- Knowing how to find the domain and the range

>> **Polynomial equations:**

- Understanding polynomial equations (especially linear, quadratic, cubic, and quartic equations)
- Graphing polynomial equations

>> **Exponential and logarithmic equations:**

- Understanding exponential and logarithmic equations
- Graphing exponential and logarithmic equations

>> **Geometry concepts for Algebra 2:**

- Finding the volume of non-prism solids (especially spheres, pyramids, and cones)
- Applying basic theorems for intersecting lines (especially vertical angles, supplementary angles, corresponding angles, and total angles in a polygon)
- Working with congruent and similar triangles
- Working with circles on the *xy*-graph

>> **Trigonometry:**

- Understanding the trigonometric ratios (especially sine, cosine, and tangent)
- Working with the special right triangles ($45° - 45° - 90°$ and $30° - 60° - 90°$)
- Understanding radian measure
- Measuring arc length
- Solving basic trig equations
- Understanding basic trig identities
- Applying the law of sines and the law of cosines

Chapter 3

Preparation and Test Strategy

As with any test, doing well on the ACCUPLACER takes some planning. You may be more worried about some parts of the test than others. And depending upon how much time you have before you have to take the test (or a particular section of it), you may have an opportunity to create a study schedule to help you do well on the test. In this chapter, I discuss both planning for and taking the ACCUPLACER from a strategic perspective.

Strategies for Taking the ACCUPLACER

If you're like most students, you're used to taking both tests in school and standardized tests. Most of these tests have a few things in common — for example, they're administered on paper (break out those #2 pencils!) and have a strictly enforced time limit. Under these circumstances, you may have your own favorite strategies for success.

The ACCUPLACER, however, is different in a variety of ways that change the game and also the strategies that work best. In this section, I explain these differences and then suggest a few possible strategies for doing your best on the test.

Getting comfy with the CAT

In most cases, the ACCUPLACER is administered over the Internet, on a computer in an office of the community college where the student is enrolled. (Less commonly, it's given as an on-paper test in its COMPANION format.)

The computer format provides a dimension to the ACCUPLACER that you won't find with an on-paper test: *computer-adaptive testing,* or CAT for short.

Computer-adaptive testing means that the computer selects each question you see from a large pool of questions, basing its selection in part on the answers that you've previously given to questions. So, when you answer a question right, the computer tends to make the next question a little tougher. Conversely, when you answer a question incorrectly, the computer often gives you a question that's a little easier. That is, the computer *adapts* the test specifically for you in response to your answers.

One reason for the CAT feature is that it makes cheating really tough. For example, even if you're sitting next to your best friend taking the same test at the same time, the two of you probably won't even receive the same first question.

But the main reason for computer-adaptive testing is that it finds out your strengths and weaknesses more quickly and effectively than a paper test. This makes sense when you think about it. It's like the difference between answering a questionnaire on paper versus being interviewed. An interviewer can shift the focus of the next question they ask based on your answer to the last question.

REMEMBER

The good news here is that the ACCUPLACER doesn't have to be as long as, say, the SAT or the ACT. On the flipside, however, try not to get spooked knowing that the computer is, in a sense, grading the test as you're taking it. Try to remember that computers aren't sentient beings (yet!), and that the computer-adaptive feature isn't really "grading" or "judging" you; it's just pulling each question from a pool based upon a program written by the test designers.

And most of all, I encourage you not to dwell on what it means when you get what feels like an easy question ("Does that mean I messed up the previous question?") versus a hard question ("Did I get the previous question right?"). More than likely, if you fall down this particular rabbit hole (or should I say CAT hole?), you'll just distract yourself from the question you're trying to answer.

Multiple-choice or guess

Every question on the ACCUPLACER is a multiple-choice question with four possible answers, A through D. Unlike tests that are administered on paper (like the SAT), once you see a question, you're stuck with that question until you answer it. The computer doesn't provide a way to leave a question blank and then come back to it later. On the bright side, you don't lose points for choosing a wrong answer, so if you're really and completely stuck with a question, make your best guess and move on.

Time and time again

REMEMBER

The ACCUPLACER isn't a timed test. Once more, with feeling: The ACCUPLACER is NOT a timed test.

You can take as long as you like (within reason!) to complete each section of the ACCUPLACER. From a strategy perspective, this fact makes the ACCUPLACER virtually unique among all the tests you've ever taken and ever will take in your life.

How do you conquer a test that isn't timed? How do you use this feature to your best advantage? That's what I want to talk about now.

In a timed test, the pressure is on, and you have to work fast to get as many right answers as you can before the clock runs out.

In contrast, a non-timed test presents a different sort of opportunity — and challenge. The opportunity is to *take your time* with the test. For example,

>> Read every question very carefully so you don't "answer a question they didn't ask."

>> In the Reading and Writing tests, read each passage thoroughly so you really understand it.

>> In the math tests, work each step of the problem carefully, and check your solution *before* clicking the answer.

The challenge here is the same as the opportunity: to *take your time*.

Taking a test is uncomfortable, like wading belly deep through a swamp full of snakes. The temptation is to hurry through it as quickly as possible to avoid the discomfort.

So I understand that you may well prefer to blow through the ACCUPLACER at top speed. But if you do that, you'll probably make careless mistakes on questions that, with a little time and thought, you could have answered right. And for every section of the ACCUPLACER that you fail, you'll be spending 15 to 30 weeks in a no-credit class.

TIP

Here's what I recommend:

>> Take just one, or at most two, sections of the ACCUPLACER on a single day. You want to be fresh as a daisy for each section, right?

>> Before you begin a section, make sure that you'll have at least an hour and a half to complete it. So, if the office where they give the test is getting ready to close in 45 minutes, you probably want to reschedule!

>> Take your time answering each question.

>> If you really don't know the answer, guess! On the ACCUPLACER, there's no penalty for guessing, so answer every question.

>> Check to make sure you selected the answer you meant before clicking the SUBMIT button. The computer interface requires you to select an answer and then submit it. So as you answer each question, think of the show *Who Wants to Be a Millionaire?* and ask yourself, "Is that your final answer?"

To calculate or not to calculate?

The ACCUPLACER math tests include some questions that permit you to use a calculator, and other questions that don't. But remember that because the ACCUPLACER is a computer-based test, you don't need to bring a calculator to the test.

Instead, when a math question permits you to use a calculator, it will appear on the screen for you to use. In this section, I explore both no-calculator and calculator questions.

No-calculator questions

No-calculator questions usually require arithmetic that's not too complex — the kind of calculations that you can do either mentally or by hand. The person who sits you down to do the test should provide you with some scratch paper to work on — especially for the math tests. Please feel free to ask them to provide it if they forget.

The Arithmetic Test — the first of the three math tests, covering the most basic math — typically includes only no-calculator questions. This feature makes sense, because the test makers are specifically trying to see whether you can do basic math without the help of a calculator.

Calculator questions

Apart from the Arithmetic Test, the ACCUPLACER includes two, more difficult math tests: the Quantitative Reasoning, Algebra, and Statistics Test (QAS) and the Advanced Algebra and Functions Test (AAF). These tests contain more difficult math than the Arithmetic Test, so they include a mix of questions that may or may not require the on-screen calculator.

Calculator questions tend to include long numbers that are difficult to calculate by hand, or may include one or more calculations (such as finding a square root) that can't easily be done without a calculator.

When a question allows for the use of a calculator, you'll see an icon at the top-right corner of the screen. In some cases, more than one calculator — including a graphing calculator — may be made available.

Plan Your Work and Work Your Plan

A lot of tests — such as the SAT and the ACT — occur on dates that are set in stone many months in advance. They occur at the convenience of the test makers, and when the day arrives, you just have to show up and do your best.

But as I discuss both in Chapter 1 and earlier in this chapter, the ACCUPLACER is a test that you can take when you're ready. Most community colleges are fairly flexible about when you can take it. Furthermore, you don't have to take the entire ACCUPLACER in one sitting. Instead, you can take one or two sections in a single sitting, and push off the other sections for another time.

In this section, I suggest a step-by-step plan for setting up a study routine so that you can use this book in the way that works for *you* on your path to ACCUPLACER success.

1. Make sure you know which sections of the ACCUPLACER your school requires.

The ACCUPLACER includes five sections, which I include here for easy reference:

>> Reading Test

>> Writing Test

>> Three Math Tests:

- Arithmetic

- Quantitative Reasoning, Algebra, and Statistics (QAS)

- Advanced Algebra and Functions (AAF)

However, some community colleges don't require you to take all five sections of the ACCUPLACER. Be sure you know which ACCUPLACER sections your school requires, so you don't study for a test you don't have to take!

2. Decide which section of the ACCUPLACER will be easiest for you and take that first.

Here's a question I almost always ask each student on the first day of my ACCUPLACER prep class: Are you more worried about the English tests (Reading and Writing), or about the math tests?

Generally speaking, most students are more worried about the math tests. If you feel that way too, then I suggest that you plan to take both the Reading and Writing tests *before* you take the math tests. Through experience, I've found that students tend to pass the Writing Test more easily than the Reading Test. So, I recommend that you take the Writing Test first and the Reading Test second.

Flipside, if math is your thing and you're more worried about the English tests, then take the math tests first and save Reading and Writing for later. Because the three math tests progress in order of difficulty, you'll probably do best to take the Arithmetic Test first, the QAS second, and the AAF third.

I sum up these recommendations as follows.

If you're worried about math, follow this order:	If you're worried about English, follow this order:
1. Writing	1. Arithmetic
2. Reading	2. QAS
3. Arithmetic	3. AAF
4. QAS	4. Writing
5. AAF	5. Reading

3. Decide roughly when you want to take your first ACCUPLACER section.

When you know which section of the ACCUPLACER you plan to take first, the next question to address is when you want (or need) to take this section. That is, how long do you think you'll need to study for it?

If you're planning to take the Writing Test first, I suggest that you flip to Part 3 of this book and leaf through the three chapters you find there. The first two chapters are meant to be read through, and the third is full of practice problems. Additionally, Part 7 also includes two practice tests, so you may want to try out the two Writing sections of these tests.

Alternatively, if you're planning to take the Arithmetic Test first, take a quick look at Part 4. Again, the first two chapters are to be read through, and the third includes practice problems that will strengthen your arithmetic skills. And again, Part 7 includes two practice tests, so your study plan can also include the two Arithmetic sections.

I sum these tasks up as follows.

Writing Test Practice Steps	Arithmetic Test Practice Steps
1. Read Chapter 7	1. Read Chapter 10
2. Read Chapter 8	2. Read Chapter 11
3. First half of Chapter 9 questions	3. First half of Chapter 12 questions
4. Second half of Chapter 9 questions	4. Second half of Chapter 12 questions
5. Complete Practice Test 1 — Writing	5. Complete Practice Test 1 — Arithmetic
6. Complete Practice Test 2 — Writing	6. Complete Practice Test 2 — Arithmetic

I estimate that each of these six items will take you roughly the same amount of time. (That's why I split up the practice questions into two parts, numbered 3 and 4.)

4. Plan out a study schedule for your first ACCUPLACER section.

When you know how much time you'll have to study for a section of the ACCUPLACER, you can split up the six study steps I mention in the previous section into a schedule.

For example, here's how you break up the steps if you plan to take the Writing Test or the Arithmetic Test in three weeks:

Writing Test	Arithmetic Test
Week 1:	**Week 1:**
1. Read Chapter 7	1. Read Chapter 10
2. Read Chapter 8	2. Read Chapter 11
Week 2:	**Week 2:**
3. First half of Chapter 9 questions	3. First half of Chapter 12 questions
4. Second half of Chapter 9 questions	4. Second half of Chapter 12 questions
Week 3:	**Week 3:**
5. Complete Practice Test 1 — Writing	5. Complete Practice Test 1 — Arithmetic
6. Complete Practice Test 2 — Writing	6. Complete Practice Test 2 — Arithmetic

I think that both of these schedules are fairly doable, even if you're already busy with school and work.

5. Repeat Steps 2 through 4 for each of the other ACCUPLACER sections.

When you've worked through a schedule for one of the ACCUPLACER sections, set a date to take your next section of the test, and then set up a new study schedule.

The Reading Test should take just about as much time to study for as either the Writing Test or the Arithmetic Test. However, both the QAS and AAF Parts each contain four chapters, and the chapters are pretty long. So, depending on how much comfort you have with this material, you probably want to plan at least one full week to read each of these chapters, plus two weeks to work through the practice problems.

6. Retake any ACCUPLACER sections as needed.

Remember that most community colleges give you two tries to take the ACCUPLACER. So if you don't pass a section on your first try, don't despair! When you finish taking all the sections of the ACCUPLACER, if you need to take one or more over again, set up a new schedule and work through the book again. You can also find an additional practice test online; read the Introduction for instructions on how to access this practice test.

Good luck!

2

The Reading Test

IN THIS PART . . .

Know the types of reading passages that the ACCUPLACER uses.

Get familiar with the types of reading questions most often asked on the ACCUPLACER.

Practice your reading skills with questions designed to prepare you for the test.

» Looking at passage length and singular versus paired passages

» Anticipating the type of content the ACCUPLACER includes

» Identifying narrative, explanatory, and persuasive passages

» Understanding the specific reading challenges you'll face on the ACCUPLACER

Chapter 4

Reading Passages

In this chapter, I give you a quick overview of the types of reading passages you'll see on the ACCUPLACER. First, I discuss the form of the passages, including passage length and the number of questions that go with each passage length. I also fill you in on paired passages, which test your ability to compare and contrast two passages on a common theme. After that, I focus on the content of the passages.

By the end of this chapter, you'll have a general sense of the kind of reading the ACCUPLACER folks expect you to know.

Outline of the ACCUPLACER Reading Test

The ACCUPLACER Reading Test includes 20 questions, organized in a specific way. Here, I pull all of this information together in a single outline so you can see exactly what your ACCUPLACER reading test will look like.

For easy reference, Table 4-1 contains this outline, breaking down the 20-question reading test into separate sections.

One long literary passage with 4 questions

The test begins with a single long literary passage that includes a four-question set. This is the only literary reading passage you face on the ACCUPLACER. As I discuss later in this chapter, this passage is a selection from a fictional narrative work, such as a novel or a short story.

TABLE 4-1 **Overview of ACCUPLACER Reading Passages and Questions**

Type of passage	Number and length	Number of questions
1 literary set	1 long passage	4 questions
1 informational set	2 medium passages	4 questions
12 informational discrete	10 short or very short passages	12 questions
	2 fill-in-the-blank questions	
		Total questions = 20

TIP

As you read a literary passage, determine as soon as you can whether the narrator is first-person (a character in the story talking from the "I" perspective) or third-person (not a character in the story). Next, get clear on the names of the main characters — usually, there will be two or three at the most — how they relate to each other, and especially how they differ.

After you understand the basics of who the main characters are, try to figure out "what's at stake" — that is, what each character wants or is trying to achieve. Here's an example:

> "The narrator (Blair) is a daughter who wants to help her mother, but the mother (Arla) becomes defensive and keeps pushing her daughter away."

> "The father (unnamed) and son (Kevin) are talking about Kevin's older brother (Stephen), who is planning to join the Army. The father objects to this plan, and gets angry when Kevin points out that the father is usually overcritical of Stephen."

> "The narrator (Aaron) is a 22-year-old Jewish man thinking back on his recent trip to Israel, where he met his girlfriend (Fadilah), a Muslim woman whom he plans to ask to marry him."

One paired set of 2 medium-length informational passages with 4 questions

Next, you'll read a paired set of medium-length passages on an informational topic, and answer a set of four related questions. Typically, the two passages in a paired set relate distinct viewpoints on a single topic.

For example, the first passage in a paired set might focus on an advance in submarine technology, and the second might discuss a particular application of this technology for studying deep ocean plant life. Or the first might present the positive aspects of a proposed United Nations resolution, and the second could present the downsides.

TIP

In most cases, two of these questions will test you on the finer points that distinguish the two passages. So, as you read paired passages, try to get a sense of the differences between the two authors' viewpoints. And if possible, see if you can find actual words to describe these differences. Here's an example:

> "Passage 1 talks about a possible threat to the Amazon rainforest, and Passage 2 talks about a possible solution."

> "The author of Passage 1 urged Abraham Lincoln to sign the Emancipation Proclamation, and the author of Passage 2 criticized him for signing it."

> "Passage 1 discusses Billie Holliday's early years, and Passage 2 discusses her musical stardom."

Ten short or very short informational passages with 1 question each

Finally, the test presents with 10 short or very short passages, each of which includes a single question for you to answer. These passages are always informational, and they tend to be a grab bag. (Later in this chapter, I briefly discuss the range of topics and styles for these shorter passages.)

TIP

When a passage has only one question, read the question first. This can help you focus so you know what to look for before you read the passage.

The test ends with two fill-in-the-blank questions that do not include a reading passage.

Passage Form — What the Passages Look Like

Reading passages for the ACCUPLACER are classified according to four different lengths:

>> Very short: 75 to 100 words

>> Short: 150 to 200 words

>> Medium: 250 to 300 words

>> Long: 350 to 400 words

Each very short and short passage is followed by a single (discrete) question. Each long passage and pair of medium passages is followed by a set of four questions based on that passage.

Paired passages are labeled as "Passage 1" and "Passage 2" and linked by a common topic or theme, each with a distinct position or point of view on that topic. For example, a pair of passages might both discuss an innovation in video game design, one written by a game designer, and the other from the perspective of a game blogger.

Passage Content — What the Passages Are About

In this section, I describe the content of the passages you'll read on the ACCUPLACER. I first touch upon the genre, which falls into two categories: literary content (fiction) and informational content (non-fiction). Next, I discuss the type of texts, which tells you about the purpose and organizational form of the passage, and falls into three categories: narrative, explanatory, and persuasive.

Understanding the genre of the passage

On the ACCUPLACER, passage genre falls into two basic categories: literary (fiction) and informational (non-fiction). In this section, I discuss both of these types of passages.

Literary

On the ACCUPLACER, you'll read one literary passage and answer four questions about this passage. *Literary passages* are selected from fictional works of prose, such as short stories and novels. (*Note:* The reading passages on the ACCUPLACER don't include works of poetry.) A literary passage is typically a narrative storyline depicting one or more characters.

Informational

Informational passages are selected from factual works, and break down into the following three topic areas: careers/history/social studies, humanities, and science. Here, you get a quick take on all three types.

» **Careers/history/social studies:** One informational topic area for reading passages on the ACCUPLACER is *careers/history/social studies*. These passages focus on career-based information, the lives of people and civilizations in other times, and texts in social studies (such as anthropology, cultural anthropology, economics, psychology, sociology).

» **Humanities:** *Humanities* passages are typically about the fine arts (visual art, music, dance, theater, and literature), architecture, movies, and other forms of creative work discussed from an academic perspective.

» **Science:** Common topics in *science* passages are biology, chemistry, physics, astronomy, earth science, and medicine. Information often includes important scientific discoveries, how theories developed, important experiments, and technological advances.

Informational text types

The informational reading passages on the ACCUPLACER include texts whose organization and purpose fall into three different types: narrative, explanatory, and persuasive.

TIP

» **Narrative passages:** A *narrative passage* describes the progression of events as they occur in time. Most literary works — and all the literary passages on the ACCUPLACER — are organized as narratives, for the purpose of telling a story that unfolds over a period of time. Additionally, a few of the informational passages you read here may be organized as narratives, although this is somewhat uncommon.

» **Explanatory passages:** An *explanatory passage* explains an event or process, or otherwise provides information. This is the most common form of informational passage on the ACCUPLACER, and most likely the most common form of reading you've done in classes such as history, social studies, and the sciences.

» **Persuasive passages:** A *persuasive passage* attempts to convince the reader that a particular viewpoint is correct, and in some cases to take a specific action based on this viewpoint (for example, to vote for a certain candidate or to begin recycling plastic bottles).

As you read an informational passage, see if you can tell whether the author's main purpose is simply to explain something, or if he or she is attempting to persuade. Remember, every informational passage will have explanatory elements. However, a persuasive passage will use these elements to craft an argument.

IN THIS CHAPTER

» **Identifying the main idea of a passage**

» **Reading between the lines**

» **Understanding questions that focus on rhetoric — the *how* rather than the *what* of a passage**

» **Analyzing a pair of passages that provide two perspectives on a single topic**

» **Concentrating on vocabulary in the context of a passage**

Chapter **5**

Working through Reading Questions

The ACCUPLACER Reading Test includes 20 questions, which break down into categories as follows:

>> Information and Ideas: 7 to 11 questions

>> Rhetoric: 7 to 11 questions

>> Synthesis (comparing texts): 2 questions

>> Vocabulary: 2 to 4 questions

In this chapter, you focus on the types of questions that you will most likely see on the test.

Questions about Information and Ideas

About half of the questions you'll be asked on the ACCUPLACER (7 to 11 questions out of 20) concern factual content — that is, information and ideas from the passage. Generally speaking, these questions fall into four types: main idea, details, summary, and inference. In this section, you work with these four basic types of questions.

TIP

Remember that shorter passages are paired with just one question. In these cases, read the question first, then the passage, and finally the four answer choices. Reading the question before you read the passage can help you to find the information you're looking for in the passage.

Looking at the big picture: Determining central ideas and themes

One of the most common types of questions on any reading test asks you to identify the *main idea* or *central idea* of a passage — that is, the most important point that the passage conveys. On the ACCUPLACER, you'll most likely answer a variety of such questions.

TIP

When faced with a question that asks you to identify the main or central idea of a passage, read the whole passage and consider it from start to finish. The correct answer should capture the entire passage without getting bogged down in details.

The *state university* system in the United States is a group of public universities supported by an individual state. These systems constitute the majority of public-funded universities and each state supports at least one such system. The amount of the financial subsidy from the state varies from university to university and state to state, but the effect is to lower tuition costs below that of private universities for students from that state or district. On the other hand, a private university is not operated by a (state) government, although many receive favorable tax credits, public student loans, and grants. Depending on their location, private universities may be subject to government regulation. (*Ethics in the University*, by James G. Speight — Page 2)

EXAMPLE

The main purpose of the passage is to

(A) Provide an example of a state university and briefly describe it.

(B) Explain how state universities provide lower-cost tuition to students from that state or district.

(C) Discuss a few key differences between state universities and private universities.

(D) Persuade the reader that state universities are often superior to private universities.

Although the first part of the passage focuses on state universities, the subsequent discussion of private universities is equally important. Thus, Answers A and B are both ruled out. However, the passage is explanatory rather than persuasive — that is, it doesn't take a position on which type of university is superior — which rules out Answer D; therefore, Answer C is correct.

In the first few weeks babies mostly cry but very quickly they can make their first real sounds. These often resemble vowels and at first they are produced more or less by chance. Later on, these sounds begin to be made on purpose and they become more varied. The children are discovering what they can do with their speech apparatus through constant experimenta-tion. Around eight months, they will produce sequences of constantly repeated syllables such as *babababa* or *gagaga*, known as *babbling*. Although these babbles are beginning to sound like real language, having intonational patterns and such, they are not yet real words. They do not have a constant reference to the same objects every time they are produced. Only when a sound sequence is produced consistently in association with an object or event do we really speak of a word, for example, when daddy is always referred to with *baba*, or when a child says *eat* every time they want something to eat. (*Linguistics*, by Anne E. Baker and Kees Hengeveld — Page 63)

EXAMPLE

The main idea of the passage is to

(A) Outline how language develops in children from birth until they speak their first words.

(B) Explain what *babbling* is and how it relates to language acquisition in children less than 1 year old.

(C) Precisely define what distinguishes a baby's first word and provide a few simple examples.

(D) Illuminate why 8 months old is a critical time when babies make a significant leap for-ward toward learning language.

The passage walks the reader through a sequence of typical stages in infant language development, from the first sounds they make until they are able to repeat actual words for communication. Answers B, C, and D focus on individual stages in this process. Only Answer A recognizes the overall structure of the passage as an outline, with no single stage identified as more important than the others, so Answer A is correct.

Focusing on the details: Reading closely

In contrast to main idea questions, *detail questions* ask you to correctly identify a specific fact or piece of information mentioned in the passage.

TIP

When answering a detail question, reading the question first is usually helpful, so you have an idea of what you're looking for in the passage.

Any object made by humans is an *artifact*. Usually you think of ancient ceramic pots or arrowheads, and indeed these items are everywhere at archaeological sites. But a temple or a palace is an artifact too — one made up of individual artifacts such as bricks or stone blocks. Most often, artifacts are portable objects excavated and brought back to a laboratory for study.

Ecofact is a term archaeologists invented to classify natural objects used by humans without modification. Animal bones left from dinner or pollen from gathered plants are ecofacts. But if a bone has been modified to become a harpoon point, that modification makes it an artifact. Even phosphates or other chemicals in the soil are ecofacts showing that people threw their organic waste on the ground. (*Archaeology For Dummies* by Nancy Marie White — Page 12)

EXAMPLE

According to the passage, which of the following is an example of an ecofact?

(A) A stone tablet with words in an ancient language written on it.

(B) The leg bone of a large animal fashioned into a weapon.

(C) A human skull placed at a gravesite to frighten strangers away.

(D) A bowl made of bronze that in its day was used to hold food.

The passage states that an ecofact is "a natural object used by humans without modification." This definition rules out Answers A and D, because both are human-made objects. The leg bone of a large animal would only be an ecofact if unmodified, but it becomes an artifact when fashioned into a weapon, so Answer B is incorrect. In contrast, a human skull placed at a gravesite is a natural object without modification, so Answer C is correct.

In some cases, a more difficult detail question may require you to identify more than one detail. This is especially true if the question includes a capitalized word such as NOT or EXCEPT. In these types of questions, you may need to verify three details — one for each incorrect answer — and answer the question by identifying information that is NOT contained in the passage. Here is an example:

The path to an academic career can be demanding — mentally, physically, and emotionally — and not everyone has the perseverance to complete years of concentrated study. But the experience of doing post-baccalaureate work is exhilarating for those with sufficient interest and determination. There may be many teachers, mentors, or colleagues who are willing to help the emerging faculty member and assist with overcoming difficult hurdles that are needed to gain confidence and the ability to think and work independently. However, no one can foresee the winding of the path along which adventurous observation and experiment may lead, and what boundaries can be set to the possibilities of interpretation. (*Ethics in the University*, by James G. Speight — Page 7)

EXAMPLE

The passage states that all of the following may be difficult aspects of pursuing an academic career EXCEPT

(A) An advanced degree typically takes years to complete.

(B) Other academics may be unhelpful or even damaging.

(C) The work required may take a toll on the body, mind, and emotions.

(D) The ultimate outcome of research can be uncertain.

Answer this question by ruling out wrong answers one by one. The passage states that an academic career requires a person to complete "years of concentrated study," which rules out Answer A. It states that "an academic career can be demanding — mentally, physically, and emotionally," which rules out Answer C. And "no one can foresee the winding of the path along which adventurous observation and experiment may lead," which rules out Answer D. In contrast, the passage mentions that "there may be many teachers, mentors, or colleagues who are willing to help the emerging faculty member and assist with overcoming difficult hurdles," which is the opposite of what Answer B states, so this is the correct answer.

Getting to the point: Summarizing

A *summary* of a passage briefly restates the most important points made in that passage. Another type of question that you will face on the ACCUPLACER requires you to read a passage and then identify a reasonably accurate summary. Here is an example:

> For much of the 20th century, our knowledge of the history of life on Earth went no further back than the dawn of the Cambrian period — "only" 550 million years ago. Fossils of quite sophisticated marine eukaryotes have been dated to that time and, during the Cambrian period itself, a very wide range of new lifeforms appeared. This flourishing of diversity in this period is known as the *Cambrian explosion.* Intense searches in pre-Cambrian rocks were conducted from the mid-1960s onward, but for many years failed to yield any fossils. However, one of those pivotal moments in science came in 1987 when the American paleobiologist William Schopf identified fossil micro-organisms dating back 3.5 billion years. (*Functional Biology of Plants*, by Martin J. Hodgson and John A. Bryant — Page 1)

EXAMPLE

Which of the following summarizes the four events discussed in the passage in order from earliest to latest:

(A) The Cambrian explosion, the development of micro-organisms, fossils of micro-organisms are discovered, fossils from the Cambrian explosion are discovered.

(B) The Cambrian explosion, the development of micro-organisms, fossils from the Cambrian explosion are discovered, fossils of micro-organisms are discovered.

(C) The development of micro-organisms, the Cambrian explosion, fossils of micro-organisms are discovered, fossils from the Cambrian explosion are discovered.

(D) The development of micro-organisms, the Cambrian explosion, fossils from the Cambrian explosion are discovered, fossils of micro-organisms are discovered.

The passage states that the Cambrian explosion occurred 550 million years ago, but that fossils of micro-organisms date back to 3.5 billion years ago, so this rules out Answers A and B. It also states that for much of the 20th century, fossils dating back to the Cambrian explosion were studied, but that the first fossils of micro-organisms were discovered in 1987, which rules out Answer C; therefore, Answer D is correct.

If Mars had an ocean in the past, the planet had to have been much warmer than it is today. If the carbon dioxide atmosphere was much thicker then, it would have trapped heat due to the greenhouse effect. If Mars once had a warm atmosphere and an ocean, then carbon dioxide from the atmosphere would have dissolved in the water. Chemical reactions then would have produced carbonates (minerals composed of carbon and oxygen). This theory predicts that there are carbonate rocks on Mars. NASA's Mars Exploration Rover, Spirit, discovered carbonate rocks in 2010!

Many experts have a variety of opinions, but I think the case is closed: Mars once had a warm climate and a whole lot of liquid water. (*Astronomy For Dummies,* by Stephen P. Maran — Page 125)

EXAMPLE

Which of the following provides the best summary of the passage?

(A) A testable prediction is made, a theory is proposed, the expected result is verified, and the passage states that the theory is most likely true.

(B) A theory is proposed, a testable prediction is made, the expected result is verified, and the passage states that the theory is most likely true.

(C) A theory is proposed, the passage states that the theory is most likely true, a testable prediction is made, and the expected result is verified.

(D) A testable prediction is made, the expected result is verified, a theory is proposed, and the passage states that the theory is most likely true.

The passage opens with the theory that Mars may have had an ocean in the past. Later in the passage, it predicts the presence of carbonate rocks on Mars. The confirmation of this prediction in 2010 is mentioned next, and the passage concludes with the writer's opinion that the theory is most likely true. Thus, Answer B is correct.

Reading between the lines: Understanding relationships and making inferences

To answer some questions on the ACCUPLACER, you need to *make inferences* — that is, use the information given explicitly in the passage to form reasonable conclusions about what the passage is saying.

In some cases, you need to make relatively straightforward inferences to answer the question. A careful reading of the passage, so that you clearly understand what it says, should be enough to answer correctly. Here is an example:

In a study performed by psychologist Paul Slovic of the University of Oregon, Slovic and his team of researchers told a group of volunteers that a young girl was suffering from starvation and malnutrition. The story of the young girl suffering was followed with a request to see if they would be willing to help her. His research team then told a different group the same story of the girl who was suffering from starvation and malnutrition but changed the request a bit. This time, his team told the group that there are millions of others suffering just like her and asked them for support because a population needs them. Through the use of statistics and images, the second group of volunteers was told of the immense breadth of the issue.

The result? The second group of volunteers, the one shown information about the population suffering from starvation and malnutrition, along with statistics and a request to support the millions of people affected, gave half as much as the volunteer group that was exposed to the message of just the individual girl. (*Social Movements for Good,* by Derrick Feldmann — Page 22)

EXAMPLE

Which of the following inferences did the experimenters most likely make as a result of their findings?

(A) The fact that the first group gave more money supports the conclusion that people give more when the welfare of a large number of people unknown to the giver is at stake.

(B) The fact that the first group gave more money supports the conclusion that people give more when the welfare of a single person known to the giver is at stake.

(C) The fact that the second group gave more money supports the conclusion that people give more when the welfare of a large number of people unknown to the giver is at stake.

(D) The fact that the second group gave more money supports the conclusion that people give more when the welfare of a single person known to the giver is at stake.

This question simply requires you to follow the narrative and draw the logical conclusion based on what is presented there: The second group gave half as much money, so Answers C and D are ruled out. And the first group was told only about one girl, which rules out Answer A; therefore, Answer B is correct.

A more difficult question may require you to carefully thread through the argument that the passage is making.

> At the core of a hotel revenue manager's job are the data and analytics in the revenue management system that form the foundation for day-to day pricing decisions. The better the data and the more accurate and robust the analytics, the better the pricing decisions will be. Better pricing decisions lead to a more effective revenue management strategy and, of course, to more revenue. This is why the industry must be concerned about the latest and greatest analytic techniques to support effective pricing in a changing marketplace. (*Hotel Pricing in a Social World*, by Kelly A. McGuire — Page 18)

EXAMPLE

Which of the following statements is NOT implied in the passage?

(A) Day-to-day pricing decisions are important among a hotel revenue manager's duties.

(B) Good pricing decisions result in increased revenue.

(C) Higher revenue leads to increased quality of data and analytics.

(D) The quality of data and analytics affects revenue management strategy.

When answering this question, be very careful to understand what the argument is saying — particularly, what the author is calling causes and what she sees as effects. The first sentence tells you analytics about pricing are "at the core of a hotel revenue manager's job," so Answer A is ruled out. And "better pricing decisions lead to more revenue," which rules out Answer B. Also, notice that the second and third sentences say that better data and analytics lead to better pricing decisions, which in turn lead to more revenue; therefore, Answer D is ruled out. In contrast, the argument tells us that increased quality of data and analytics leads to higher revenue, rather than the reverse, so Answer C is correct.

Questions Focusing on Rhetoric

Rhetoric is the study of effective writing. ACCUPLACER questions that focus on rhetoric (which number 7 to 11 questions out of 20) ask you to identify not so much *what* the passage is telling you, but rather *how* it's telling you. In this section, you work with a variety of rhetorical issues.

Analyzing word choice rhetorically

Sometimes, it's not what you say, but how you say it. ACCUPLACER Reading questions focusing on word choice require you to decide why the author chose the specific words that they used. Let's return to an earlier passage:

> For much of the 20th century, our knowledge of the history of life on Earth went no further back than the dawn of the Cambrian period — "only" 550 million years ago. Fossils of quite sophisticated marine eukaryotes have been dated to that time and, during the Cambrian period itself, a very wide range of new lifeforms appeared. This flourishing of diversity in this period is known as the *Cambrian explosion*. Intense searches in pre-Cambrian rocks were conducted from the mid-1960s onward, but for many years failed to yield any fossils. However, one of those pivotal moments in science came in 1987, when the American paleobiologist William Schopf identified fossil micro-organisms dating back 3.5 billion years. (*Functional Biology of Plants* by Martin J. Hodgson and John A. Bryant — Page 1)

EXAMPLE

Which of the following is the most plausible explanation why the authors place quotes around the word "only" in the first sentence?

(A) To acknowledge that 550 million years is an approximation, while implying their confidence in it.

(B) To indicate that they, personally, dispute the accuracy of this estimate.

(C) To underscore the irony that 550 million years could be considered a short length of time.

(D) To downplay the extraordinary achievement inherent in this discovery.

There is no indication that the dates in the passage are inaccurate or even questionable, so Answers A and B are ruled out. And the passage is simply factual, with no indication that the author wishes to downplay the discovery, so Answer D is ruled out; Answer C is correct.

Analyzing text structure

A passage is always more than just a random collection of facts and opinions. *Text structure* describes the overall organization of a passage, shedding light on how information in the passage is revealed to the reader. Here's a short list of common organizational strategies for passages.

> » *Anecdote:* The passage opens with a story and then relates this story to the topic that the author wants to discuss.
>
> » *Argument:* The passage presents a set of premises leading to a conclusion.
>
> » *Cause and Effect:* The passage describes a cause and then its resultant effect.
>
> » *Compare and Contrast:* The passage compares two or more ideas or theories, potentially settling on one as superior.
>
> » *Narrative:* The passage recounts a series of events in time, from earliest to latest.
>
> » *Question and Answer:* The passage opens with a question and then explores this question, proposing one or more answers and potentially settling on one as correct.

TIP

When analyzing the structure of a text, try to step back from it and view it as a whole rather than getting lost in the details. What's the big picture? Is the author presenting a problem and then proposing a solution? Are they posing a question and then attempting to answer it? Are two or more ideas being compared or contrasted? Does the passage reach a conclusion, or just explore ideas without taking a side?

Saint Augustine believes that everything that exists in time — everything we can see with our eyes or touch with our hands — is insubstantial and necessarily unsatisfying. For him what is here today and gone tomorrow is not fully real. He offers a brilliant analysis of what is now known as "time consciousness" and argues that a life implicated in the past-present-future flow of time is, by itself, no more than a disappearing wisp of smoke. As a result he thinks it is altogether reasonable for us to orient ourselves toward what is immune to the flow of time; to, in other words, the Eternal; to God.

By contrast, Nietzsche finds Augustine's denigration of temporal flow despicable. For him, to deny the reality and goodness of the passage of time is a kind of self-hatred. Yes, time flies, but that's all we've got, and so Nietzsche challenges us to have the courage to affirm our lives for what they are rather than to escape into an imaginary and lifeless beyond. (*Thinking Philosophically — An Introduction,* by David Roochnik — Page 13)

EXAMPLE

Which of the following best describes the relationship of the second paragraph to the first?

(A) The first paragraph poses a question and the second paragraph answers it.

(B) The first paragraph states a problem and the second paragraph solves it.

(C) The first paragraph presents a valid approach to a problem and the second paragraph presents another valid approach.

(D) The first paragraph presents a flawed approach to a problem and the second paragraph presents an improved approach that the author explicitly endorses.

The two paragraphs both discuss how to think about the passage of time. More specifically, they present two contrasting approaches to this question by two different thinkers: Saint Augustine and Nietzsche. Their approaches differ greatly, but there is no indication that the author is arguing that one is superior to the other. Rather, the two paragraphs describe two approaches that are presented as equally valid ways of thinking about time, so Answer C is correct.

Analyzing point of view

The *point of view* of a passage is the author's personal view of the topic that they are discussing. Generally speaking, the point of view of an author can range from positive through neutral (no strong opinion) to negative.

TIP

Point-of-view questions tend to use a relatively short list of words to describe an author's attitude. Words like *favorable* and *enthusiastic* describe an author's positive stance. Words like *impartial* and *journalistic* imply a neutral position. And words like *critical* and *hostile* denote a negative opinion.

Additionally, positive and negative opinions can be further characterized as strong or mild. For example, words like *forcefully, utterly, wholeheartedly, candidly,* and *unreservedly* indicate a strong positive or negative stance. In contrast, words like *warily, thoughtfully, prudently, guardedly,* and *cautiously* suggest a milder opinion.

Our sports and our athletes represent us; they embody our identities and aspirations. In playing and watching sports, individuals, groups, and nations lay bare their characters and social values, as well as our common human nature. Situations of physical effort, stress, and rivalry show more about the character of people than is revealed in superficial and formal social settings. In the intensity of sports, we drop our veneer of socialization or civilization; we show our true natures, a human condition somewhere between animals and angels. We learn much about ourselves in examining how we prepare, compete, strain, and sweat — how we handle our greatest feats and defeats. (*Sport and Spectacle in the Ancient World,* by Donald G. Kyle — Page 5)

EXAMPLE

The author's attitude toward sports could best be described as

(A) Candidly enthusiastic

(B) Guardedly optimistic

(C) Journalistically neutral

(D) Unreservedly hostile

The author's point of view toward sports is positive rather than negative or neutral, so you can rule out Answers C and D. Additionally, his opinion is strong rather than mild, which rules out Answer B; therefore, Answer A is correct.

Analyzing purpose

Each sentence in a passage provides information that has a *purpose* when understood in the context of the entire passage. Some questions on the ACCUPLACER require you to uncover the purpose of a given sentence, explaining why the author wrote this sentence. Returning to an earlier passage,

In the first few weeks babies mostly cry but very quickly they can make their first real sounds. These often resemble vowels and at first they are produced more or less by chance. Later on, these sounds begin to be made on purpose and they become more varied. The children are discovering what they can do with their speech apparatus through constant experimentation. Around eight months, they will produce sequences of constantly repeated syllables such as *bababababa* or *gagaga*, known as *babbling*. Although these babbles are beginning to sound like real language, having intonational patterns and such, they are not yet real words. ***They do not have a constant reference to the same objects every time they are produced.*** Only when a sound sequence is produced consistently in association with an object or event do we really speak of a word, for example, when daddy is always referred to with *baba*, or when a child says *eat* every time it wants something to eat. (*Linguistics*, by Anne E. Baker and Kees Hengeveld — Page 63)

EXAMPLE

What is the purpose of the sentence in boldface, reproduced here?

They do not have a constant reference to the same objects every time they are produced.

(A) To provide an example of a word.

(B) To provide an example of babbling.

(C) To provide an explanation of why certain sounds are not words.

(D) To provide an explanation of why some children do not speak until they are close to 1 year old.

The boldface sentence provides an explanation rather than an example, so Answers A and B are ruled out. This explanation sheds light on why repeated sounds such as *gagaga* are not words, so Answer C is correct.

Analyzing arguments

Some questions focus on the argument that the writer is making — that is, what the writer wants to convince their reader to be true. To help you answer questions that relate directly to the writer's argument, here are a few useful terms:

>> **Conclusion:** A statement that the writer wants to convince the reader is true

>> **Premise:** A statement that supports the conclusion

>> *Assumption:* A premise that isn't stated directly in the argument

>> *Counter-premise:* A statement that opposes the conclusion, which the argument must overcome

Let's return to an earlier passage:

The path to an academic career can be demanding — mentally, physically, and emotionally — and not everyone has the perseverance to complete years of concentrated study. But the experience of doing post-baccalaureate work is exhilarating for those with sufficient interest and determination. There may be many teachers, mentors, or colleagues who are willing to help the emerging faculty member and assist with overcoming difficult hurdles that are needed to gain confidence and the ability to think and work independently. However, no one can foresee the winding of the path along which adventurous observation and experiment may lead, and what boundaries can be set to the possibilities of interpretation. (*Ethics in the University,* by James G. Speight — Page 7)

EXAMPLE

How does the sentence in boldface function in the writer's argument?

(A) As the conclusion that the writer is arguing for.

(B) As a premise that leads to the writer's conclusion.

(C) As a fact that the writer must counter in order to strengthen the argument.

(D) As a common misconception that the author dismisses as irrelevant to the argument.

The writer is arguing *for* rather than *against* the rewards of an academic career. In this context, the boldface sentence works in opposition to the argument, which rules out Answers A and B. The statement that an academic career is difficult and not for everyone is in opposition to what the writer is arguing. The writer doesn't dismiss this concern, so Answer D is incorrect. Instead, the writer argues that for people who are determined enough, the difficulties can be surmounted; therefore, Answer C is correct.

Questions That Compare Two Texts

Recall from Chapter 4 that the ACCUPLACER Reading Test will include four questions that quiz you on a pair of related passages. Of these questions, at least two will require you to compare both texts. Many students find questions that compare passages to be among the most difficult on the test.

To get a feel for these types of questions, begin by reading the following two passages. As you do so, think about these two questions:

What do these two passages have in common? (Usually, some aspect of the topic under discussion is common to both passages.)

What does each passage have that makes it distinct from the other passage?

Passage 1

The Mars atmosphere is mostly carbon dioxide, but the Martian atmosphere is much thinner than the atmospheres of Earth. Mars also has clouds of water-ice crystals, which resemble the cirrus clouds on Earth. In winter, some carbon dioxide from the Martian atmosphere freezes on the surface, leaving thin deposits of dry ice (frozen carbon dioxide). The South Polar Cap is always covered in dry ice (water ice may lie below it). But water ice is seen at the North Pole when it's summer there and the dry ice has evaporated for the season.

If Mars had an ocean in the past, the planet had to have been much warmer than it is today. If the carbon dioxide atmosphere was much thicker then, it would have trapped heat. If Mars once had a warm atmosphere and an ocean, then carbon dioxide from the atmosphere would have dissolved in the water. Chemical reactions then would have produced carbonates (minerals composed of carbon and oxygen). This theory predicts that there are carbonate rocks on Mars. NASA's Mars Exploration Rover, Spirit, discovered carbonate rocks in 2010!

Many experts have a variety of opinions, but I think the case is closed: Mars once had a warm climate and a whole lot of liquid water. (*Astronomy For Dummies*, by Stephen P. Maran — Page 125)

Passage 2

One of the most persuasive pieces of evidence for prolonged surface drainage is meandering channels, which are created where streams flow across a floodplain that is resistant to erosion. Erosion takes place on the outside of the bend, where the flow is faster, with deposition on the inside, where it is slower. Over time, the meander loops increase in size and migrate down-stream, widening the flood plain. On Mars, meandering river sediments have been found in many places, suggesting that much of the planet experienced lengthy periods of surface run off. Low features within the bends, known as scroll bars, show where the meanders migrated over time.

Rivers on Earth often enter lakes or the sea through deltas. The river splits into numerous distributaries as it crosses the flat, low-lying area which is created by deposition of a large sediment load when the rate of flow decreases. Deltas are also found on Mars, often at the margins of craters. In these cases, the channels probably disgorged into a lake which filled the impact basin.

The most famous example is a fossil delta in the Eberswalde crater. After its formation, the delta material was further buried by other materials — probably sediments — that are no longer present. The entire area became cemented and hardened to form rock. Subsequent erosion stripped away the overlying rock, re-exposing the delta. Today, the hardened delta sediments are preserved as ridges. (*Exploring the Solar System*, by Peter Bond — Page 181)

Before proceeding to the test questions, spend some time with the two questions you were asked before you read the passages:

What do these two passages have in common? (Usually, some aspect of the topic under discussion is common to both passages. Be as specific as possible.)

Both passages discuss Mars — more specifically, evidence for the presence of water on the surface of Mars, probably a long time ago.

What does each passage have that makes it distinct from the other passage?

Passage 1 discusses the atmosphere of Mars. It suggests that if a great deal of water had been present on Mars at one time, the atmosphere would have been warmer than it currently is. It also mentions the presence of carbonates as evidence that this theory is correct.

Passage 2, in contrast, focuses on patterns of erosion found on Mars that are common to dried-up riverbeds and river deltas on Earth. These patterns suggest that water was probably flowing on Mars sometime in the past.

Now, proceed to the questions:

EXAMPLE

Which of the following provides the best summary of the two passages?

(A) Passage 1 discusses evidence suggesting there was once water on Mars, but Passage 2 disputes this evidence.

(B) Passage 1 discusses evidence suggesting there was once water on Mars, and Passage 2 discusses different evidence suggesting this.

(C) Passage 1 discusses evidence suggesting there was once liquid water on Mars, but Passage 2 discusses evidence suggesting that this water was frozen.

(D) Passage 1 discusses evidence suggesting Mars had liquid water in the past, and Passage 2 discusses evidence of it in the present.

Both passages discuss evidence that there was once water on Mars, so Answer A is incorrect. And both passages discuss evidence for water in the distant past, which rules out Answer D. In Passage 2, the evidence for the earlier presence of water implies that it had been flowing rather than frozen, so Answer C is ruled out. Thus, Answer B is correct.

EXAMPLE

Which passage or passages cite(s) a testable prediction whose later confirmation suggests the presence of water on Mars?

(A) Passage 1 only

(B) Passage 2 only

(C) Passage 1 and Passage 2

(D) Neither Passage 1 nor Passage 2

Both passages mention evidence of water on Mars. However, the only testable prediction mentioned is made in Passage 1: that the theory implies the presence of carbonate rocks on Mars, which were found in 2010. Therefore, Answer A is correct.

Questions That Test Vocabulary

The previous version of the ACCUPLACER included a separate section for vocabulary questions. In the new ACCUPLACER, vocabulary questions (2 to 4 questions out of 20) are included on the Reading Test. Some of these questions place vocabulary words in the context of an entire passage, and others are discrete fill-in-the-blank questions. In this section, I discuss both types of questions.

Passage-based vocabulary questions

Passage-based vocabulary questions are presented in the same way as the questions you've already looked at in this chapter: a reading passage followed by a question based on that passage.

TIP

Even if you're not entirely sure what a word means, you can use the passage to help you make sense of it. Try replacing the vocabulary word with each of the words that are provided as answer choices. In this way, you may be able to eliminate one, two, or even all three wrong answers.

For much of the 20th century, our knowledge of the history of life on Earth went no further back than the dawn of the Cambrian period — "only" 550 million years ago. Fossils of quite sophisticated marine eukaryotes have been dated to that time and, during the Cambrian period itself, a very wide range of new lifeforms appeared. This flourishing of diversity in this period is known as the *Cambrian explosion*. Intense searches in pre-Cambrian rocks were conducted from the mid-1960s onward, but for many years failed to yield any fossils. However, one of those **pivotal** moments in science came in 1987 when the American paleobiologist William Schopf identified fossil micro-organisms dating back 3.5 billion years. (*Functional Biology of Plants,* by Martin J. Hodgson and John A. Bryant — Page 1)

EXAMPLE

In context, which of the following best captures the meaning of the word *pivotal*?

(A) Challenging

(B) Disputed

(C) Outrageous

(D) Significant

The passage describes years of failure from the mid-1960s onward, then signals a shift in 1987 with the word *however*. It describes that year as *pivotal* because of Shopf's discovery of fossil micro-organisms. The best synonym for this important discovery is *significant*, so Answer D is correct.

In the first few weeks babies mostly cry but very quickly they can make their first real sounds. These often resemble vowels and at first they are produced more or less by chance. Later on, these sounds begin to be made on purpose and they become more varied. The children are discovering what they can do with their **speech apparatus** through constant experimentation. Around eight months, they will produce sequences of constantly repeated syllables such as *babababa* or *gagaga*, known as *babbling*. Although these babbles are beginning to sound like real language, having intonational patterns and such, they are not yet real words. They do not have a constant reference to the same objects every time they are produced. Only when a sound sequence is produced consistently in association with an object or event do we really speak of a word, for example, when daddy is always referred to with *baba*, or when a child says *eat* every time it wants something to eat. (*Linguistics,* by Anne E. Baker and Kees Hengeveld — Page 63)

EXAMPLE

In context, the words *speech apparatus* refer to

(A) A baby's throat and mouth.

(B) The repetitive sounds a baby makes.

(C) The unintentional sounds a baby makes.

(D) A baby's first successful words.

The word *apparatus* refers to something physical that performs a function, so *speech apparatus* means the physical portion of the body that produces speech — that is, the throat and mouth. Answer A is correct.

Fill-in-the-blank vocabulary questions

The last two questions on the Reading Test test your vocabulary by asking you to fill in the blank to complete a sentence.

As with passage-based vocabulary questions, the challenge with vocabulary questions comes when you're faced with one or more words you don't know. Your best bet is to rule out wrong answers by reading the sentence to yourself, switching in each possible answer.

For example:

EXAMPLE

Concerned that his supervisor would not accept his excuse for being late, Jason chose to _____ rather than simply say that he overslept.

(A) ameliorate

(B) equivocate

(C) speculate

(D) terminate

Glancing over this list of answers, you may be sketchy about words *ameliorate* and *equivocate*. However, you may know that *speculate* means "take a guess" and that *terminate* means "finish." Neither of these meanings works when placed in the blank, so you can rule out Answers C and D. At this point, if you don't know the other two words, use your spider sense to help you guess. The word *ameliorate* means "improve," so Answer A is incorrect. The word *equivocate* means "avoid being honest," which fits well with the sentence, so Answer B is correct.

EXAMPLE

After a 12-hour shift at work followed by a bachelorette party for her best friend, Brie was so _____ that she fell asleep in her clothes on the couch.

(A) demented

(B) enervated

(C) incentivized

(D) traumatized

Again, this question includes words that you may not know. However, if you know that *demented* means "crazy or irrational" and that *traumatized* means "deeply disturbed," you can see that both Answers A and D are wrong. The more difficult words are *enervated* ("exhausted") and *incentivized* ("motivated"), so Answer B is correct.

Chapter 6

Reading Practice Questions

n this chapter, you put your reading skills to the test. First up are five reading passages with a total of 20 questions. After that, you try out 10 fill-in-the-blank questions to test your vocabulary in the context of sentences. When you're finished, or if you need help answering any of the questions, you can flip to the end for a detailed explanation of each answer.

Reading Passage Questions

Line   What then is "Greek" about Greek art? And how much of it is "art"? For the Greeks, "art" was craft and artists were by and large thought of as artisans: good with their hands and not much else (though famous ones, like Pheidias, came to be respected for their political power and the money that it made them.) Much of what we appreciate as "Greek
(5)   art" today, or have done so in the past, has been elaborated, embellished, and reinvented. In short, it has been translated by the crucial intervention of Rome and the Middle Ages, not to mention the systematic efforts of Western European elites in early modernity. (*A Companion to Greek Art*, edited by Tyler Jo Smith and Dimitris Plantzos — Page 9)

1. Which of the following best describes the central idea of the passage?

(A) Modern people perceive Greek art quite differently from how the Greeks of the time perceived it.

(B) Greek art was, in fact, not crafted by the Greeks themselves, but by artists in other places such as Rome.

(C) Greek artists did not think of themselves as artists, but rather as artisans and craftspeople.

(D) The effort of modern historians to reinvent the concept of Greek art has done irreparable damage to our understanding of it.

2. According to the passage, the Greeks themselves thought of their artists much as we today think of

   (A) A famous singer or actor.

   (B) A reclusive artist or writer.

   (C) A successful entrepreneur or founder of a company.

   (D) A skilled plumber or electrician.

3. Which of the following can be inferred about Pheidias?

   (A) He was obscure in his own time, but later recognized as an artist of great talent.

   (B) He was successful as an artist and craftsman among Greeks of his own time.

   (C) He founded an important movement in Greek art that is still appreciated today.

   (D) He was respected, even lionized, in his day but sought scrupulously to avoid the public eye.

4. Which of the following is the best explanation of why the writer places the words "art" and "Greek art" in quotes?

   (A) He wants to bring into focus common assumptions and fallacies in the very use of these terms.

   (B) He is attempting to persuade the reader that Greek artists were, in fact, not particularly good artists in comparison with artists from other regions.

   (C) He hopes to underscore his opinion that no definition of art has ever been agreed upon among scholars.

   (D) He is confessing the fact that he, himself, is unclear about the meaning of these terms as they relate to Greek art.

Line   In addition to attempting to provide basic food supplies, water, and sanitation, governments are expected to ensure that biomedical technologies are readily available, including immunization and indispensable medication including antibiotics and painkillers. Increasingly, however, as the media reminds us every day, money is spent on weapons as a result (5) of local conflicts, leaving ever fewer resources for medical care. Furthermore, regulations implemented in the name of security as a response to threats of terrorism, real and imagined, have brought about restrictions by local governments on the movement of peoples as they attempt to flee from violence and abject poverty, resulting in a phenomenal rise in refugee and squatter populations. Epidemics of infectious disease thrive in conditions of (10) poverty and instability, and today have the potential to wreak widespread havoc in a matter of hours, striking even the world's wealthiest — as demonstrated by the case of Atlanta lawyer Andrew Speaker in 2007, who was infected with a lethal strain of extremely drug-resistant tuberculosis (XDR-TB), leading to a frantic search for other exposed passengers on transatlantic flights. (*An Anthropology of Biomedicine* by Margaret Lock and Vinh-Kim (15) Nguyen — Page 4)

5. Which of the following best sums up the main idea of the passage?

   (A) A variety of current geopolitical realities make epidemics of infectious diseases an appreciable risk to people everywhere.

   (B) Money that governments spend on the military could be put to better use fighting the spread of infectious diseases.

   (C) Infectious diseases tend to thrive in conditions such as those often found where people fleeing from violence and poverty are forced to gather.

   (D) Generally speaking, world leaders tend to worry about the spread of disease only when wealthy people are infected.

6. The writer includes the words, "as the media reminds us every day" (line 4), potentially for any of the following reasons EXCEPT

(A) To encourage the reader to acknowledge that some of what she is saying is easily verifiable.

(B) To hint at the fact that the trend she is citing is not likely to change anytime soon.

(C) To call into question the validity of the media as a reputable source of information.

(D) To underscore that the diversion of funds from healthcare to the military is commonplace.

7. Which choice best characterizes the overall structure of the passage?

(A) A problem is stated explicitly, followed by a supporting example, and then some potential causes of the problem are enumerated.

(B) A problem is stated explicitly, and then some potential causes of the problem are enumerated, followed by a supporting example.

(C) Some potential causes of a problem are enumerated, and then the problem is stated explicitly, followed by a supporting example.

(D) A supporting example of a problem is provided, then the problem is stated explicitly, followed by its potential causes.

8. The author's purpose in writing the passage is most likely

(A) To advocate for a radical solution to a problem that she anticipates may not be popular.

(B) To expose the hidden source of a problem and bring its perpetrators to justice.

(C) To explain multiple sources of a problem and its implications.

(D) To discredit an explanation of a problem that she believes to be untenable.

Line   More has been written about *Hamlet* than about any other single piece of literature. Not only has it been commented upon by poets and thinkers such as Coleridge, Goethe, and Freud, but it even has its own journal, *Hamlet Studies*, and every year dozens of articles and books are published on it. Almost every literary movement in some way co-opts the play, and every
(5)   school of criticism undertakes an interpretation of it. *Hamlet* functions as a touchstone: To interpret it convincingly is to validate one's literary theory or approach. Even people who have read little or no Shakespeare know about the play by hearsay, and the character of Hamlet has so much apparent substance that his name signifies a certain kind of person. For many young people he functions as a literary liberator, because he seems so much like their secret
(10)   selves — the person whom they feel themselves to be, unknown to their families and friends.

    While Hamlet is one of the most compelling of literary creations, he is also one of the most elusive. Therefore, it is worth the effort to consider what it is about the play that makes it at once so popular, so compelling, and so puzzling. The play generates its power by staging the characters' struggle with these questions.

(15)   What delays Hamlet in getting his revenge on Claudius? What is Hamlet's tragic flaw, the one quality that leads to the nearly total destruction at the end of the play? Is he crafty, insane, or a little of both? (*Thinking About Shakespeare*, by Kay Stockholder — Page 74)

9. Which of the following quotations from the passage best substantiates the specific claim that *Hamlet* has attained tremendous fame?

(A) "Hamlet functions as a touchstone: To interpret it convincingly is to validate one's literary theory or approach."

(B) "Even people who have read little or no Shakespeare know about the play by hearsay."

(C) "For many young people he functions as a literary liberator, because he seems so much like their secret selves."

(D) "The play generates its power by staging the characters' struggle with these questions."

10. Which of the following statements would the author be LEAST likely to agree with?

(A) *Hamlet* has attained a prominence that outstrips virtually any other work of literature.

(B) In the centuries since it was written, *Hamlet* has inspired interpretation by a wide variety of literary critics.

(C) No single interpretation of *Hamlet,* however convincing, will be likely to satisfy all or even most of the play's audience.

(D) The character of Hamlet is likely to appeal to individuals more as they get older.

11. When the writer describes Hamlet as "elusive," she most likely means

(A) Hard to pin down

(B) Easily led by others

(C) Volatile in his moods

(D) Quick to take offense

12. Which of the following does the author state explicitly about the three questions that end the passage?

(A) These questions are rhetorical and, thus, they are not meant to be answered.

(B) She will attempt to provide satisfactory answers to these questions later in the essay.

(C) The play is so captivating because its characters are forced to wrestle with these questions.

(D) Virtually all previous attempts by literary critics to answer these questions have ended in relative failure.

Line   The Great Depression of the 1930s transformed attitudes. Unemployment rocketed to 25 percent. People's life savings vanished in a tsunami of bank failures. More than half of older Americans were poor. In desperation, people turned to Washington for help, and Washington looked overseas for ideas. President Franklin D. Roosevelt and his advisors,
(5)   including Labor Secretary Frances Perkins, considered an idea known as *social insurance,* which had gained popularity in Europe. The idea was that governments could adapt insurance principles to protect their populations from economic risks. Unlike private insurance arrangements, which are supposed to protect individuals, social insurance programs are supposed to help *all* of society.

(10)   In 1889, German Chancellor Otto von Bismarck pioneered the idea with a system of old-age insurance that required contributions from workers and employers. By the time of the Great Depression, dozens of nations had launched some sort of social insurance effort. U.S. leaders, eager to ease the economic pain engulfing the nation, took a more serious look at social insurance from Europe. Others viewed social insurance as radical and un-American.

(15)   After a lengthy debate, Congress passed the Social Security Act, and President Roosevelt signed it into law on August 14, 1935. The law provided unemployment insurance as well as help for seniors and needy children. Title II of the act, "Federal Old-Age Benefits," created the retirement benefits that many people now see as the essence of Social Security. (*Social Security For Dummies,* by Jonathan Peterson — Page 10)

13. The main purpose of the passage is

(A) To provide a brief history of Social Security, showing where this idea originated, as well as when and why it was implemented.

(B) To convince a potentially skeptical reader that although Social Security was necessary in its time, it is now outdated and should be eliminated.

(C) To critique Social Security as essentially un-American, having emerged from a variety of foreign governments.

(D) To demonstrate the way in which German Chancellor Otto von Bismarck was influential in creating Social Security as we know it today.

14. According to the passage, what distinguishes Social Security and other such programs from private insurance?

(A) Private insurance originated in the U.S.; Social Security originated outside the U.S.

(B) Private insurance is intended to help individuals; Social Security is intended to help society as a whole.

(C) Private insurance dates back to the 1800s; Social Security dates back to the 1900s.

(D) Private insurance provides benefits to people of all ages; Social Security provides benefits only to children and senior citizens.

15. Which of the following, if any, would most likely have objected to President Roosevelt's plan for Social Security?

(A) Both Frances Perkins and Otto von Bismarck

(B) Frances Perkins but not Otto von Bismarck

(C) Otto von Bismarck but not Frances Perkins

(D) Neither Frances Perkins nor Otto von Bismarck

16. Why does the author place the words "Federal Old-Age Benefits" in quotation marks?

(A) To cast doubts that these benefits actually exist within Title II of the Social Security Act.

(B) To imply that even President Roosevelt himself questioned the validity of these benefits.

(C) To indicate a non-literal usage of these words.

(D) To clarify that these precise words are found in Title II of the Social Security Act.

Line   Once, Samanas had travelled through Siddhartha's town, ascetics on a pilgrimage, three skinny, withered men, neither old nor young, with dusty and bloody shoulders, almost naked, scorched by the sun, surrounded by loneliness, strangers and enemies to the world, strangers and lank jackals in the realm of humans. Behind them blew a hot scent of
(5)   quiet passion, of destructive service, of merciless self-denial.

In the evening, after the hour of contemplation, Siddhartha spoke to Govinda: "Early tomorrow morning, my friend, Siddhartha will go to the Samanas. He will become a Samana."

Govinda turned pale, when he heard these words and read the decision in the motionless face of his friend, unstoppable like the arrow shot from the bow. Soon and with the first
(10)   glance, Govinda realized: Now it is beginning, now Siddhartha is taking his own way, now his fate is beginning to sprout, and with his, my own. And he turned pale like a dry banana-skin.

"O Siddhartha," he exclaimed, "will your father permit you to do that?"

Siddhartha looked over as if he was just waking up. Arrow-fast he read in Govinda's soul, read the fear, read the submission.

(15)   "O Govinda," he spoke quietly, "let's not waste words. Tomorrow, at daybreak I will begin the life of the Samanas. Speak no more of it."

Siddhartha entered the chamber, where his father was sitting on a mat of bast, and stepped behind his father and remained standing there, until his father felt that someone was standing behind him. Quoth the Brahman: "Is that you, Siddhartha? Then say what you (20) came to say."

Quoth Siddhartha: "With your permission, my father. I came to tell you that it is my longing to leave your house tomorrow and go to the ascetics. My desire is to become a Samana. May my father not oppose this."

The Brahman fell silent, and remained silent for so long that the stars in the small (25) window wandered and changed their relative positions, 'ere the silence was broken. Silent and motionless stood the son with his arms folded, silent and motionless sat the father on the mat, and the stars traced their paths in the sky. Then spoke the father: "Not proper it is for a Brahman to speak harsh and angry words. But indignation is in my heart. I wish not to hear this request for a second time from your mouth."

(30)   Slowly, the Brahman rose; Siddhartha stood silently, his arms folded.

"What are you waiting for?" asked the father.

Quoth Siddhartha: "You know what." (*Siddhartha*, by Hermann Hesse — Pages 9–10)

17. Which of the following best summarizes the passage?

(A) A young man gives his friend information about himself which changes the friend's opinion of him, and then the friend tells his father about it.

(B) A young man and his friend have a difference of opinion, and then the young man seeks the counsel of his father to help resolve it.

(C) A young man informs his friend of a momentous decision, and then requests that his father support that decision.

(D) A young man hears distressing news from his friend, and then tells his father the news.

18. Which of the following quotations from the passage provides the best evidence that Govinda is less ready than his friend for his life to change?

(A) Lines 6–7: "Early tomorrow morning, my friend, Siddhartha will go to the Samanas. He will become a Samana."

(B) Lines 10–11: "Now it is beginning, now Siddhartha is taking his own way, now his fate is beginning to sprout, and with his, my own. And he turned pale like a dry banana-skin."

(C) Line 12: "'O Siddhartha,' he exclaimed, 'will your father permit you to do that?'"

(D) Lines 24–25: "The Brahman fell silent, and remained silent for so long that the stars in the small window wandered and changed their relative positions, 'ere the silence was broken."

19. Siddhartha's announcements could best be described as

(A) Modest

(B) Hasty

(C) Conflicted

(D) Determined

20. Which of the following best paraphrases Siddhartha's apparent meaning when he says, "You know what"?

    (A) "You have asked me not to speak of something, and I am not speaking about it."

    (B) "You have asked me not to make the same request twice, but I still want what I asked for."

    (C) "You have forced me to request your permission, but I do not need it."

    (D) "You have told me something that I don't want to hear, and I am pretending that I didn't hear it."

Fill-in-the-Blank Questions

1. Surprisingly, Rebecca did not _____ her professor's criticism; instead, she took it in the most constructive possible way.

    (A) adapt to

    (B) insist upon

    (C) react with

    (D) recoil at

2. Finding neither a diplomatic nor a military solution _____, the Pentagon committee sought another resolution to their dilemma.

    (A) crucial

    (B) feasible

    (C) lamentable

    (D) nominal

3. They describe his uncle as a _____ man, not normally given to small talk, and even less inclined to express personal thoughts or feelings.

    (A) confrontational

    (B) figurative

    (C) suggestive

    (D) taciturn

4. Although the band was dedicated and sincere, some of the audience experienced their music as _____, and found it necessary to leave before the show was over.

    (A) cacophony

    (B) deterrence

    (C) fortitude

    (D) obstinacy

5. The vice president concurred that our firing of Caldwell was justified, but added that our most _____ error had been hiring him in the first place.

(A) auspicious

(B) egregious

(C) impenetrable

(D) obligatory

6. Because his promotion from copywriter to journalist took more than 15 years, he often joked that success was more a matter of _____ than talent.

(A) leniency

(B) novelty

(C) scholarship

(D) tenacity

7. While doing her best to avoid _____, she said that her experience as a nun had led her to believe that religious people of all faiths tended to have certain characteristics in common.

(A) castigating

(B) generalizing

(C) obstruction

(D) unruliness

8. The astronomer _____ her students toward the profession not with guarantees of great monetary gain but rather with tantalizing accounts of a boundless universe waiting to be explored.

(A) derided

(B) enticed

(C) infiltrated

(D) permeated

9. The speaker was known to _____ his words with hand gestures, at times almost pounding the lectern as he spoke.

(A) accentuate

(B) excoriate

(C) pummel

(D) rectify

10. While Washington D.C. may appear to be quite _____ by day — especially for tourists who choose to focus exclusively on its monuments and museums — by night it's just as vibrant and eclectic as any other American city.

(A) benevolent

(B) dissipated

(C) malleable

(D) venerable

Answers to Reading Passage Questions

1. **A.** The passage states that the Greeks thought of artists principally as craftsmen, which is different from how modern people think of them. It also states that the modern understanding of Greek art has been influenced by centuries of thought about it by other cultures. Both of these statements reinforce the central idea that modern people perceive Greek art differently from how it was understood at the time, so Answer A is correct.

2. **D.** The passage states that the Greeks thought of artists as "artisans: good with their hands and not much else." This way of thinking most closely tracks with how modern people think of a skilled plumber or electrician, so Answer D is correct.

3. **B.** According to the passage, Pheidias became successful at his craft, and this led to fame, fortune, and political power. Thus, Answer B is correct.

4. **A.** The writer is calling into question not only what Greek art is, but also whether it is specifically Greek and specifically art in the way the reader assumes. Thus, he's focusing on the possible assumptions and fallacies that the reader brings to the passage, hoping to adjust this understanding, so Answer A is correct.

5. **A.** According to the passage, one expectation of governments is to help fight disease. However, pressures on the government to spend money in other ways limit resources available for this purpose. Also, the existence of refugee populations creates conditions that allow diseases to flourish. These two political realities increase the danger of epidemics; therefore, Answer A is correct.

6. **C.** The writer's use of the words, "as the media reminds us every day," encourages the reader to verify what she is saying, so Answer A is incorrect. It tells the reader that what she is describing is commonplace, so Answer D is incorrect. And it also hints at the fact that the trend is not likely to change, so Answer B is incorrect. In contrast, she doesn't call the validity of the media into question, so Answer C is correct.

7. **C.** The passage opens with two explanations of why government can fall short of providing adequate conditions for populations to remain healthy. After this, the danger of epidemics of infectious diseases is explicitly stated, followed by an example of such an emergency in 2007. Thus, the passage gives some potential causes of a problem, then states the problem, and finally provides a supporting example, so Answer C is correct.

8. **C.** The passage gives at least two separate sources of the problem — why epidemics of infectious diseases can overwhelm the ability of governments to prevent them. It also discusses the implications of this problem and provides an example. Answer C is correct.

9. **B.** The fame that *Hamlet* has attained is best substantiated by the quote, "Even people who have read little or no Shakespeare know about the play by hearsay." That is, the play is so famous that even if you haven't read it, you've probably heard about it. Thus, Answer B is correct.

10. **D.** In the last sentence of the first paragraph, the author states that, "For many young people [Hamlet] functions as a literary liberator, the person whom they feel themselves to be." Thus, he states directly that young people identify with the character of Hamlet, so Answer D is correct.

11. A. In the second paragraph, the writer is discussing the effect of the character of Hamlet on his audience, rather than his character as seen by others in the play. Thus, when he describes Hamlet as "elusive," he means that he is hard for his audience to pin down; therefore, Answer A is correct.

12. C. The last sentence before the three questions is, "The play generates its power by staging the characters' struggle with these questions." Here, the writer explicitly states that the play is so captivating because its characters are forced to wrestle with these questions, so Answer C is correct.

13. A. The passage is an informational narrative, giving the history of the origins of Social Security. There is little to indicate that the passage is persuasive, ruling out Answers B and C. And the mention of Bismarck is a supporting detail rather than central to the topic, so Answer D is incorrect. Thus, Answer A is correct.

14. B. The passage identifies Social Security as a form of social insurance, and states, "Unlike private insurance arrangements, which are supposed to protect individuals, social insurance programs are supposed to help *all* of society." Answer B is correct.

15. D. The passage states that Frances Perkins considered an idea for social insurance, which became known as Social Security, so she wouldn't have objected. It also states that Otto von Bismarck pioneered the idea in Germany, so he wouldn't have objected. Therefore, Answer D is correct.

16. D. Quotation marks have a variety of possible uses, but in this case they are used to clarify precise wording; therefore, Answer D is correct.

17. C. The key aspect of the passage is Siddhartha's decision to join the Samanas. He tells his friend Govinda of his decision, and then requests permission from his father. Therefore, Answer C is correct.

18. B. "Siddhartha was going his own way; his destiny was beginning to unfold itself, and with his destiny, his own. And he became as pale as a dried banana skin." When Govinda hears Siddhartha's decision, he realizes that his own life will now change, and becomes pale at the realization. Answer B is correct.

19. D. Throughout conversations with both Govinda and his own father, Siddhartha remains unwavering in his determination to follow through with his decision, so Answer D is correct.

20. B. Siddhartha asks for his father's permission out of respect, but when he says, "I trust my father will not object," he implies that he will not take no for an answer. When his father replies, "I should not like to hear you make this request a second time," Siddhartha honors the words but in his silence makes clear that his request still stands. Answer B is correct.

Answers to Fill-in-the-Blank Questions

1. **D.** In the second half of the sentence, Rebecca's reaction is positive. In contrast, she didn't respond negatively, so Answer D is correct.

2. **B.** The sentence requires a word like *practical* or *workable*. The word *feasible* is a synonym for these words, so Answer B is correct.

3. **D.** The sentence requires a word that means *quiet* or *understated*. The word *taciturn* captures this meaning, so Answer D is correct.

4. **A.** The sentence implies that the music was loud and annoying. The noun *cacophony* implies these characteristics, so Answer A is correct.

5. **B.** Apparently, hiring Caldwell was a very bad idea, so this error was *severe* or *outrageous*. The word *egregious* captures this meaning, so Answer B is correct.

6. **D.** Working for 15 years requires *perseverance* or *fortitude*. The word *tenacity* is a synonym for these words, so Answer D is correct.

7. **B.** The speaker is trying to avoid *oversimplifying* or *prejudging*. The word *generalizing* captures this meaning more closely than any of the other words, so Answer B is correct.

8. **B.** The astronomer's "tantalizing accounts" of astronomy *tempted* or *allured* her students. The word *enticed* captures this essence, so Answer B is correct.

9. **A.** The speaker seems to *stress* or *emphasize* his words when he speaks. The word *accentuate* is close in meaning, so Answer A is correct.

10. **D.** The second part of the sentence describes Washington D.C. as fun and exciting. In contrast, the monuments and museums are implied to be *stuffy* or *boring*. The closest meaning here is *venerable*, which means "*respectable*" and "*time-honored*," so Answer D is correct.

3

The Writing Test

IN THIS PART . . .

Clarify the rules of grammar most often tested on the ACCUPLACER.

Understand usage questions that test word choice, style, and tone.

Use the skills in this part of the book to answer a variety of practice questions.

Chapter **7**
Standard English Grammar Conventions

Grammar is a set of rules that allow words to function as sentences that convey ideas.

If you already speak and read English (going out on a limb here, but I'm guessing you do), then you've already absorbed a ton of basic grammatical rules, even if you can't quite explain how those rules work. I call this *inner ear* knowledge — the way a sentence either does or doesn't sound right when you read it to yourself.

On the Writing section of the ACCUPLACER, 9 to 11 of the 25 questions you'll face test your knowledge of grammar. To do well, it helps to translate your inner ear knowledge to an explicit understanding of the when, how, and why of grammar. That's what this chapter is all about.

Review of English Sentence Grammar

Grammar describes the way that sentences function or, in some cases, break down. This understanding is as important to writing as the study of human anatomy and physiology are to medicine. If you're not sure how the body is supposed to function when it's healthy, you won't be able to treat it when it's ill or injured. In a similar way, if you're not quite sure what a good sentence looks like, you may have trouble correcting bad sentences.

In this section, I give you a brief introduction to English grammar. Here, I identify the elements that every sentence must have — namely, the subject and the verb. I focus on how sentences are built from smaller components such as words, phrases, and clauses. I also discuss the ways in which words and phrases are used in sentences — that is, as nouns, pronouns, adjectives, and so forth.

Simple sentences

The simplest form of sentence is conveniently called a *simple sentence*. What makes a simple sentence simple is that it has just one subject. In this section, I show you some simple and then a few not-so-simple simple sentences.

The subject and the verb

Every simple sentence has a *subject* and a *verb*.

> **The subject tells who or what the sentence is about.**

> **The verb tells what the subject is doing (or in some cases, how it's being).**

For example,

> *Geoffrey* **reads.** *Caitlin* **dances.** *Children* **play.**

Here, the word in *italics* is the *subject*, and the word in **bold** is the **verb.** The subject and the verb are the core of every sentence. When you get good at finding them, you can begin to untangle a variety of grammatical problems that the ACCUPLACER will throw at you.

The subject of a simple sentence can take three forms: a noun, a pronoun, or a noun phrase.

A *noun* is a word that represents a person, a place, or a thing — for example, Ralph, New York, or chair. A *proper noun* is the name of a specific person or place, and is always capitalized. In the preceding example sentences, the subject is a noun.

A *pronoun* can replace a noun as the subject of a sentence. For example,

> *He* **reads.** *She* **dances.** *They* **play.**

A *noun phrase* is a string of two or more words that can also replace a noun as the subject of a simple sentence. For example,

> The <u>boy</u> **reads.** That talented <u>girl</u> **dances.** The two happy <u>children</u> on the rollercoaster **play.**

Every noun phrase includes a noun at its core, which is the most important word in that phrase. This word is called the <u>head</u> of the phrase, which I have <u>underlined</u> in each of the example sentences.

Modifiers in noun phrases

In the previous section, you worked with the noun phrase *the two happy children on the rollercoaster*:

> the two happy <u>children</u> on the rollercoaster

Here, I underline the head of this phrase, *children*, which is the most important word in the phrase. Everything else in the phrase helps to *modify* this noun:

> *The* and *two* are determiners. *Determiners* give non-descriptive information about the noun.

> *Happy* is an adjective. *Adjectives* give descriptive information about the noun.

> *On the rollercoaster* is a prepositional phrase beginning with the preposition *on*. *Prepositional phrases* can provide information about the location of a noun.

Notice that these three types of modifiers are all clustered around the noun: first the determiners, then the adjective, and finally the prepositional phrase. Here's another example of a noun phrase, again as the subject of a sentence:

That big beautiful red <u>car</u> in the driveway of my dad's house **is** mine.

In this case, the noun *car* is the head of the phrase. All the other words help to modify this noun:

Determiner: *that*

Adjectives: *big, beautiful, red*

Prepositional phrase: *in the driveway of my dad's house*

Here's another example:

That big beautiful red <u>car</u> in the driveway of my dad's house and the shiny black <u>motorcycle</u> on the street **are** mine.

In this case, the *compound subject* now includes two key nouns — *car* and *motorcycle*. Notice here how each of the modifiers — the determiners (*that* and *the*), the adjectives (*big, beautiful, red, shiny,* and *black*), and the prepositional phrases (*in the driveway of my dad's house* and *on the street*) — clusters around the noun that it modifies.

Compound sentences

A *compound sentence* has more than one subject, and each subject has its own verb. Another way to say this is that a compound sentence has more than one clause. If you're like most students, your blood pressure rose significantly the moment you read the word *clause*. Breathe deeply for a moment as I reassure you that this concept isn't as frightening as you may believe. In this section, I clarify these anxiety-producing and eye-glazing concepts.

What's a clause and why do I care?

A compound sentence has more than one subject, each with its own verb. This isn't difficult to understand. For example,

I like apples, and you like oranges.

You can probably see that this sentence is made up of two simple sentences ("I like apples." and "You like oranges.") that are attached by a comma plus the conjunction *and*. These are two examples of clauses. **A clause is a string of words that can (sometimes with minor adjustments) stand on its own as a sentence.**

That's it! So now that you know that clauses don't bite, here's another example of a compound sentence made up of two longer clauses:

My wife and kids wanted to spend the entire summer at the lake, but I surprised them with a trip to Aruba.

This time, the two clauses are longer, but the form of the sentence is the same: two clauses, connected by a conjunction (*but*) plus a comma to make the sentence more readable.

Building compound sentences using the FANBOYS conjunctions

English has seven conjunctions that are commonly used to build compound sentences: *for, and, nor, but, or, yet,* and *so*. You can remember these seven words using the mnemonic *FANBOYS*. For example, in the previous section, I show you two examples of compound sentences that include the conjunctions *and* and *but*. Here are examples using the other five:

> I believe in this prophecy, <u>for</u> it is so written.

> Claude certainly isn't rich, <u>nor</u> is he really French.

> We can drive you to the airport, <u>or</u> you can take the train.

> The dinner was delicious, <u>yet</u> the recipe was very easy to follow.

> Jack didn't want to tell Lisa the truth, <u>so</u> he gave her a series of excuses.

As you can see, a compound sentence may or may not include a comma. This is mostly a matter of taste — if you feel the need to pause, place a comma before the conjunction; otherwise, leave it out.

For example, shorter compound sentences usually don't need a comma:

> I love you <u>and</u> you love me.

However, longer sentences often benefit from a comma to let the reader know when to mentally pause and take a breath:

> I'm rebuilding the engine of a 1965 Ford Mustang in my best friend's garage, <u>and</u> we're looking online for a steering wheel and some engine parts to use for the project.

Using a semicolon to build compound sentences

You can remove the comma and conjunction from a compound sentence and replace it with a semicolon (;). For example,

> I gave you my word; let's try to make the best of a difficult situation.

Using a semicolon in this way is completely grammatical; it's just a bit formal. See what I mean? It forces the reader to stop mid-sentence and then start again, which breaks the flow of the writing. In some cases — especially when you're discussing a serious topic or making an important point — this can be a good stylistic choice.

Complex sentences

A *complex sentence* has at least one subordinate clause. (Stay with me now.) In this section, I explain the entire subordinate clause thing.

Main and subordinate clauses

So far, all the clauses that you've seen in this section have been main clauses (or independent clauses). A *main clause* can stand on its own as a sentence. For example,

> <u>I like apples</u>, and <u>you like oranges</u>.

You can easily rewrite this sentence as a pair of sentences:

I like apples. You like oranges.

The seven short *FANBOYS* conjunctions (*for, and, nor, but, or, yet,* and *so*) are used to connect main clauses. For this reason, they are also called *coordinating conjunctions*.

In contrast, longer conjunctions are called *subordinating conjunctions*. For example,

I like apples; <u>however</u>, you like oranges.

Here, the conjunction *however* connects the two clauses, allowing them to function within a single sentence. But even though this sentence is short, the longer conjunction requires a semicolon to end the first clause, and a comma after the conjunction.

This is a typical construction required in formal English. And on the ACCUPLACER, you'll probably see at least one question that tests you on it. Here are a few more words that require this type of punctuation:

The parties have signed the contract; <u>therefore</u>, I believe we should honor it.

We're surprising our parents with a party; <u>consequently</u>, my sister sent out invitations.

You don't have to read the textbook; <u>nevertheless</u>, you do have to pass the exam.

If you like skiing, we could go to Vermont; <u>otherwise</u>, maybe Florida would be better.

Subordinate clauses with relative pronouns

Relative pronouns are words that allow you to connect a pair of related sentences together. For example, consider these two sentences:

The boss gave the promotion to Elaine. Elaine has been running the main office for three years.

You can avoid repeating the name *Elaine* by blending the sentences together using the relative pronoun *who*:

The boss gave the promotion to Elaine, <u>*who*</u> has been running the main office for three years.

Here, the relative pronoun *who* refers back to the noun *Elaine* and replaces it. As you can see, when connecting two sentences using a relative pronoun, a comma is needed to replace the period at the end of the first sentence.

When the noun you want to replace is a thing rather than a person, use *which* instead of *who*:

I spent five minutes looking for my glasses, <u>*which*</u> were on my head the entire time.

In this case, the relative pronoun *which* refers back to the noun phrase *my glasses*. Again, a comma is needed just before the relative pronoun.

Two other common relative pronouns are *where* and *when*, which are used to refer back to a place and a time, respectively:

My grandparents went to Niagara Falls, <u>*where*</u> they had spent their honeymoon.

We hiked up the trail until noon, <u>*when*</u> we stopped to rest and have lunch.

Nouns, Pronouns, and Possessives

In the previous section, you discover how a pronoun can take the place of a noun or a noun phrase. Pronouns have a variety of grammatical forms that you need to be aware of. Additionally, both nouns and pronouns have possessive forms (such as *John's* and *its*) to describe relationships and ownership. In this section, you focus on ACCUPLACER questions that address these issues.

Pronoun-antecedent agreement

Recall that a *pronoun* replaces a noun (or a noun phrase or noun clause) in a sentence. For example,

Geoffrey was surprised.	*He* was surprised.
The boy on the train showed us a magic trick.	*He* showed us a magic trick.
The boy who showed us that trick is from Sweden.	*He* is from Sweden.

In each of these examples, the pronoun *he* replaces a longer, more complicated element of a sentence.

The element that the pronoun refers back to is called the *antecedent*. In many cases, the antecedent affects your choice of pronoun. This issue is called *pronoun-antecedent agreement*.

On the ACCUPLACER, you'll most likely be tested on five varieties of pronouns:

She/her/hers — Antecedent is singular and female:

My *grandmother* is visiting from Boston, and *she* doesn't have a car, so I lent *her* mine.

He/him/his — Antecedent is singular and male:

The *man* at the hotel said *he* wasn't sure but promised to ask *his* supervisor.

It/its — Antecedent is a singular inanimate (an object or an idea):

That *book* is old, so *its* cover is missing, but *it* still contains useful information.

He or she / him or her / his or her / his or hers — Antecedent is singular and could be either male or female:

Next semester, I hope the *professor* makes *his* or *her* exams a little easier to pass.

They/them/their/theirs — Antecedent is plural:

Those *kids* can leave *their* lunches in the refrigerator while *they* go swimming.

As you can see, when sentences are relatively uncomplicated, choosing the correct pronoun isn't too difficult. On the ACCUPLACER, your first task is to figure out exactly what the antecedent is, so you can pick the right answer. For example,

EXAMPLE

The miner panned for any traces of gold that might be there, even though no gold had been found in that region of California, because <u>he</u> could be sold for the greatest price.

(A) (as it is now)

(B) he or she

(C) it

(D) they

As it stands, the antecedent of the pronoun *he* is the miner; however, this makes no sense in the context of the sentence, so Answer A is incorrect. Answer B is incorrect for similar reasons. The pronoun *they* could refer to *traces* of gold; however, what could be sold for the greatest price isn't *traces*, but rather *gold*, which rules out Answer D. Answer C is correct, because the antecedent of the pronoun is *gold*, so the pronoun *it* makes the most sense.

Pronoun clarity

A sentence lacking in *pronoun clarity* includes a pronoun whose antecedent is uncertain or ambiguous. (For a quick review of pronouns and antecedents, flip to the previous section, "Pronoun-antecedent agreement.") For example,

EXAMPLE

Grace told her best friend, Eleanor, the unfortunate news that <u>her son</u> had been suspended from school.

(A) (as it is now)

(B) Graces son

(C) Grace's son

(D) they're son

This sentence is confusing because the pronoun *her* could refer to either Grace or Eleanor. Answers B and C both supply this information, but Answer B incorrectly uses a plural (*Graces*) instead of a possessive (*Grace's*), as in Answer C. Answer D incorrectly uses the contraction *they're* instead of the possessive *their*. The correct answer is C.

Distinguishing plural and possessive nouns

In spoken English, adding an −*s* sound to the end of a noun can change that noun in a variety of ways, making it either plural or possessive, or both. In written English, these forms are all distinguished by the placement of the letter *s* and, in some cases, the appearance of an apostrophe (').

For example, here's how the noun *boy* is changed in these various forms:

Singular noun: *boy*

Plural noun: *boys*

Singular possessive: *boy's*

Plural possessive: *boys'*

Most students are fairly comfortable using the singular and plural forms of nouns:

The **boy** owns a dog. (one boy)

The **boys** own a dog. (more than one boy)

The possessive form includes an apostrophe *before* the *s*:

The **boy's** dog is named Samson. (dog belonging to one boy)

The possessive plural form can get confusing, however:

The **boys'** dog is named Samson. (dog belonging to more than one boy)

In this case, the apostrophe appears *after* the s, signaling that the dog belongs to more than one boy.

EXAMPLE

Sheri was allowed to use <u>her parents car</u> to drive to school while they were in Israel.

(A) (as it is now)

(B) her parent's car

(C) her parents' car

(D) her parents's car

The sentence refers to the parents as *they*, which is plural, so you need to use the plural possessive to refer to the car that belongs to them; this rules out Answers A and B. Answer D, however, is simply an incorrect form, so Answer C is correct.

Using possessive determiners

Earlier in this section, I discuss how pronouns (such as *you, it,* and *they*) can be used to replace nouns in a sentence.

Like nouns, pronouns also have a possessive form. For example,

> You own a car. *Your* car is a Toyota.
>
> The cat has a collar. *Its* collar is blue.
>
> The teachers had a meeting. *Their* meeting ran late.

Words like *your, its,* and *their* are called *possessive determiners*. As you can see, unlike possessive nouns, which include an apostrophe, possessive determiners *never include an apostrophe.*

WARNING

Although none of these three words has an apostrophe, each has a sound-alike that does have an apostrophe:

> You're means *you are.*
>
> It's means *it is.*
>
> They're means *they are.* (Additionally, *there* means *that place.*)

Be sure to keep these words clear — this is an area that the ACCUPLACER will probably test you on.

Practicing Perfect Punctuation

I'll bet you already know that *punctuation marks* are used to make writing clearer and more effective. Moreover, in many situations, the use of punctuation marks is explicitly defined by a variety of grammatical rules. In this section, I set you up for success on the ACCUPLACER by explaining how, when, and why to use punctuation.

End-of sentence punctuation

End-of-sentence punctuation marks include the period (.), question mark (?), and exclamation point (!). Most students feel pretty comfortable with them, but here's what they're used for:

A *period* (.) ends a statement (a sentence that provides information) or a command (a sentence that gives an instruction):

George Washington crossed the Delaware in 1776.

Don't go in the garage.

A *question mark* (?) ends a question (a sentence that requests information):

What is the capital of Wisconsin?

An *exclamation point* (!) ends an exclamation (a sentence that expresses a strong emotion):

The dog is eating your birthday cake!

Presenting information using a colon

The *colon (:)* is used to present information that follows in two key ways. First, the colon can present a list of items. For example,

Before we vote on the proposal, we should consult with three people: our bookkeeper, our accountant, and our lawyer.

Another way the colon is often used is to present a single piece of information in a dramatic fashion. For example,

Although his life in Los Angeles appeared ideal, there was just one problem: He was miserable.

These two uses for the colon are essentially the *only* ways that this punctuation mark gets used.

Separating clauses with a semicolon

Earlier in this chapter, in the section, "Using a semicolon to build compound sentences," I show you how a semicolon is used to separate a pair of clauses in two distinct ways:

Separating two clauses *without* a conjunction (usually, two related ideas, creating a dramatic effect). For example,

The letter arrived on Friday; I opened it with trembling fingers.

Separating two clauses *with* a subordinating (long) conjunction (note the presence of a comma after the conjunction). For example,

The letter arrived on Friday; however, I was too nervous to open it.

In contemporary writing, these are by far the most common uses for a semicolon, and the only ones that you should expect to see on the ACCUPLACER.

Using commas to separate items in a series

When you list more than two items in a phrase, you need to place commas after all but the last item. For example, consider these two sentences:

Calvin and Andrew are on the basketball team.

David, Calvin, and Andrew are on the basketball team.

In the first sentence, no commas are needed because the conjunction *and* connects the two nouns *Calvin* and *Andrew*. In the second sentence, however, commas are placed after the names *David* and *Calvin*, making the sentence easier to read.

WARNING

Some older texts omit the second comma. This is an outdated convention, and is no longer used in standard English writing. To see why, look at this sentence:

David, Calvin and Andrew are on the basketball team.   WRONG!

You could read this sentence incorrectly, as if you are talking to David and telling him that Calvin and Andrew are on the basketball team.

Separating items in a series with commas is even more important when the items are individual phrases rather than words. For example,

EXAMPLE

The senator from Wyoming the two representatives from New York and the advisors from the United Nations met to discuss the proposal.

(A) (as it is now)

(B) The senator from Wyoming, the two representatives from New York and the advisors from the United Nations met to discuss the proposal.

(C) The senator from Wyoming, the two representatives from New York, and the advisors from the United Nations met to discuss the proposal.

(D) The senator from Wyoming, the two representatives from New York, and the advisors from the United Nations, met to discuss the proposal.

Without commas, this long sentence is confusing, not to mention incorrect, which rules out Answer A. Answer B is missing the second comma, so this is also incorrect. And in Answer D, the extra comma after *Nations* is not needed. Answer C is correct.

Setting off parenthetical elements

A *parenthetical element* in a sentence (also called a *non-restrictive element*) is a part of the sentence that could be removed without changing the essential grammatical structure of the sentence. Parenthetical elements are set off from the rest of the sentence by a pair of punctuation marks.

To see how this works, consider the following pair of sentences:

Judge Patricia Garcia, *who was appointed by the governor almost twelve years ago*, is now intending to retire from the bench.

Judge Patricia Garcia is now intending to retire from the bench.

The first sentence includes a parenthetical element (in *italics*) that's set off from the rest of the sentence by a pair of commas. The second sentence excludes this element, but still makes grammatical sense, even though it lacks some information.

Parenthetical elements are usually set off from the rest of a sentence by a pair of commas. You can think of parenthetical elements as a detour: The first comma directs you off the main highway and the second comma directs you back onto it.

Two other relatively common ways to punctuate parentheticals are with dashes or parentheses:

> **Judge Patricia Garcia — *who was appointed by the governor almost twelve years ago* — is now intending to retire from the bench.**

> **Judge Patricia Garcia (*who was appointed by the governor almost twelve years ago*) is now intending to retire from the bench.**

English grammar very rarely has rules that can never be broken, but here's one you can take as virtually inviolable: A parenthetical element *must* begin and end with the same type of separator — that is, a pair of commas, dashes, or parentheses.

For example,

EXAMPLE

Like many similar large animals, the Bengal tiger — although technically the most numerous <u>subspecies of tiger in Asia, is threatened</u> by a variety of factors and currently considered endangered.

(A) (as it is now)

(B) subspecies of tiger in Asia; is threatened

(C) subspecies of tiger in Asia — is threatened

(D) subspecies of tiger in Asia is threatened

Here, the parenthetical element separates the subject of the sentence (*the Bengal tiger*) from the verb (*is threatened*). The first separator is a dash, so the second must also be a dash, and so you can rule out Answers A, B, and D; therefore, the correct answer is C.

Setting off dangling modifiers with a comma

A *dangling modifier* is a set of words at the beginning of a sentence that appears before the subject and modifies that subject. The sentence and modifies a noun that appears later in the sentence. Problems with dangling modifiers take a little practice to see, but once you're used to them, you'll see them coming a mile away.

For example, can you see what's wrong with this sentence?

> **Waiting for the bus, the bell in the tower rang three times.**

As it stands, the sentence seems to be implying that the bell was waiting for the bus when it rang. Yet you can probably understand that what the sentence is trying to say is more like this:

> **Waiting for the bus, I heard the bell in the tower ring three times.**

The phrase *waiting for the bus* is a modifier, providing information about the subject of the sentence (*I*). When a modifier appears at the beginning of the sentence, two things usually need to be done to make the sentence grammatical:

> Separate off the modifier with a comma.

> Introduce the subject *immediately* after the comma.

EXAMPLE

While baking a cake for her son's birthday, it turned out that Alissa didn't have enough eggs.

(A) (as it is now)

(B) the eggs turned out to be not enough.

(C) Alissa's eggs turned out to be insufficient.

(D) Alissa found that she didn't have enough eggs.

The opening of the sentence, *While baking a cake for her son's birthday,* is a dangling modifier that provides information about a woman who is baking a cake. To be correct, this construction needs to end with a comma, and then immediately identify this woman. Answers A, B, and C all fail to do this, so Answer D is correct.

Setting off adverbial adjuncts with a comma at the beginning of a sentence

An *adverbial adjunct* is a word or short phrase that helps to identify where or when an action takes place. For example:

> *In springtime,* we *often* eat our lunch *in the park.*

This sentence has three adverbial adjuncts: *in springtime*, *often*, and *in the park*.

When a sentence begins with adverbial adjunct, use a comma to set it off from the rest of the sentence; otherwise, don't use a comma.

EXAMPLE

After the holidays were <u>finished Tom promised to study more diligently in the weeks ahead</u>.

(A) (as it is now)

(B) finished, Tom promised to study more diligently in the weeks ahead

(C) finished Tom promised to study more diligently, in the weeks ahead

(D) finished, Tom promised to study more diligently, in the weeks ahead

This sentence has two adverbial adjuncts related to time: *After the holidays were finished* and *in the weeks ahead*. The first adverbial adjunct begins the sentence, so it needs to be set off by a comma; in contrast, the second adverbial adjunct does not. Therefore, Answer B is correct.

Other Key Grammar Issues

In this section, you round out your grammar skills with a few more important topics that are likely to appear on your ACCUPLACER.

Subject-verb agreement

The subject of every sentence is either singular or plural. This distinction is important on the ACCUPLACER because the subject dictates *subject-verb agreement*:

> A *singular* subject
>
> Denotes just one person, place, or thing.
>
> Requires a *singular* verb.

A *plural* subject

Denotes more than one person, place, or thing.

Requires a *plural* verb.

For example, consider these two sentences:

That talented <u>girl</u> **dances.**   Those talented <u>girls</u> **dance.**

In the first sentence, the subject is singular, so the verb *dances* is singular. In the second sentence, the subject is plural, so the verb *dance* is also plural.

WARNING

One peculiarity of English is that in all but rare exceptions, the singular verb includes an −s or −es that gets dropped from the plural form. (You'd think it would be the reverse, right?)

But don't let this confuse you by overthinking it. With relatively uncomplicated subjects, you shouldn't have any trouble spotting errors in subject-verb agreement. For example, read this sentence to yourself and decide which word is best:

The two happy children on the rollercoaster **(is/are) laughing.**

If you find this problem fairly easy, don't worry about the following explanation:

The subject here is the entire first part of the sentence in italics. This subject includes two nouns: *children* and *rollercoaster*. One of these is the head of the phrase, and to find this, you need to ask yourself: "Who or what is laughing?" The <u>children</u> are laughing, and this is a plural noun, so the subject is plural; therefore, the plural verb **are** is needed for subject-verb agreement.

One trick the ACCUPLACER folks like to play on their unsuspecting victims is to complicate a sentence by inserting a parenthetical element between the subject and the verb. A *parenthetical element* is a part of the sentence that could be removed without changing the essential grammatical structure of the sentence. For example,

The two <u>children</u> on the rollercoaster — Jacob, the boy in the blue sweater, and his younger brother, Benjamin, in the white T-shirt — **are laughing.**

Here, the part of the sentence between the two dashes (—) is a parenthetical element. As you can see, the fundamental structure of the sentence isn't altered by the parenthetical element: The head of the subject phrase is still <u>children,</u> and the verb is still *are laughing.* But the inclusion of the parenthetical element may, in practice, make it harder to identify subject-verb agreement issues.

EXAMPLE

Empathy, the ability to intuit the emotional states of other people in different circumstances, <u>are some of the most</u> important qualities that a good social worker possesses.

(A) (as it is now)

(B) are among of the most

(C) is some of the most

(D) is among the most

The subject of the sentence is *empathy*, which is singular, so subject-verb agreement requires a singular verb; this rules out Answers A and B. Answer C incorrectly uses *some* to refer to this singular noun, so Answer D is correct.

Parallel structure

In some cases, a simple sentence — that is, a sentence with a single subject — has more than one verb. For example,

> *Their babysitter* **carries** the children upstairs, **tucks** them into their beds, **reads** them a story, **sings** to them, and then **turns off** the light.

In this sentence, the subject is simply *the babysitter*, but there are five verbs: *carries, tucks, reads, sings,* and *turns off*. As you can see, subject-verb agreement is necessary for all five of these verbs, and if you change the subject from singular to plural, the verbs must also change:

> *Their older brothers* **carry** the children upstairs, **tuck** them into their beds, **read** them a story, **sing** to them, and then **turn off** the light.

This form is called *parallel construction*, which provides a structural framework that help make long sentences like these easier to read.

In some cases, parallel structure allows you to write the first part of a compound verb (a verb consisting of more than one word) only once rather than repeating it. For example,

> They ***have been waiting*** for over an hour and ***growing*** more impatient every minute.

In this sentence, the two verbs are *have been waiting* and *have been growing*. As you can see, however, parallel construction allows you to drop the words *have been* in the second instance.

EXAMPLE

Our grandparents had stopped at a restaurant on the road and <u>eating lunch</u>, so they weren't very hungry when they arrived.

(A) (as it is now)

(B) are eating lunch

(C) eaten lunch

(D) would have eaten lunch

The parallel construction begins with the verb *had stopped* and needs to continue in a similar way, with a verb form that also fits after the word *had* and makes sense with the rest of the sentence. The words *eaten lunch* fit, because *had eaten lunch* is correct, and this also makes sense to explain why they weren't hungry, so Answer C is correct.

Understanding shifts in verb forms

The verb in a sentence tells the action that takes place. Depending on the *verb form*, the time frame of this action can shift. Officially, English has a total of 12 verb forms. For example, consider these three sentences:

> Past: The party ***was*** fun.

> Present: The party ***is*** fun.

> Future: The party ***will be*** fun.

In each case, changing the verb form shifts the time that the sentence (and the party) takes place.

Progressive verbs use forms of the auxiliary verb *to be* plus a present participle (for example, am talking, was eating, were studying). These verbs talk about ongoing action that is already in progress. For example,

Past Progressive: The band ***was playing*** when I arrived.

Present Progressive: The band ***is playing*** when I arrive.

Future Progressive: The band ***will be playing*** when I arrive.

Perfect verbs use forms of the auxiliary verb *to have* plus a past participle (have stolen, had taken, will have made). These verbs denote action that is completed. For example,

Past Perfect: The guests ***had eaten*** when she made a toast.

Present Perfect: The guests ***have eaten*** when she makes a toast.

Future Perfect: The guests ***will have eaten*** when she makes a toast.

Finally, *perfect progressive verbs* use both auxiliary verbs plus a past present participle (have been looking, had been reading, will have been singing). These verbs denote a subtle mixture of meaning from both progressive and perfect verbs. For example,

Past Perfect Progressive: We ***had been dancing*** for an hour when the lights were dimmed.

Present Perfect Progressive: We ***have been dancing*** for an hour when the lights are dimmed.

Future Perfect Progressive: We ***will have been dancing*** for an hour when the lights are dimmed.

EXAMPLE

James <u>has talked</u> with his parents for several hours when his sister finally arrived.

(A) (as it is now)

(B) has been talking

(C) had been talking

(D) will have been talk

The second part of the sentence occurs in the past, and the first part refers to a ongoing action that occurred before this, so Answer C is correct.

Beyond these 12 verb forms, other *modal verbs* (can, could, should, would, and so forth) can be used to express action that is possible, desirable, or conditional. For example,

I *would bake* a cake if you paid for the ingredients.

Elton *could do* his homework, but he doesn't want to.

Miriam *should have taken* an umbrella.

EXAMPLE

Bonnie <u>would have been</u> able to attend the graduation if she had gotten time off from work.

(A) (as it is now)

(B) will have been

(C) wasn't being

(D) wasn't

The action of the sentence occurs in the past, so Answer B is incorrect. Neither Answer C nor Answer D works in the context of the sentence. Answer A includes the word *would,* which implies the conditional nature of Bonnie's attendance, so this is the correct answer.

Logical comparison

An ACCUPLACER question on *logical comparison* tests to see whether you understand how standard English requires comparisons to be made. As a quick example, see if you can tell what's wrong with this sentence:

Amanda's score was higher than Brittany.

The problem here is that *Amanda's score* is illogically being compared to *Brittany herself*. Here's a quick fix:

Amanda's score was higher than *Brittany's*.

Now, *Amanda's score* is being compared to *Brittany's (score),* which is logical. If that makes sense, you're ready for an ACCUPLACER-level example:

EXAMPLE

Which is the best answer choice for the underlined section of the following sentence?

Unlike the novels of Charles Dickens — which are usually set in cities like London or Paris — Thomas Hardy most often set his novels in less urban locales.

(A) (as it is now)

(B) Thomas Hardy most often was setting his novels

(C) Thomas Hardy's novel is most often set in

(D) those of Thomas Hardy are most often set in

As it is now, the sentence compares the *novels* of Charles Dickens to the *author* Thomas Hardy, which is an illogical comparison, so Answer A is wrong. Answer B also fails to fix this problem. In Answer C, only one Thomas Hardy novel is being compared, which is again an illogical comparison. Answer D correctly compares the *novels* of Charles Dickens to *those of* Thomas Hardy, so this is the correct answer.

Conventional expression

Conventional expression (sometimes called *diction*) is sort of a catch-all category that means you should try to avoid writing that reads awkwardly.

EXAMPLE

Which version of the following sentence would NOT be acceptable?

I thought about Chase for many years after our first meeting.

(A) (as it is now)

(B) thought of

(C) remembered

(D) remembered about

Standard English conventions allow you to write *thought about, thought of,* and *remembered* relatively interchangeably. But for whatever reason, *remembered about* is awkward and should be avoided, so Answer D is correct.

Fortunately or unfortunately, there is often no logical reason why one phrasing is conventional and another is unconventional. Your best friend for these kinds of questions is your inner ear — that little voice that speaks to you as you read. When answering this type of question, read all the choices to yourself in turn and try to "hear" whether each one sounds right or wrong.

Commonly confused words

Some ACCUPLACER questions test your knowledge of words that sound somewhat alike and tend to be mistaken for each other. Here are a few examples of such words:

» **Accept and except:** *Accept* is a verb that means "receive"; *except* is a preposition that means "excluding."

- Wendy *accepted* the award graciously, but gave the prize money to charity.
- Gina works every day *except* Sundays and holidays.

» **Affect and effect:** *Affect* is a verb that means "influence"; *effect* is a noun that means "result."

- Get some sleep, so that exhaustion doesn't *affect* your mood.
- As an *effect* of the economic slump, we didn't receive bonuses this year.

» **Complement and compliment:** *Complement* is a verb that means "provide balance"; *compliment* is both a noun and a verb that means "praise."

- The blue curtains *complemented* the mahogany floors.
- When she said you were authoritarian, she didn't mean it as a *compliment*.

» **Discreet and discrete:** The adjective *discreet* means "tactful"; the adjective *discrete* means "separate."

- Please be *discreet* about Marcia's plans for the surprise party.
- In our company, the sales and marketing departments are two *discrete* entities.

» **Eminent and imminent:** The adjective *eminent* means "distinguished"; the adjective *imminent* means "about to happen."

- The *eminent* economist was the keynote speaker at the conference.
- She warned that a global recession was *imminent*.

» **Flaunt and flout:** *Flaunt* is a verb that means "show off"; *flout* is a verb that means "not follow a rule."

- Evan shouldn't *flaunt* the fact that he's a better pitcher than Ryan.
- If we don't make a turkey for Thanksgiving, we'll be *flouting* tradition.

» **Loose and lose:** *Loose* is an adjective that means "not tight"; *lose* is a verb that means "fail to win."

- When you plan to exercise, be sure to wear *loose* clothing.
- I don't like to *lose* at tennis.

» **Imply and infer:** *Imply* is a verb that means "hint at"; *infer* is a verb that means "draw a conclusion."

- The writer *implied* that the story wasn't true.
- The reader *inferred* that the writer may have been biased.

» **Passed and past:** The verb *passed* means "went by"; the word *past* can be used as a noun meaning "an earlier time" or as an adjective meaning "earlier."

- When Henry was bored, time *passed* very slowly.
- In the *past*, the Dutch and the Spanish were naval enemies.
- Lonnie's *past* attempts to find a job were unsuccessful.

>> **Precede and proceed:** *Precede* is a verb that means "happen earlier"; *proceed* is a verb that means "go forward."

- The appetizer course *precedes* the main course.

- When the theater doors opened, we *proceeded* to our seats.

>> **Than and then:** *Than* is always used for making comparisons; *then* means "at that time" or "in that case."

- My brother is taller *than* I am.

- Well, *then*, if the movie starts at 2:00, let's see it *then*.

Chapter **8**

Expression of Ideas

More than half of the ACCUPLACER questions you'll face (about 60 percent) focus on the expression of ideas — that is, the meaning of what's being expressed rather than its grammatical form.

Your goal here is not to search with a magnifying glass for violations of the fine points of grammar, but rather to adopt a wider focus and see if you can figure out how to improve the flow of the paragraph you're reading.

In this section, I show you how to do this by addressing three major topics: development, organization, and effective language use.

Development

ACCUPLACER questions that focus on *development* test your understanding of how the writer structures their essay.

The structure of an essay resembles the structure of a tree: The *proposition* (or *main idea*) is like the roots and trunk of the tree — without them, the tree would fall over. *Supporting details* are like the branches and leaves — they help the tree to function, but each by itself could be removed without killing the tree.

Generally speaking, if a sentence doesn't contain information that either states or supports the main idea, it should be cut from the essay because it blurs the *focus* of the essay. ACCUPLACER questions that test you on focus ask you to state whether and why a sentence should be included or removed.

In this section, you get a handle on all these important issues.

Proposition

ACCUPLACER questions that cover *proposition* require you to understand the central theme — the *main idea* — of a text or paragraph, and to understand how revisions might clarify or detract from this theme.

Finding the main idea of a paragraph is more of an art than a science. Often, the main idea is stated up front in the first sentence, and the rest of the paragraph fleshes out this idea with a variety of supporting details. For example,

EXAMPLE

(1) The tragic fire at the Triangle Shirtwaist Factory in 1911, where 146 workers perished in less than an hour, was an accident waiting to happen. (2) The factory was located on the eighth, ninth, and tenth floors, just above the reach of rescue ladders and too high for trapped workers to jump safely from windows. (3) These floors were overcrowded with hundreds of workers, most of them women in their teens and early twenties. (4) While the building itself was made of stone and concrete, it was filled with cloth as well as completed dresses and shirts hanging from the ceiling, all of which were highly flammable. (5) Furthermore, to prevent theft, the factory owners kept some doors to the stairwells locked, and limited access to others by blocking hallways with large boxes.

Which sentence contains the main idea of the paragraph?

(A) Sentence 1

(B) Sentence 2

(C) Sentence 4

(D) Sentence 5

Here, the first sentence contains the main idea that conditions at the Triangle Shirtwaist Factory were ripe for a deadly fire. Then, each of the remaining sentences enumerates a different condition that contributed to the tragedy. Thus, Sentences 2, 4, and 5 all contain supporting details, and Sentence 1 contains the main idea, so Answer A is correct.

In other cases, the main idea of a paragraph appears in the last sentence, with the supporting details all building up to this idea, sentence by sentence. For example,

EXAMPLE

(1) The radio in the cockpit was playing Led Zeppelin's "Stairway to Heaven." (2) I remember hoping that this song wasn't a bad omen. (3) My two friends, Brad and Roy, were joking with each other, high on adrenalin — but then, they'd done this before. (4) A few of the others with us made conversation back and forth until the instructor broke in and told us, (5) "Okay, everybody, we've cleared 13,000 feet. Once we've cleared the edge of the canyon in about two minutes, it's showtime." (6) There was no turning back: I was about to jump out of my first airplane.

Which sentence contains the main idea of the paragraph?

(A) Sentence 2

(B) Sentence 3

(C) Sentence 5

(D) Sentence 6

Here, the writer builds tension by purposely holding back the key fact that makes sense of the paragraph: that he's about to jump out of an airplane. This sentence explains why, for example, the song "Stairway to Heaven" might be a bad omen, or why Brad and Roy might be high on adrenalin. Answer D is correct.

Perhaps the most difficult case is when the main idea is buried someplace in the middle of the paragraph. For example,

EXAMPLE

(1) Every year, Americans spend untold millions on a wide variety of over-the-counter cold and flu medications. (2) Some of these are meant to relieve symptoms, and they can be helpful. (3) Others purport to prevent colds or at least reduce the chances of infection. (4) Believe me, I've tried them all, and zinc lozenges are the only such remedy I've ever found to be effective. (5) Allowing a zinc lozenge to melt on your tongue seems to make the human rhinovirus — the most common cause of the common cold — less likely to take hold in one of its favorite breeding grounds, the human throat. (6) The trick, however, is to take one at the onset of a cold, when you feel that first telltale tickle of soreness in your throat. (7) If you wait long enough for the virus to get a foothold, the positive effect of zinc lozenges is sharply reduced.

Which sentence contains the main idea of the paragraph?

(A) Sentence 1

(B) Sentence 3

(C) Sentence 4

(D) Sentence 6

This paragraph begins by offering a background on cold remedies, before getting to the point in Sentence 4: that zinc lozenges are the only useful remedy the author has ever found. Sentences 5 through 7 are supporting details that tell how zinc lozenges work and under what circumstances they're most effective. Therefore, Answer C is correct.

Support

Some ACCUPLACER questions ask you to identify, add, or revise a *supporting detail* in a way that best serves the development of the passage. For example,

EXAMPLE

(1) When my daughter, Sara, asked me, "Was it hard growing up in the days before people had cellphones and the Internet?" I had to chuckle a bit. (2) The question seemed to imply that she thought of life before these relatively recent conveniences as some sort of Dark Ages. (3) But then, I realized that I, too, had always felt a similar unexamined *pity* for people who had lived a generation before I was born. (4) For example, when I discovered as a boy that my parents had grown up a decade before television was available, I remember thinking, "Wow, radio must have been really popular back then!"

The quotation in Sentence 4 is reproduced here.

"Wow, radio must have been really popular back then!"

Which of the following provides the best detail to support the author's emotional reaction to his parents' experiences?

(A) (as it is now)

(B) "Hmmm, I wonder if they had advertising in those days?"

(C) "Nice, there must have been fewer distractions in those days."

(D) "Ugh, what did they do for fun in those days!"

Here, the writer identifies "pity" as a common thread running through both his daughter's and his own reactions. Answers A, B, and C all reflect varying degrees of curiosity about what life was like at that time. Only Answer D expresses pity, so that is the correct answer.

EXAMPLE

(1) When deciding to buy a house, your first and most important consideration should be what you can really afford. (2) Obviously, if possible, you want to find a house you love in the perfect neighborhood for you. (3) But struggling each month to meet your mortgage can turn any dream home into a nightmare. (4) That's why I advise all my clients to <u>visit each of the schools that their children will attend before signing on the dotted line.</u>

Which of the following best completes the sentence?

(A) (as it is now)

(B) choose a new home location not just for its attractiveness, but also for a variety of other considerations, such as safety and proximity to their workplace.

(C) write down a maximum dollar amount their budget can handle, post it on the refrigerator, and not spend one penny more.

(D) try to meet the people next door and ask them for the pros and cons of living in that particular neighborhood.

Here, the writer makes her main point in the first sentence, that homebuyers should not spend more than they can afford. Thus, her advice should match this consideration, so Answer C is correct.

Focus

ACCUPLACER questions that test for *focus* usually ask you to determine whether a given sentence (or portion of a sentence) should be included, and the reason why or why not. For example, consider the next two questions as they relate to this paragraph:

EXAMPLE

(1) In her book *Silences,* author Tillie Olsen discusses the reasons why gifted writers often tragically fail to fully realize their potential. (2) In particular, she examines the lives of many women writers who — repeatedly derailed in their efforts to write by the economic or societal demands placed specifically upon women — can only write sporadically, sometimes producing little to no work for years, even decades. (3) In her opening chapter, Olsen states that she considers herself to be such a "silence." (4) As such, her voice is a vital addition to the literary criticism that emerged concurrently with the Women's Movement of the 1970s.

Sentence 3 is reproduced here.

In her opening chapter, Olsen states that she considers herself to be such a "silence."

The writer is considering adding the following text after this sentence.

Although Olsen was born in Nebraska, she lived most of her adult life in San Francisco.

Should the writer make this addition?

(A) Yes, because the text provides a clarifying detail about why Olsen considers herself to be a "silence."

(B) Yes, because the text gives the reader a clue as to why Olsen may have embraced the Women's Movement.

(C) No, because the text interrupts the narrative flow of the paragraph with an unnecessary biographical detail about Olsen.

(D) No, because the text unnecessarily repeats information that is available elsewhere in the paragraph.

Although this detail about Tillie Olsen's life may be interesting, it's not pertinent to the focus of this paragraph, so it should not be added. Therefore, you can rule out Answers A and B. However, this text doesn't repeat information, so you can also rule out Answer D. The correct answer is C.

Now, however, consider the following question, which refers to the same passage:

Sentence 3 is reproduced here.

EXAMPLE

In her opening chapter, Olsen states that she considers herself to be such a "silence."

The writer is considering adding the following text after this sentence.

The publication of her first and second books was separated by a gap of more than 20 years.

Should the writer make this addition?

(A) Yes, because the text provides a clarifying detail about why Olsen considers herself to be a "silence."

(B) Yes, because the text provides a clue foreshadowing why Olsen may not herself have embraced the Women's Movement.

(C) No, because the text interrupts the narrative flow of the paragraph with an unnecessary biographical detail about Olsen.

(D) No, because the text unnecessarily repeats information that is available elsewhere in the paragraph.

This time, the detail relates directly to the focus of the paragraph, and in fact clarifies Tillie Olsen's perspective on herself as a writer. Thus, you can rule out Answers C and D. The text provides no clue about Olsen's perspective on the Women's Movement, so you can rule out Answer B. The correct answer is A.

Organization

The *organization* of a passage refers to the flow of its structure. In some cases, organization is based on time, so that the reader can clearly understand the order in which key events take place. In other cases, it's based on the unfolding of an argument, so that the reader can grasp the point that the author is trying to make.

Organizational elements also include an introduction, in which the writer establishes the reader's expectations, and a conclusion, where the writer appropriately sums up their main points. Additionally, transitional sentences and words help the reader to navigate an essay, underscoring when a narrative or argument is about to change direction.

In this section, you see how the ACCUPLACER commonly measures your understanding of such organizational elements.

Logical sequence

Questions that address *logical sequence* usually require you to consider whether the meaning and flow of a paragraph could be improved if a given sentence were moved to a different location. For example,

EXAMPLE

(1) I visited Tokyo for the first time in 2017, and was immediately charmed by the city and its people. (2) I made plans to return there six months later, and on my second trip decided to consider living there. (3) With this understanding of Japanese immigration law, I planned my third trip, intending to stay about two months, during which time I hoped to find an apartment and investigate work options. (4) I found an apartment without too much trouble, but my relative lack of fluency with the Japanese language closed a lot of doors to otherwise attractive jobs. (5) So I turned my attention to improving my Japanese, attending daily language classes for four hours per day. (6) I did some research and found that as a U.S. citizen, I could use my current passport to stay in Japan for up to 90 days.

Which is the most logical placement for Sentence 6 (reproduced here)?

I did some research and found that as a U.S. citizen, I could use my current passport to stay in Japan for up to 90 days.

(A) Where it is now
(B) After Sentence 1
(C) After Sentence 2
(D) After Sentence 4

This paragraph is in narrative form, because it gives a sequence of events in the order that they occurred in time. Where it is now, Sentence 6 seems like an afterthought. Furthermore, Sentence 3 begins, "With this understanding of Japanese immigration law," implying some previous information that appears to be lacking. Placing Sentence 6 after Sentence 2 fills in this gap, so the correct answer is C.

Introductions and conclusions

Introductions and conclusions provide information that helps the reader to navigate the entire text:

> » The *introduction* begins a passage, orienting the reader by giving them a sense of the writer's purpose and why they should be interested.

> » The *conclusion* ends a passage, providing closure and giving the writer a chance to provide the reader with a wider view of the topic.

The ACCUPLACER tests you on your understanding of both.

TIP

Although a question about an introduction is usually the first question related to a passage, you may need to read a few sentences or even a couple of paragraphs into the text before answering it. For example, consider this opening sentence and the question that follows:

EXAMPLE

(1) In 1960, Stanford University psychologists Walter Mischel and Ebbe Ebbesen were looking for a clever way to observe and measure "delayed gratification": the ability to wait to receive something pleasurable. (2) They designed an experiment in which children were given an interesting choice. (3) One by one, each child was placed alone in a small room with only a table, a chair, and a plate containing a delicious treat, such as a marshmallow. (4) Each child was told that the marshmallow was theirs to eat, but if they could wait for 15 minutes, they would be given two marshmallows to enjoy. (5) Some children ate the marshmallow immediately or managed to hold out for a few minutes before giving into temptation; others endured the whole 15 minutes and received a second treat.

(6) Long-term studies of the children showed that, generally speaking, the kids who were able to delay this small gratification tended to be more successful in a variety of ways. (7) As a group, they had better grades in school, higher SAT scores, and when they entered the workforce earned more money than their counterparts who succumbed to temptation.

Which of the following, if placed before Sentence 1, would work best as an introductory sentence for the passage?

(A) Is experimental psychology really useful, or is it just hype?

(B) Can a marshmallow really predict your child's future?

(C) How do children behave when they're alone and no one is watching?

(D) Is your child a natural born optimist or pessimist?

If you try to answer this question without reading further down the page, you'll likely get confused. However, after you read the first paragraph or two, the point that the author is making gets clearer: An experiment that uses a marshmallow seems to have some value in predicting the behavior of children, so Answer B is correct.

In contrast, a conclusion question is usually the last question related to a passage. So, by the time you reach it, you'll probably already be familiar with the entire passage, and will need to draw on this understanding to answer the question. In the following example, I give you the last paragraph of the essay and ask you to choose the best concluding sentence:

EXAMPLE

(8) Follow-up studies by Mischel and Ebbesen, as well as other researchers, have duplicated the results of the original experiment. (9) But outside the psych lab, the results of the Marshmallow Experiment, as it is now known, have appeared in countless self-help and business books. (10) For these authors and their readers, it stands as <u>an object lesson in the rewards of not allowing oneself to be swayed by shiny objects, however appealing they might be.</u>

Which of the following best completes Sentence 10 as the concluding sentence of the essay?

(A) (as it is now)

(B) a reminder that children are only as honest and trustworthy as the adults who guide them — their parents, relatives, teachers, and coaches.

(C) an inspirational tale that children need not only food for their minds but also nourishment for their souls.

(D) a clarion call for those in experimental research fields to continue to reach out to the greater public.

This question tests whether you can get the main takeaway from the essay — the connection between delaying self-gratification and subsequent success — and apply it in a slightly new context. Answer A is correct.

Transitions

Transitions help the reader anticipate important shifts in perspective or focus. A *transitional sentence* is often used in the middle of a text to signal to the reader that the focus of the text is changing, for example, from the advantages of an idea to its potential drawbacks. *Transitional words* within a paragraph help the reader navigate the complexity of thought that the writer is trying to convey.

In this section, I show you how to answer ACCUPLACER questions that test your understanding of transitional words and sentences.

Transitional words

Transitional words are often conjunctions and short phrases that give the reader cues about the direction in which the writer is taking the narrative or argument. As a simple example, consider the following two sentence fragments:

Fragment #1: **I love you, and**

Fragment #2: **I love you, but**

Both fragments start with good news. The word *and* implies that Fragment #1 will end equally well — maybe even with a proposal of marriage. However, the word *but* warns you that Fragment #2 is not likely to be so much fun, and may even end with a breakup.

This small example illustrates the power of transitional words to help the reader understand what the writer is trying to say.

Transitional words tend to fall into a few major categories.

Words that continue an argument with a *similar* statement:

and	also	additionally	similarly
indeed	likewise	moreover	furthermore
besides	in addition	by the same token	in the same way

Words that pivot to a *contrasting* statement:

but	yet	however	nevertheless
though	although	since	nonetheless
instead	regardless	in contrast	on the other hand
even so	despite	in spite of	unfortunately

Words that sum up an argument or set of facts with a *concluding* statement:

so	therefore	thus	because
consequently	hence	accordingly	as such
in conclusion	this is why	as a result	for these reasons
to these ends			

Beyond these big categories, a few other words and phrases also function as conjunctions, with a variety of meanings:

For example, for instance — Provide a specific case of a general principle.

In other words, in broad terms, in short, in sum, in effect — Restate something that has already been discussed.

Normally, usually, generally, as a rule, in most cases — Describe how things are most of the time.

Naturally — Introduces an expected event.

Obviously — Of course.

Surprisingly — Introduces an unusual event.

Alternatively — Introduces a possibility or speculation.

In any case — Returns an argument from speculation to fact.

Previously — Discusses an event that took place earlier in time.

Subsequently — Discusses an event that takes place later in time.

Finally — Discusses an event that takes place last or after a long interval of time.

In the following examples, see if you can find the transitional word that best fits the context.

EXAMPLE

Which is the best version of the underlined portion of the following sentence?

The charter plane left on schedule; <u>naturally</u>, owing to severe weather conditions, the team arrived in Dallas later than expected.

(A) (as it is now)
(B) consequently
(C) however
(D) moreover

In this example, the sentence pivots from a positive state of affairs to a negative one. Thus, the transitional word *however* best handles this contrast; Answer C is correct.

EXAMPLE

(1) Although most mammals do not lay eggs, the order of Monotremata are a rare exception. (2) <u>In sum</u>, female duck-billed platypuses, the most widely known of the monotremes, lay eggs but also produce milk to provide food for their young.

Which of the following is the best version of the underlined portion of the sentence?

(A) (as it is now)
(B) for instance
(C) nonetheless
(D) accordingly

This time, Sentence 1 states a generalization about mammals that includes exceptions. Sentence 2 introduces a specific case, so the transitional words *for instance* work best. Therefore, Answer B is correct.

EXAMPLE

(1) Nursing is not only a well-paid profession, but also one that is unlikely to be outsourced to workers in other parts of the world. (2) While other types of work, such as computer programming, can be done remotely from anywhere with access to the Internet, in most cases nurses must be in the same room as their patients. (3) Furthermore, not everybody is suited to the particular demands of this type of work, particularly the physical and psychological stresses. (4) <u>In any case</u>, students who believe they have what it takes to succeed in this field would be well advised to look further into nursing as a career choice.

Which of the following is the best version of the underlined portion of the sentence?

(A) (as it is now)
(B) Thus
(C) Additionally
(D) Subsequently

Here, the passage is examining the pros and cons of nursing as a career path. The final sentence reaches the conclusion that those who are suited for the demands of the field should consider it. The concluding transition word *thus* fits the context best, so Answer B is correct.

Transitional sentences

A *transitional sentence* provides a logical connection between contrasting parts of an essay. It anticipates and helps to reduce potential confusion on the part of the reader. For example,

EXAMPLE

(1) The best boss I ever had was at a small Italian restaurant in Cambridge, Massachusetts. (2) His name was Arlo, and he was the evening manager, overseeing the wait staff (I was a waiter) as well as the kitchen crew during the dinner rush. (3) Restaurant work is hard, and ongoing squabbles between these two groups of workers are pretty common. (4) But because Arlo had personally gained the trust and respect of every employee, we tended to bring him our disputes before they got out of hand. (5) Like most good managers, Arlo was measured and tempered in his style of behavior. (6) When an argument broke out between the maître d' and the head chef in front of the customers, Arlo lost it. (7) He brought both men up to his office and gave them both a taste of his wrath. (8) It probably lasted only ten minutes, but it was epic, and months later we were still talking about it.

Which of the following sentences, if placed after Sentence 5, creates the best transition to unify the paragraph?

(A) Yet he knew how to have a good time, and when a particularly busy shift was done, he was often the first person to suggest that we go out as a group to blow off some steam.

(B) In fact, in the two years I worked there, I can remember only one time when I ever heard him raise his voice.

(C) In that way, he was quite different from Sheila, who usually managed our lunchtime shift.

(D) Arlo was from Texas, and spoke with a thick drawl that contrasted sharply with the New England regional accents we were used to hearing.

Without a transitional sentence, this paragraph could be confusing. Up to Sentence 5, the writer describes Arlo as "measured and tempered," having "gained the trust and respect of every employee." But Sentence 6 tells you that "Arlo lost it," which appears to contradict the first part of the paragraph. To bridge this difference, the essay requires a transitional sentence to put this incident into perspective. Answer B provides this perspective, explaining that this was a single incident over a two-year period.

Effective Language Use

On the ACCUPLACER, *effective language use* covers a variety of issues related to word choice and sentence structure. In this section, I cover the most common issues that you'll run across when taking the test.

Precision

A *precision* question tests whether word choice provides a specific shade of meaning needed for clear communication. For example,

EXAMPLE

Which is the best version of the underlined portion of the following sentence?

When her supervisor scuttled the project without informing her, Sharon was <u>volatile</u> and immediately sent out a heated email in protest.

(A) (as it is now)

(B) irate

(C) precarious

(D) hazardous

Here, you're looking for a word that describes Sharon's immediate state as "angry." *Volatile* means "unstable" as a general rather than immediate state of being. *Hazardous* means "dangerous" and also implies a general state. And *precarious* means "risky" as applied to a situation rather than a person. Only *irate* means "angry," so the correct answer is B.

Conciseness

Conciseness — or *concision* — refers to brief, clear writing that avoids the following problems:

>> **Extra words** — using five words when two words are sufficient

>> **Redundancy** — unnecessarily repeating an idea

>> **Awkwardness** — non-standard phrasing that feels clumsy to the reader

For example, here's a question that tests for conciseness:

EXAMPLE

Which is the best version of the underlined portion of the following sentence?

Jack's <u>not being able to handle</u> his new responsibilities as manager led to his demotion to his former position.

(A) (as it is now)

(B) ineffectiveness at the handling of

(C) inability to handle

(D) inability to handle or cope with

As it is now, you can probably understand what the writer is trying to convey. The problem is awkwardness: the phrasing "Jack's not being able to handle" is clumsy. Answer B is a little better, but the words "at the handling of" are still a little awkward and could be trimmed. Using only three words, Answer C provides the same information as Answer B without awkwardness. Answer D may appear to be an improvement on Answer C; however, "handle" and "cope with" mean essentially the same thing, so they are redundant. The correct answer is C.

Style

Style refers to the writer's way of expressing himself, especially with regard to the level of formality. Very formal writing — such as in legal briefs, medical journals, and scientific papers — is usually dry and difficult to read, full of long words and opaque sentence structure. In contrast, very informal writing — such as text messages between friends or posts to social media — often permits broad liberties with regard to grammar, such as no capital letters, minimal punctuation, many abbreviated words, and so on.

On the ACCUPLACER, the writing samples tend to run from the moderately formal to the moderately informal. Questions about the author's style gauge how perceptive you are at picking up cues about the author's level of formality.

For example, a quick way to identify informal writing in a given sample is to notice whether the author uses first-person words, such as *I, me, my,* and *mine.* A second clue is the presence of words such as *you* and *your* that address the reader directly.

In more formal writing, the writer usually avoids the use of first person, preferring to keep the writing impersonal. This convention is appropriate to academic writing, technical and science writing, and news journalism (although in recent years, more writers in these fields are incorporating first-person expression). Here is an example of formal writing:

EXAMPLE

(1) While contemporary legal scholars were united in their estimation of *Brown v. Board of Education* as a landmark moment in jurisprudence, not all viewed the decision favorably. (2) For example, Learned Hand, Chief Judge of the Second Circuit Court of Appeals, considered the ruling to be <u>a boneheaded</u> example of judicial overreach.

Which is the best phrase for the underlined portion of the last sentence?

(A) (as it is now)

(B) an egregious

(C) a morbid

(D) a whacky

Here, the formal writing style is in keeping with the topic: a historically significant Supreme Court decision. The informality of the word *boneheaded* is inconsistent with this style, as is *whacky,* so Answers A and D are incorrect. The word *morbid* (depressing) is simply incorrect in this context, so Answer C is ruled out. The word *egregious* (strongly negative) is correct in both style and meaning, so Answer B is correct.

In contrast, less formal writing that includes the first person allows writers to speak about themselves casually, including personal anecdotes and interjecting their own opinions more freely. Feature and travel journalism, and in recent years blogging, are examples of literary arenas that generally permit the author to step in and speak about their own experiences directly. For example,

EXAMPLE

(1) If you went to camp like I did, you probably remember singing songs like "Do Your Ears Hang Low?", "On Top of Spaghetti," and "Baby Bumble Bee" around the campfire at night. (2) <u>The sillier a song was, the more fun we had singing it.</u> And now, when my kids come home singing and laughing their heads off, I get to enjoy them all again!

Which of the following versions of Sentence 2 fits best stylistically with the rest of the paragraph.

(A) (as it is now)

(B) Our enjoyment increased exponentially with the drollness of each song.

(C) How utterly pithy they all were!

(D) Humor mixed with pathos is, indeed, a winning combination.

Here, the topic of this first-person passage is far from serious. It addresses the reader directly ("If you went to camp"), and includes informal phrases such as "like I did" and "laughing their heads off." As it is now, Sentence 2 matches this informal style, while Answers B, C, and D all include vocabulary and syntax that push beyond formal to stuffy. Answer A is correct.

Tone

Tone refers to the author's feeling about their topic, from strongly negative to relatively neutral to strongly positive. For example,

EXAMPLE

(1) For more than 15 years, Space Exploration Technologies Corporation — better known to the world as SpaceX — has been reimagining the possibility of human space travel. (2) Founded by visionary Elon Musk, SpaceX has made a name for itself as the first privately funded company to duplicate and now surpass a variety of accomplishments previously achieved only by government-funded entities, such as NASA in the U.S. and Roscosmos in Russia. (3) And the company's greatest achievements may well lie ahead: Musk's ambitious timeline calls for an excursion to Mars by 2024, with human colonization of that planet projected not far beyond.

The author's tone here can be best described as

(A) Sharply critical

(B) Journalistically reserved

(C) Guardedly optimistic

(D) Proudly enthusiastic

Answer A is strongly negative, Answer B is neutral, Answer C is somewhat positive, and Answer D is strongly positive. Which of these most accurately represents the author's tone? Scanning the passage, phrases such as "reimagining the possibility," "visionary Elon Musk," "company's greatest achievements," and "Musk's ambitious timeline" all point to a strongly positive tone: one that is proudly enthusiastic. The correct answer is D.

Syntax

In linguistics, *syntax* refers to the arrangement of elements in a sentence in accordance with the rules of grammar for a specific language. On the ACCUPLACER, questions that test your understanding of syntax are of several varieties. In some cases, you'll need to revise a single sentence into a more elegant form. For example,

EXAMPLE

(1) To finish my homework, my mom told me before TV watching or video game playing.

Which of the following is the best version of this sentence?

(A) (as it is now)

(B) Before watching TV or playing video games, my mom told me, to finish my homework.

(C) To finish my homework, my mom told me before watching TV or playing video games.

(D) My mom told me to finish my homework before watching TV or playing video games.

As it is now, the sentence is awkward and unclear. Answer B has non-standard grammar and incorrectly implies that the writer's mom rather than the writer wants to watch TV or play video games. Answer C is also awkward, implying that the writer's mom will finish the writer's homework. Answer D clearly states what the writer has in mind, so this is the correct answer.

In other cases, an ACCUPLACER question will ask you to combine a pair of sentences into a single sentence, often to eliminate the need for repeated words or phrases. For example,

(1) For centuries, Japanese *kabuki* theater has been the strict province of male actors.
(2) Kabuki was actually originated by a woman, Izumo no Okuni.

Which of the following best combines Sentences 1 and 2 into a single sentence?

(A) Japanese *kabuki* theater has for centuries been the strict province of male actors, and Izumo no Okuni, a woman, actually originated the form.

(B) Although Japanese *kabuki* theater has for centuries been the strict province of male actors, a woman — Izumo no Okuni — actually originated the form.

(C) For centuries, Japanese *kabuki* theater has been the strict province of male actors; therefore, the form was actually originated by a woman, Izumo no Okuni.

(D) While a woman, Izumo no Okuni, was actually originating Japanese *kabuki* theater, for centuries the form has been the strict province of male actors.

Combining these two sentences might allow you to eliminate the repetition of the word *kabuki*. Additionally, the flow can be improved by highlighting the irony that a woman invented a form of theater that is now only practiced by men. Answers A and C both fail to highlight this contrast, using the conjunctions *and* and *therefore* to link the two sentences. Answer D correctly includes *while* as a contrasting conjunction, but implies that male actors were practicing at the same time as Izumo no Okuni was originating it. Answer B correctly highlights the contrast between the origins and subsequent practice of *kabuki*.

Chapter **9**

Writing Practice Questions

Ready to practice your skills from Chapters 7 and 8? This chapter includes 30 questions that focus on both grammar and the expression of ideas. If you get stuck, flip to the end of the chapter, where I provide a detailed explanation for each question.

Questions on Grammar

For each of the following questions, decide whether the underlined portion of the sentence is correct as written or can be improved.

1. Jonathan worried about traveling <u>in the snowstorm, however by the time</u> he left the house, the weather had improved considerably.

    (A) (as it is now)

    (B) in the snowstorm, however, by the time

    (C) in the snowstorm; however by the time

    (D) in the snowstorm; however, by the time

2. The male peacock displays a dazzling array of colors, but <u>it's</u> female counterpart is, by comparison, rather drab.

    (A) (as it is now)

    (B) its

    (C) their

    (D) they're

3. Even if Simon completes the assignment — which is doubtful, <u>because it is due on Tuesday, to avoid summer school,</u> he still needs to do well on the exam.

    (A) (as it is now)

    (B) because it is due on Tuesday; to avoid summer school,

    (C) because it is due on Tuesday — to avoid summer school,

    (D) because it is due on Tuesday) to avoid summer school,

4. Every morning, Alexandra used to arrive at the shop at 6:00 a.m., start the coffee brewing, and then <u>opens</u> the doors at around 6:15.

   (A) (as it is now)
   (B) open
   (C) opening
   (D) was opening

5. The magician showed us a series of rope <u>tricks these</u> my nephews really enjoyed.

   (A) (as it is now)
   (B) tricks, these
   (C) tricks, which
   (D) tricks, who

6. The accent wall is painted in a deep orange, and <u>its affect</u> on the rest of the room is striking.

   (A) (as it is now)
   (B) its effect
   (C) it's affect
   (D) it's effect

7. <u>Jane's husband's car</u> is being repaired, so he is using her car.

   (A) (as it is now)
   (B) Janes husband's car
   (C) Janes husbands' car
   (D) Jane's husbands' car

8. Fortunately, the new job offered Jennifer an important <u>benefit: the opportunity</u> to travel to Spain just about every month.

   (A) (as it is now)
   (B) benefit, the opportunity
   (C) benefit; the opportunity
   (D) benefit this being the opportunity

9. By the end of the First World War, many soldiers had been out of the country for well over a year and <u>was</u> looking forward to being reunited with their families.

   (A) (as it is now)
   (B) were
   (C) were being
   (D) will be

10. Unlike her mother, <u>Bethany's height was nearly as tall as her father</u> by the time she was 16 years old.

   (A) (as it is now)
   (B) Bethany was nearly as tall as her father's height
   (C) Bethany's height was nearly as great as her father's
   (D) Bethany was nearly as tall as her father

11. When they <u>have been walking</u> for about two hours, Sandra and Chuck sat down in a shaded area and had a picnic.

   (A) (as it is now)
   (B) were walking
   (C) had been walking
   (D) will have been

12. Elissa doesn't just own <u>pugs because she enjoys them,</u> she also earns a portion of her income by breeding them and selling them as pets.

   (A) (as it is now)
   (B) pugs, because she enjoys them,
   (C) pugs because she enjoys them;
   (D) pugs, because she enjoys them;

13. <u>Last summer, David, Laura, and Raymond</u> all visited the Grand Canyon together.

   (A) (as it is now)
   (B) Last summer David, Laura and Raymond
   (C) Last summer David, Laura, and Raymond
   (D) Last summer, David, Laura and Raymond

14. It is usually more polite to <u>accept a complement</u> when one is given than to explain why you don't deserve it.

   (A) (as it is now)
   (B) accept a compliment
   (C) except a compliment
   (D) except a complement

15. Upon returning to Manhattan, <u>the front door had been unlocked the whole time that they were gone was what the Woolcotts discovered.</u>

   (A) (as it is now)
   (B) the front door had been unlocked the whole time that the Woolcotts had been gone, as they discovered.
   (C) the Woolcotts discovered that the front door had been unlocked the whole time that they were gone.
   (D) the Woolcotts discovered the front door's having been unlocked the whole time that they were gone.

16. The two-week trip to Paris took Kevin away from his <u>studies; consequently he</u> had a stack of assignments waiting for him upon his return.

   (A) (as it is now)
   (B) studies; consequently, he
   (C) studies: consequently he
   (D) studies: consequently, he

17. When the lawyers left the hearing, we could tell from the looks on <u>they're faces that they</u> were not pleased.

   (A) (as it is now)

   (B) their faces that they

   (C) their faces that he or she

   (D) they're faces that he or she

18. Jacob told his brother, Isaac, the good news: <u>his daughter's</u> essay had won first prize, which included a $5,000 scholarship.

   (A) (as it is now)

   (B) his daughters

   (C) Isaacs' daughter's

   (D) Isaac's daughter's

19. Sarah's principal concern was for her daughter <u>Claire had neglected to bring</u> a jacket warm enough for a long day of ice skating.

   (A) (as it is now)

   (B) Claire: that had neglected to bring

   (C) Claire, who had neglected to bring

   (D) Claire; who had neglected to bring

20. Torn in several places and missing its front cover, <u>I noticed that David's copy of *The Catcher in the Rye* showed that he had read it multiple times.</u>

   (A) (as it is now)

   (B) David had read his copy of *The Catcher in the Rye* multiple times.

   (C) David's copy of *The Catcher in the Rye* showed that he had read it multiple times.

   (D) the condition of David's copy of *The Catcher in the Rye* showed that he had read it multiple times.

Questions on Expression of Ideas

For each of the following passages, read the early draft of an essay and then choose the best answer to each question that follows.

(1) The *Star Trek* television franchise has proved to be one of the most popular and durable such ventures in the history of Hollywood. (2) In the half century since its premier in 1966, the original *Star Trek* series has spawned a surfeit of offshoots in a wide variety of media forms: films, spin-offs, animated series, prequels, video games, and even fan-created series.

(3) Although its opening monologue famously touts the "five-year mission" of the Starship Enterprise, the show was cancelled in its third season. (4) Over the next few years, however, the syndication of *Star Trek* episodes outside of prime-time television hours sparked a relatively small but fiercely loyal cult following. (5) In an age predating the Internet by a generation, fans managed to find each other, often at *Star Trek* conventions that occurred with increasing frequency throughout the 1970s. (6) These fans formed and then steadfastly maintained these notorious connections, further extending the *Star Trek* "community."

(7) With a viewership eager for more, *Star Trek* was finally ready for its next phase, and so was Roddenberry. (8) In 1979, the first full length feature film, *Star Trek — The Motion Picture —* was released. (9) Five more movies followed until Roddenberry's death in 1991. (10) Also during that time period, Roddenberry launched *Star Trek — The Next Generation*, the first of a variety of equally successful *Star Trek* series.

1. Should the writer delete the underlined portion of Sentence 2 (reproduced here)?

   In the half century since its premier in 1966, the original Star Trek *series has spawned a surfeit of offshoots in a wide variety of media forms: films, spin-offs, animated series, prequels, video games, and even fan-created series.*

   (A) Yes, because the underlined portion blurs the focus of the essay with unnecessary information.
   (B) Yes, because the underlined portion unintentionally weakens the argument that the show was originally unsuccessful.
   (C) No, because the underlined portion provides clarifying information.
   (D) No, because the underlined portion provides evidence that the show was originally unsuccessful.

2. Which of the following sentences, if placed at the beginning of the second paragraph, provides the best transition to the information that follows?

   (A) Before he created *Star Trek*, Gene Roddenberry got his start in Hollywood as a freelance scriptwriter for popular 1960s shows such as *Have Gun — Will Travel* and *Highway Patrol*.
   (B) Notwithstanding its ultimate success, the series initially struggled to find an audience and did not attain the popularity that its creator, Gene Roddenberry, had hoped for.
   (C) Consequently, series creator Gene Roddenberry and his heirs have earned an estimated $500 million in royalties from the successful franchise.
   (D) From the inception of the original series, its creator Gene Roddenberry envisioned the *Star Trek* universe as not just an escape from our world, but a vantage point from which to explore it.

3. Which is the best version of the underlined portion of Sentence 6 (reproduced here)?

   These fans formed and then steadfastly maintained these notorious *connections, further extending the* Star Trek *"community."*

   (A) (as it is now)
   (B) cherished
   (C) formidable
   (D) prosperous

4. Why does the writer place the words *Star Trek* "community" in quotation marks?

   (A) To identify the use of a term in a non-standard way.
   (B) To identify the introduction of a term that the reader may not understand without a definition.
   (C) To indicate the use of a direct quotation by Gene Roddenberry.
   (D) To convey the writer's subtle contempt for that community.

5. Which of the following best characterizes the writer's overall stance on the *Star Trek* franchise?

(A) Journalistic impartiality

(B) Unmasked antipathy

(C) Mild praise

(D) Overall admiration

(1) I have to admit that when I first noticed the entrance to The World Bags and Luggage Museum in Tokyo, I thought it was a joke. (2) "What's it filled with?" I wondered, "A lot of luggage on display behind glass?" (3) Because of my reservations, I walked inside and found, happily, that admission to the museum was free.

(4) Inside, I found pretty much what I'd been expecting: luggage behind glass. (5) What I wasn't expecting, though, was how surprising and artistic the exhibits actually were.

(6) Most of the pieces were from the period between the late 19th century and the late 20th century. (7) For example, a collection of steamer trunks of various sizes and shapes were displayed in front of a large black-and-white photograph of the *Titanic*. (8) Although I'd seen such items before, I'd never connected them with a historical event of such magnitude. (9) Trunks like these, I discovered, would be typical of those that affluent travelers took with them when crossing the ocean. (10) One was even open and set on its side, showing off that it doubled as a chest of drawers, a useful feature given the relatively confined cabin space that would have been available aboard a ship.

(11) Another display presented a variety of mid-20th-century luggage made from the hides of exotic animals: alligators, crocodiles, snakes, plus others that I didn't recognize. (12) The pieces themselves were well-preserved and ornately beautiful. (13) And with changes in fashion — not to mention regulations prohibiting the hunting of animals for such uses — examples of this kind were most likely never again to be reproduced.

(14) Oddly, some of the less outlandish pieces held more emotional interest for me. (15) A row of rather workaday looking travel suitcases from the 1960s and 70s reminded me of the one that my dad would carry with him on business trips when I was growing up. (16) At the time, I never thought twice about it; now, it brought to mind sitting on his bed talking with him as he packed, and knowing that I would miss him while he was away. (17) I imagine that a woman might have a similar moment of wistful recognition among the wide collection of vintage purses and handbags.

6. Which of the following is the best version of the underlined portion of Sentence 3 (reproduced here)?

Because of my reservations, I walked inside and found, happily, that admission to the museum was free.

(A) (as it is now)

(B) With equal enthusiasm,

(C) Despite my initial skepticism,

(D) Not surprisingly,

7. Which of the following sentences, if placed between Sentences 6 and 7 (reproduced here), creates the best transition between these two sentences?

(6) Most of the pieces were from the period between the late 19th century and the late 20th century.
(7) For example, a collection of steamer trunks of various sizes and shapes were displayed in front of a large black-and-white photograph of the Titanic.

(A) A few even had multiple uses, which explains perhaps why they were so popular in their day.

(B) Some were set in a historical context that gave them greater interest.

(C) The pieces looked so old that they now appeared to need special temperature and humidity-controlled conditions.

(D) Although they might have been expensive in their time, you might well pick up something like this for five dollars at a garage sale.

8. Which decision is best regarding the underlined portion of Sentence 11 (reproduced here)?

Another display presented a variety of mid-20th-century luggage made from the hides of exotic animals: alligators, crocodiles, snakes, plus others that I didn't recognize.

(A) Leave it as it is now

(B) Revise it to "the types of animals you might normally expect to find in the zoo."

(C) Revise it to "many dangerous varieties that can only be hunted by exceedingly courageous stalkers."

(D) DELETE it and end the sentence with a period.

9. Which of the following is the best decision for the underlined portion of Sentence 17 (reproduced here)?

I imagine that a woman might have a similar moment of wistful recognition among the wide collection of vintage purses and handbags.

(A) (as it is now)

(B) shocked disdain

(C) bottomless depression

(D) ironic musing

10. Which of the following would provide a concluding sentence that best fits the tone and content of the essay?

(A) If you're in the area, please don't pass over this unusual gallery!

(B) One often finds in such unlikely venues an opportunity not only for edification but also for delight.

(C) Despite the extravagant cost, the museum is entirely worth budgeting for.

(D) In short, I believe that women are likely to enjoy the experience far more than men.

Answers to Questions on Grammar

1. **D.** The subordinating conjunction *however* requires a semicolon before it and a comma after it, so Answer D is correct.

2. **B.** A possessive determiner is required rather than a contraction, so Answers A and D are ruled out. The antecedent of this determiner is the singular noun *peacock*, which rules out Answer C, so Answer B is correct.

3. **C.** The parenthetical element begins with a dash, so it needs to end with a dash; therefore, Answer C is correct.

4. **B.** The parallel construction begins with the words *used to arrive*, continues with the verb *start*, and needs to continue with the verb *open*, so Answer B is correct.

5. **C.** The sentence as it is now is a run-on, so Answer A is incorrect. To fix this, a comma is required after the first clause — that is, after the word *tricks* — followed by the relative pronoun *which* to refer back to the noun phrase *rope tricks*. Thus, the correct answer is C.

6. **B.** The contraction *it's* means "it is," which is incorrect in the context of the sentence, ruling out Answers C and D. The verb *affect* means "influence," which is also incorrect in this context, so Answer A is ruled out. Answer B is correct.

7. **A.** In this phrase, the singular nouns *Jane* and *husband* both require singular possessive forms. Answers B and C are incorrect because *Janes* is a plural form. Answer D is incorrect because *husbands'* is a plural possessive form. Answer A is correct.

8. **A.** In this sentence, the colon is properly used to present the information that follows in a dramatic way; therefore, Answer A is correct.

9. **B.** The verb needs to agree with the subject *many soldiers*, which is plural, so Answer A is incorrect. The sentence refers to past events, so the future verb *will be* is improper; therefore, Answer D is wrong. Additionally, the verb phrase *were being* doesn't make sense when placed in front of the words *looking forward*, so Answer C is incorrect; thus, Answer B is correct.

10. **D.** The first words of the sentence set up a comparison with Bethany's mother rather than the height of Bethany's mother. Thus, the comparison must continue with references to *people* rather than to their height. Answer A refers to Bethany's height, so this is incorrect. Answer B refers to Bethany's father's height, which is incorrect. Answer C refers to both Bethany's height and her father's height, which is incorrect. Thus, Answer D is correct.

11. **C.** The second part of the sentence indicates action in the past, so the first part takes place earlier than this. Therefore, either *had walked* or *had been walking* could work in this context, so Answer C is correct.

12. **C.** The first and second parts of this sentence are both clauses, so a comma is insufficient to connect them; therefore, Answers A and B are incorrect. Additionally, a comma placed after the word *pugs* is unnecessary in this context, so Answer D is incorrect; therefore, Answer C is correct.

13. **A.** The adverbial adjunct *last summer* precede the subject, so they need to be set off from the sentence with a comma, which rules out Answers B and C. The subject of the sentence contains three elements, so a comma is required after the first two elements — that is, after *David* and *Laura* — so Answer D is incorrect; thus, Answer A is correct.

14. **B.** The preposition *except* means *excluding* and isn't correct for this context, so Answers C and D are ruled out. The verb *complement* means "provide balance" and isn't correct in this context, so Answer A is wrong; therefore, Answer B is correct.

15. **C.** The sentence begins with a dangling modifier, so the subject must appear immediately after the comma, and must state who or what returned to Manhattan; therefore, Answers A and B are incorrect. Additionally, Answer D includes awkward phrasing, so it is incorrect; thus, Answer C is correct.

16. **B.** The sentence contains two clauses, which cannot be separated by a colon, so Answers C and D are ruled out. A comma must follow the subordinating conjunction *consequently*, so Answer A is ruled out; therefore, Answer B is correct.

17. **B.** The word *they're* is a contraction that means *they are*, which isn't correct in this context, so Answers A and D are incorrect. The antecedent to the pronoun is the noun phrase *the lawyers*, which is plural and thus requires the plural pronoun *they* rather than *he or she*, so Answer C is ruled out; therefore, Answer B is correct.

18. **D.** The first part of the sentence includes the names of two men, so the antecedent of the pronoun *his* is unclear; therefore, Answer A is incorrect. Answer B has the same problem, compounded by the use of the plural noun *daughters* rather than the possessive noun *daughter's*, so it is also incorrect. Answer C incorrectly uses the plural possessive *Isaacs'*, so it is incorrect; therefore, Answer D is correct.

19. **C.** As it stands, the sentence is a run-on containing two clauses inappropriately connected by a comma, so Answer A is incorrect. The use of a colon with the word *that* is incorrect, so Answer B is incorrect. And the use of a semicolon with the relative pronoun *who* is also incorrect, so Answer D is incorrect. Answer C correctly uses a comma with the relative pronoun *who* to refer back to the words *her daughter*, so this is the correct answer.

20. **C.** The sentence begins with a dangling modifier that ends at the comma, so the subject of the sentence must indicate what is being modified — that is, David's copy of *The Catcher in the Rye*. In Answer A, the subject is *I*, which is incorrect. In Answer B, the subject is *David*, which is also incorrect. And in Answer D, the subject is *the condition of David's copy of The Catcher in the Rye*, which is also incorrect. Thus, Answer C is correct.

Answers to Questions on Expression of Ideas

1. **C.** The underlined portion includes useful information about the variety of media forms that *Star Trek* has inspired, so it should not be deleted; Answer C is correct.

2. **B.** The second paragraph tells the reader that *Star Trek* was originally unsuccessful, in contrast to the information given in the first paragraph about the show's later success. Thus, a transition is needed to bridge this contrast, so Answer B is correct.

3. **B.** The friendships among *Star Trek* fans resulted in a group of people who considered themselves a form of community, so the connections among them would be significant and valued. The word *cherished* captures this meaning best, so Answer B is correct.

4. **A.** Although many *Star Trek* fans enjoyed their familiarity as a group, they wouldn't, strictly speaking, constitute a community. The quotation marks capture this loose or non-standard usage, so Answer A is correct.

5. **D.** Throughout the essay, the writer continually chooses positive words, designating the franchise as "popular and durable," applauding its "surfeit of offshoots," and describing the fans as "eager for more." Thus, he is strongly positive toward his topic, so Answer D is correct.

6. **C.** The first two sentences describe the writer's first impression of the museum as mildly negative. Thus, a contrasting transition is needed to explain his decision to enter, so Answer C is correct.

7. **B.** Sentence 7 mentions an exhibit of steamer trunks displayed in front of a photo of the *Titanic*. This is an example of an exhibit that is set in a historical context, so Answer B is correct.

8. **A.** As it is now, the sentence includes specific examples of exotic animals, to help the reader picture the exhibit being described, so Answer D is ruled out. Answer B is less specific, and Answer C blurs the focus of the paragraph, so neither of these answers is correct. Thus, Answer A is correct.

9. **A.** The writer's story about his father's suitcase mentions "sitting on his bed talking with him as he packed, and knowing that I would miss him while he was away." This indicates a moment of a wistful recognition, so Answer A is correct.

10. **A.** The essay's tone is informal, so Answer B is incorrect. It mentions that the museum is free, so Answer C is incorrect. And the writer is a man who has clearly enjoyed the experience, so Answer D is incorrect. Answer A, which ends the essay with a positive recommendation, is correct.

4

The Arithmetic Test

IN THIS PART . . .

Work with whole numbers and fractions.

Solve problems that involve decimals, percents, and number equivalencies.

Apply these skills to a variety of practice questions.

Chapter **10**

Arithmetic: Whole Numbers and Fractions

The good news about arithmetic is that it's not as tough as algebra. The not-so-good news may be that you have not reviewed your arithmetic skills since around sixth grade. (You remember sixth grade, right? Falling asleep in homeroom, forgetting your locker combination, getting lost trying to find the cafeteria.)

In this chapter, you have a golden opportunity to refresh these skills without having to bug your guidance counselor *again* for those three pesky numbers. Here, the focus is on working with whole numbers and fractions. Then in Chapter 11, I help you round out your arithmetic knowledge with decimals, percents, and word problems.

See you at the pep rally! (Do they still have those?)

Whole Numbers

On the ACCUPLACER, questions involving whole numbers usually require you to work with a few basic math ideas that you probably already know: the order of operations (fondly remembered with the mnemonic PEMDAS); estimation and rounding numbers; and working with inequalities using the number line. In this section, you get a refresher on all these topics.

Understanding the order of operations (PEMDAS)

When working with whole numbers, you can evaluate complicated expressions using the memory device PEMDAS: Parentheses, Exponents, Multiplication and Division, Addition and Subtraction. For example, suppose you want to calculate the following:

$$4 - (19 - 3 \times 5)^2 \div 2$$

To begin calculating, start inside the parentheses, beginning with the multiplication:

$$= 4 - (19 - 15)^2 \div 2$$
$$= 4 - 4^2 \div 2$$

With the parentheses handled, evaluate the exponent:

$$= 4 - 16 \div 2$$

Now, focus on the division:

$$= 4 - 8$$

And to finish, subtract:

$$= -4$$

Rounding

To round a whole number to the nearest ten, focus on the ones digit:

>> If the ones digit is 0 through 4, change it to 0.

>> If the ones digit is 5 through 9, change it to 0 **and** add 1 to the tens digit.

For example,

$$34 \rightarrow 30 \qquad 165 \rightarrow 170 \qquad 497 \rightarrow 500$$

To round a whole number to the nearest hundred, focus on the tens digit:

>> If the tens digit is 0 through 4, change it and the ones digit to 0.

>> If the tens digit is 5 through 9, change it and the ones digit to 0 **and** add 1 to the hundreds digit.

For example,

$$148 \rightarrow 100 \qquad 756 \rightarrow 800 \qquad 1998 \rightarrow 2000$$

To round a whole number to the nearest thousand, focus on the hundreds digit:

>> If the hundreds digit is 0 through 4, change it and all digits to its right to 0.

>> If the hundreds digit is 5 through 9, change it and all digits to its right to 0 **and** add 1 to the thousands digit.

For example,

$$2378 \rightarrow 2000 \qquad 8693 \rightarrow 9000 \qquad 49{,}632 \rightarrow 50{,}000$$

Estimating

Estimation uses rounding as a fast way to find a reasonably close answer to an otherwise difficult calculation. For example, suppose you are faced with the following question:

EXAMPLE

Which of the following is closest to 8975×629?

(A) 222,000

(B) 777,000

(C) 3,330,000

(D) 5,550,000

Multiplying these two numbers would be a long process. Instead, round 8975 to 9000, and round 629 to 600. Now, multiply $9000 \times 600 = 5,400,000$, which is closest to 5,550,000, so the answer is D.

Inequalities on the number line

An *inequality* is a statement that compares a pair of different values. You have to watch for four different inequality signs on the ACCUPLACER, as shown in Table 10-1.

TABLE 10-1 **Inequalities on the Number Line**

Inequality sign	Meaning	On the number line
<	Less than	Open circle
>	Greater than	Open circle
≤	Less than or equal to	Closed circle
≥	Greater than or equal to	Closed circle

You can illustrate an inequality on a number line, as shown in Figure 10-1.

FIGURE 10-1:
The inequalities on the number line are $x < 2$, $x > 2$, $x \leq 2$, and $x \geq 2$.

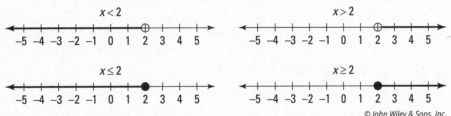

© John Wiley & Sons, Inc.

Notice that the two inequalities < and > are represented by open circles, which tell you that the value of the inequality is *not included* in the solution set. In contrast, the inequalities ≤ and ≥ are represented by closed circles, showing that this value is included in the solution set.

You can also represent the solution set for a pair of inequalities on a single number line. For example, here's how you represent the inequality $-3 < x \leq 1$, which means $x > -3$ and $x \leq 1$:

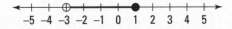

Notice here that the solution set for these two inequalities is the space on the number line between -3 and 1. Additionally, the value 1 is included in the solution set, as indicated by the closed circle. However, the value -3 is not included, as shown by the open circle.

Another way to represent a pair of inequalities is by using the word *or*. For example, here's how you represent the solution set for the inequalities $x \leq -2$ *or* $x > 3$:

This time, because the word *or* is used, the solution set includes the values that are less than –2 *or* greater than 3. Additionally, the value –2 is also included, as indicate by the closed circle, but the value 3 is excluded from the solution set.

Fractions

For many math students, fractions are a sticking point. Even solid math students often struggle to remember how to add a pair of mixed numbers. And if you're not among the small group of folks who feel confident in math, fractions may be one of your least favorite parts of the math equation.

What's also true is that fractions loom large on the ACCUPLACER. If your skills with fractions are rusty, this section will give you what you need to succeed.

Recalling a few fraction basics

Fractions are one of a variety of ways to represent values that are less than 1 — that is, parts of a larger whole.

Identifying the numerator and the denominator

Every fraction consists of two whole numbers placed one above the other. The top number is called the *numerator* and the bottom number is called the *denominator*.

For example, in the fraction $\frac{2}{9}$, the numerator is 2 and the denominator is 9.

Understanding reciprocals

The *reciprocal* of a fraction is simply that fraction turned upside-down, so that the numerator (top number) and denominator (bottom number) are exchanged. For example, the reciprocal of $\frac{3}{5}$ is $\frac{5}{3}$.

When a fraction has a 1 in the numerator, the reciprocal of that fraction is the denominator. For example, the reciprocal of $\frac{1}{7}$ is $\frac{7}{1}$, which simplifies to 7.

You can turn any whole number into a fraction by placing it in the numerator of a fraction with 1 in the denominator. For example, 14 is equal to the fraction $\frac{14}{1}$, so its reciprocal is $\frac{1}{14}$.

On the ACCUPLACER, you'll use the reciprocal of a fraction when dividing fractions.

Simplifying and increasing the terms of fractions

You can express the value of a single fraction in a variety (actually, an infinite number!) of ways by changing the values of both the numerator (top number) and denominator (bottom number). In this section, I show you how to put a fraction into its simplest form. I also show you how to increase the terms of a fraction without changing its value.

SIMPLIFYING FRACTIONS

Consider the fraction $\frac{6}{8}$. You can probably see that both 6 and 8 are divisible by 2. You can *simplify* (or *reduce*) this fraction by dividing both of these numbers by this common factor:

$$\frac{6 \div 2}{8 \div 2} = \frac{3}{4}$$

The simplified form of a fraction is usually the preferred form, and most ACCUPLACER multiple-choice answers involving fractions include only simplified fractions.

To simplify a fraction, divide both the numerator and denominator by a common factor. You may need to repeat this process until the fraction is completely simplified. For example, consider the fraction $\frac{60}{72}$. You can begin simplifying this fraction by repeatedly dividing by 2 as many times as possible:

$$\frac{60 \div 2}{72 \div 2} = \frac{30 \div 2}{36 \div 2} = \frac{15}{18}$$

Now, notice that both 15 and 18 are divisible by 3, so divide both by 3:

$$= \frac{15 \div 3}{18 \div 3} = \frac{5}{6}$$

This fraction is fully simplified because 5 and 6 have no common factor greater than 1.

INCREASING THE TERMS OF FRACTIONS

To solve some problems that involve fractions, you may need to *increase the terms* of both the numerator (top number) and denominator (bottom number) without changing the value of the fraction. To do this, multiply both the numerator and denominator by the same number.

For example, you can increase the terms of $\frac{1}{4}$ by multiplying its numerator and denominator by the same number:

$$\frac{1 \times 2}{4 \times 2} = \frac{2}{8} \qquad \frac{1 \times 3}{4 \times 3} = \frac{3}{12} \qquad \frac{1 \times 4}{4 \times 4} = \frac{4}{16}$$

The resulting fractions — $\frac{2}{8}$, $\frac{3}{12}$, and $\frac{4}{16}$ — all look different, but are all equal to $\frac{1}{4}$.

Often, the goal of changing the terms of a fraction is to change the denominator to a specific number that is a multiple of the current denominator. For example, suppose you want to change the terms of $\frac{6}{25}$ so that the denominator equals 100. To do this, consider what value you need to multiply the denominator by to achieve this goal:

$$\frac{6 \times ?}{25 \times ?} = \frac{?}{100}$$

As you can see, to change the denominator of 25 to 100, you need to multiply the denominator by 4. And to preserve the value of the fraction, you also need to multiply the numerator by 4:

$$\frac{6 \times 4}{25 \times 4} = \frac{24}{100}$$

Therefore, $\frac{24}{100}$ is an alternative way to express the value of $\frac{6}{25}$.

Working with improper fractions and mixed numbers

When the numerator (top number) of a fraction is less than its denominator, that fraction is called a *proper fraction*. For example,

$$\frac{2}{3} \qquad \frac{4}{13} \qquad \frac{81}{244}$$

Every positive proper fraction expresses a value that is between 0 and 1.

In contrast, when a fraction's numerator is greater than or equal to its denominator, it's called an *improper fraction*. For example,

$$\frac{3}{2} \qquad \frac{63}{29} \qquad \frac{100}{100}$$

Every positive improper fraction expresses a value that is greater than or equal to 1.

Like improper fractions, a *mixed number* also expresses a value of 1 or greater. A mixed number includes both a whole number and a proper fraction: for example,

$$1\frac{1}{2} \qquad 12\frac{7}{10} \qquad 139\frac{47}{83}$$

Every mixed number has an equivalent value that can be expressed as an improper fraction.

Converting mixed numbers to improper fractions

Improper fractions are often easier to calculate with, especially when multiplying and dividing fractions. So on the ACCUPLACER, you may need to change a mixed number to an improper fraction at the beginning of a problem in order to solve it.

To change a mixed number to an improper fraction:

1. **Multiply the denominator by the whole number, then add the numerator, and copy this value as the numerator of your answer.**

2. **Copy the denominator.**

For example, to change $3\frac{2}{5}$ to an improper fraction, multiply $3 \times 5 + 2$ to find the numerator of the answer, then copy the denominator of 5:

$$3\frac{2}{5} = \frac{3 \times 5 + 2}{5} = \frac{17}{5}$$

Converting improper fractions to mixed numbers

Mixed numbers tend to be easier to read and understand than improper fractions. For example, if I tell you that I put $13\frac{1}{2}$ gallons of gasoline in the tank, you can easily estimate that I bought between 13 and 14 gallons. In contrast, if I express this same value as $\frac{27}{2}$ gallons, you're likely to be confused. For this reason, multiple-choice answers on the ACCUPLACER are usually expressed as mixed numbers rather than improper fractions.

You can change an improper fraction to a mixed number in two ways, depending on whether the improper fraction is small or large. In this section, I show you both methods.

CONVERTING SMALLER IMPROPER FRACTIONS

When you perform basic operations on fractions, the result is often an improper fraction in which the numerator of the improper fraction is not much larger than its denominator. Here is a fast and easy way to change that improper fraction to a mixed number:

1. Write the number 1 (or add 1 to the previous whole number value).

2. Copy the denominator.

3. Subtract the numerator from the denominator to find the new numerator.

A few examples will help. Here's how to convert the improper fraction $\frac{8}{5}$ to an equivalent mixed number:

$$\frac{8}{5} = 1\frac{3}{5}$$

Here, I wrote the number 1, copied the denominator of 5, and subtracted $8 - 5 = 3$ to find the numerator.

Notice that the denominator of the original improper fraction and its mixed-number equivalent remain the same. This is always true when converting an improper fraction to a mixed number.

As another example, change $\frac{13}{4}$ to a mixed number by repeating this process:

$$\frac{13}{4} = 1\frac{9}{4} = 2\frac{5}{4} = 3\frac{1}{4}$$

Here, I repeatedly increased the value of the whole number by 1, and repeatedly subtracted the denominator of 4 from the numerator. The process is complete when the fractional portion of the result is a proper fraction — in this case, $\frac{1}{4}$.

CONVERTING LARGER IMPROPER FRACTIONS

When the numerator of an improper fraction is much larger than its denominator, the method I provide in "Converting smaller improper fractions" can become time-consuming.

For example, suppose you need to change $\frac{81}{2}$ to a mixed number. You may not want to go through this process:

$$\frac{81}{2} = 1\frac{79}{2} = 2\frac{77}{2} = 3\frac{75}{2} = 4\frac{73}{2} = 5\frac{71}{2} = \ldots$$

A better method in this case is to divide the numerator by the denominator. When you complete the division,

» The quotient (result of the division) is the whole number value.

» The remainder is the numerator.

» The denominator stays the same.

For example, to convert $\frac{81}{2}$ to a mixed number, you divide to get $81 \div 2 = 40r1$ — that is, 40 with a remainder of 1 — so,

$$\frac{81}{2} = 40\frac{1}{2}$$

Adding and subtracting fractions that have the same denominator

Just as with whole numbers, you can apply the Big Four operations — addition, subtraction, multiplication, and division — to fractions.

When adding and subtracting fractions, the main thing to notice first is whether the fractions have the same denominator (bottom number). In this section, I show you how to add and subtract fractions with the same denominator.

Adding fractions with the same denominator

To add fractions that have the same denominator:

1. **Add the numerators.**

2. **Keep the denominator the same.**

For example,

$$\frac{1}{5} + \frac{3}{5} = \frac{4}{5} \qquad \frac{5}{9} + \frac{2}{9} = \frac{7}{9} \qquad \frac{6}{25} + \frac{13}{25} = \frac{19}{25}$$

This rule works equally well for adding more than two fractions:

$$\frac{1}{11} + \frac{2}{11} + \frac{3}{11} + \frac{4}{11} = \frac{10}{11}$$

In some cases when adding fractions, you may need to simplify the result in one or both of the following ways:

>> Simplify the fraction (see "Simplifying fractions" earlier in this chapter).

>> Change an improper fraction to a mixed number (see "Converting improper fractions to mixed numbers").

For example:

$$\frac{7}{12} + \frac{1}{12} = \frac{8}{12}$$

You can reduce this fraction by dividing both the numerator and the denominator by 4:

$$= \frac{8 \div 4}{12 \div 4} = \frac{2}{3}$$

As another example:

$$\frac{7}{10} + \frac{9}{10} = \frac{16}{10}$$

Convert this improper fraction to a mixed number, and then simplify the result:

$$= 1\frac{6}{10} = 1\frac{3}{5}$$

Subtracting fractions with the same denominator

Subtracting fractions that have the same denominator is as simple as adding them:

1. Subtract the numerators.

2. Keep the denominator the same.

For example,

$$\frac{5}{7} - \frac{2}{7} = \frac{3}{7} \qquad \frac{10}{11} - \frac{6}{11} = \frac{4}{11} \qquad \frac{80}{99} - \frac{10}{99} = \frac{70}{99}$$

As with whole-number subtraction, when you subtract a smaller fraction minus a greater one, your result is a negative number. For example,

$$\frac{2}{9} - \frac{7}{9} = -\frac{5}{9}$$

Here, subtracting the numerators $2 - 7 = -5$ results in the negative fraction $-\frac{5}{9}$.

As when adding fractions, in some cases you will need to simplify your result. For example:

$$\frac{11}{12} - \frac{5}{12} = \frac{6}{12}$$

Here, to simplify the result, divide both the numerator and the denominator by 6:

$$= \frac{6 \div 6}{12 \div 6} = \frac{1}{2}$$

Adding and subtracting fractions that have different denominators

Adding and subtracting fractions that have different denominators can be a little complicated, and may seem absolutely daunting if you got lost in the shuffle in the third or fourth grade.

The reason for this difficulty is that most students learn to add or subtract fractions by finding a *common denominator* — that is, by getting both fractions to have the same denominator before you add or subtract. And this step is often the most confusing part of adding and subtracting fractions.

Here, I offer two simple methods for finding a common denominator:

Cross-multiplication — This method *always* works, but sometimes results in large numbers that are hard to work with.

Increasing the terms of the smaller fraction — This method only works in some cases, but it can help keep the numbers small.

I find that when I teach ACCUPLACER students how to add and subtract fractions using these two methods, they're often surprised at how straightforward it seems in comparison with their memories from grade school.

I hope you find your own new experiences of adding and subtracting fractions to be similar to theirs.

Method #1 — Using cross-multiplication

When a pair of fractions have different denominators, you can always add or subtract them using cross-multiplication — that is, multiplying the numerator of each fraction by the denominator of the other. This step gets the denominators equal to each other, which allows you to add or subtract the fractions easily, as I show you in "Adding and subtracting fractions that have the same denominator."

ADDING FRACTIONS WITH CROSS-MULTIPLICATION

To add a pair of fractions using cross-multiplication, follow these steps:

1. **Cross-multiply the two fractions to find the new numerators.**
2. **Multiply the denominators to find the new denominator.**

For example, to add $\frac{2}{5} + \frac{1}{4}$, cross-multiply the two fractions — $2 \times 4 = 8$ and $5 \times 1 = 5$ — placing these results in the numerators to add in the next step; then multiply $5 \times 4 = 20$ to get the denominator of both fractions.

$$\frac{2}{5} + \frac{1}{4} = \frac{8}{20} + \frac{5}{20}$$

The result is a pair of fractions with the same denominator. Now you can add these fractions by adding numerators and keeping the denominator the same:

$$= \frac{13}{20}$$

Not horrible, right?

In some cases, after you finish adding a pair of fractions, the result can be simplified. For example, consider the addition problem $\frac{1}{6} + \frac{3}{4}$. You can solve this problem using cross-multiplication, as follows:

$$\frac{1}{6} + \frac{3}{4} = \frac{4}{24} + \frac{18}{24} = \frac{22}{24}$$

As it stands, this problem is still incomplete, because the answer can be simplified by dividing both the numerator and the denominator by 2:

$$= \frac{22 \div 2}{24 \div 2} = \frac{11}{12}$$

This simplified version of the fraction is most likely the one that will appear as a multiple-choice answer on the ACCUPLACER.

As a final example, here's how you add $\frac{6}{7} + \frac{3}{5}$:

$$\frac{6}{7} + \frac{3}{5} = \frac{30}{35} + \frac{21}{35} = \frac{51}{35}$$

This time, the result is an improper fraction, so you need to change it to a mixed number (as I show you in "Converting improper fractions to mixed numbers"):

$$= 1\frac{16}{35}$$

SUBTRACTING FRACTIONS WITH CROSS-MULTIPLICATION

Cross-multiplying works equally well for subtracting fractions with different denominators:

1. Cross-multiply the two fractions to find the new numerators.

2. Multiply the denominators to find the new denominator.

For example, to subtract $\frac{7}{9} - \frac{3}{5}$, cross-multiply the two fractions — $7 \times 5 = 35$ and $9 \times 3 = 27$ — and place these result in the numerator to subtract in the next step. Then multiply $9 \times 5 = 45$.

$$\frac{7}{9} - \frac{3}{5} = \frac{35}{45} - \frac{27}{45}$$

Again, the result is two fractions with the same denominator. Now subtract the numerators and leave the denominator the same:

$$= \frac{8}{45}$$

As when subtracting fractions with the same denominator, when you subtract a smaller fraction from a greater one, the result is a negative fraction. For example:

$$\frac{1}{4} - \frac{5}{7} = \frac{7}{28} - \frac{20}{28} = -\frac{13}{28}$$

Here, the subtraction $7 - 20 = -13$ results in the negative fraction $-\frac{13}{28}$.

In some cases when subtracting, you may need to reduce your answer. For example:

$$\frac{3}{8} - \frac{1}{10} = \frac{30}{80} - \frac{8}{80} = \frac{22}{80}$$

To complete this problem, reduce this fraction by a factor of 2:

$$= \frac{22 \div 2}{80 \div 2} = \frac{11}{40}$$

Method #2 — Increasing the terms of one fraction

When the denominator of one fraction is divisible by the denominator of another fraction, you can get the denominators the same by increasing the terms of the smaller denominator. In this section, I show you how this method works, and why it can be preferable to cross-multiplication.

ADDING FRACTIONS BY INCREASING THE TERMS OF ONE FRACTION

Consider the problem $\frac{3}{8} + \frac{9}{16}$. If you try to add it using cross-multiplication, this is the result:

$$\frac{3}{8} + \frac{9}{16} = \frac{48}{128} + \frac{72}{128} = \frac{120}{128}$$

These numbers grew large very quickly, which increases your chance of making a calculation error. Worse yet, you're still not done, because you need to reduce this final fraction.

However, there's an easier way to do this problem. Notice that the denominator 8 is a factor of the denominator 16. This fact allows you to get the denominators the same in the following way:

1. Increase the terms of the fraction with the lower denominator so that this denominator matches that of the other fraction (see "Increasing the terms of fractions").

2. Add the fractions using the rules for fractions with the same denominator (see "Adding fractions with the same denominator").

I'll use this method to add $\frac{3}{8} + \frac{9}{16}$, increasing the terms of $\frac{3}{8}$ so that the denominator becomes 16, and then adding the results:

$$\frac{3}{8} + \frac{9}{16} = \frac{6}{16} + \frac{9}{16} = \frac{15}{16}$$

Much better, right?

Remember that when adding fractions by any method, you still may have to change an improper fraction to a mixed number. For example, to add $\frac{7}{10} + \frac{13}{20}$, increase the terms of the first fraction by a factor of 2:

$$\frac{7}{10} + \frac{13}{20} = \frac{14}{20} + \frac{13}{20} = \frac{27}{20}$$

Now, change the resulting improper fraction to a mixed number:

$$= 1\frac{7}{20}$$

TIP

Here's a quick trick that you can use in some cases when adding fractions: When a pair of fractions both have 1 in the numerator, you can add them quickly and easily as follows:

1. **Add the denominators to find the numerator of the answer.**

2. **Multiply the denominators to find the denominator.**

For example,

$$\frac{1}{2} + \frac{1}{3} = \frac{5}{6} \qquad \frac{1}{4} + \frac{1}{9} = \frac{13}{36} \qquad \frac{1}{10} + \frac{1}{13} = \frac{23}{130}$$

With a little practice, you can amaze your friends by doing this calculation in your head!

Remember, in some cases, you'll need to reduce the result. For example:

$$\frac{1}{4} + \frac{1}{10} = \frac{14}{40} = \frac{7}{20}$$

SUBTRACTING FRACTIONS BY INCREASING THE TERMS OF ONE FRACTION

As when adding fractions, in some cases when subtracting fractions, you can avoid the large numbers that cross-multiplication sometimes produces.

1. **Increase the terms of the fraction with the lower denominator so that this denominator matches that of the other fraction (see "Increasing the terms of fractions").**

2. **Subtract the fractions using the rules for fractions with the same denominator (see "Subtracting fractions with the same denominator").**

For example, you could calculate $\frac{17}{18} - \frac{7}{9}$ using cross-multiplication (try it if you dare!), but an easier way is to change the terms of $\frac{7}{9}$, changing the denominator to 18.

$$\frac{17}{18} - \frac{7}{9} = \frac{17}{18} - \frac{14}{18} = \frac{3}{18}$$

In this case, the result can be simplified:

$$= \frac{3 \div 3}{18 \div 3} = \frac{1}{6}$$

This method works equally when subtracting a smaller fraction minus a larger one. For example, to subtract $\frac{3}{14} - \frac{5}{7}$, increase the terms of the second fraction, changing its denominator to 14:

$$\frac{3}{14} - \frac{5}{7} = \frac{3}{14} - \frac{10}{14} = -\frac{7}{14}$$

This time, simplify by dividing both the numerator and the denominator by 7:

$$-\frac{1}{2}$$

Multiplying and dividing fractions

Here's some good news: Multiplying and dividing fractions is usually a lot easier than adding or subtracting them. In this section, I discuss both of these operations.

Multiplying fractions

You can multiply any pair of fractions without worrying whether their denominators are the same or different. And to sweeten the deal, you can often simplify the problem *before* you multiply to make the numbers smaller and easier to calculate.

To multiply a pair of fractions, multiply the two numerators to get the numerator of the answer, and multiply the two denominators to get the denominator. For example,

$$\frac{2}{5} \times \frac{1}{3} = \frac{2}{15} \qquad \frac{7}{8} \times \frac{3}{4} = \frac{21}{32} \qquad \frac{8}{9} \times \frac{7}{11} = \frac{56}{99}$$

In some cases, you can *cancel factors* in the numerator of one fraction and the denominator of the other fraction before you multiply. For example, to multiply $\frac{8}{9} \times \frac{9}{14}$, begin by canceling out a factor of 9 in the numerator of the second fraction and the denominator of the first fraction, replacing both with the number 1:

$$\frac{8}{9} \times \frac{9}{14} = \frac{8}{\cancel{9}^{1}} \times \frac{\cancel{9}^{1}}{14}$$

Next, divide both 8 and 14 by a factor of 2.

$$= \frac{\cancel{8}^{4}}{\cancel{9}^{1}} \times \frac{\cancel{9}^{1}}{\cancel{14}^{7}}$$

Now, finish up by multiplying across as usual, $4 \times 1 = 4$ and $1 \times 7 = 7$:

$$= \frac{4}{7}$$

Dividing fractions

To divide one fraction by another, turn the problem into fraction multiplication using the mnemonic (memory trick) **Keep-Change-Flip:**

1. *Keep* **the first fraction just as it is.**

2. *Change* **the division sign (÷) to a multiplication sign (×).**

3. *Flip* **the second fraction — that is, change it to its reciprocal (for more on reciprocals, flip to "Understanding reciprocals" earlier in this chapter.)**

From there, find the answer by multiplying the two resulting fractions. An example will help make this idea clear. Suppose you want to divide $\frac{5}{8} \div \frac{2}{3}$. Turn the problem into multiplication by applying **Keep-Change-Flip**, then multiply fractions across. (For more on fraction multiplication, see "Multiplying fractions" earlier in this section.)

$$\frac{5}{8} \div \frac{2}{3} = \frac{5}{8} \times \frac{3}{2} = \frac{15}{16}$$

In some cases when dividing fractions, you will have an opportunity to simplify the problem before you multiply by canceling factors. For example,

$$\frac{7}{10} \div \frac{4}{5} = \frac{7}{10} \times \frac{5}{4}$$

Now, cancel a factor of 5 in both the numerator and denominator, and complete the multiplication:

$$= \frac{7}{\cancel{10}^2} \times \frac{\cancel{5}^1}{4} = \frac{7}{8}$$

WARNING

Be careful not to cancel factors *before* changing the problem to multiplication. For example, suppose you want to divide $\frac{1}{6} \div \frac{6}{11}$. Here, you may be tempted to cancel factors of 6. However, watch what happens when you change the division to multiplication:

$$\frac{1}{6} \div \frac{6}{11} = \frac{1}{6} \times \frac{11}{6} = \frac{11}{36}$$

As you can see, when you change $\frac{6}{11}$ to $\frac{11}{6}$, the opportunity to cancel a factor of 6 disappears. *Never* try to cancel factors in a fraction division problem until you have changed it to multiplication.

Operations on mixed numbers

Just a guess here, but working with mixed numbers is probably not your favorite weekend activity when compared with, say, snowboarding or beach volleyball. This section probably won't change your mind, but I hope it will convince you that these problems now feel at least a little easier than they did when you were in middle school.

Adding mixed numbers — without carrying!

Adding mixed numbers isn't much more difficult than adding fractions. The method I outline next includes three steps that allow you to *avoid carrying,* which many students find confusing.

To add a pair of mixed numbers, follow these steps:

1. **Add the whole number values.**

2. **Add the fractional values.**

3. **Add the results of Steps 1 and 2.**

The only difficult step here is Step 2, where you add fractions (see "Adding fractions" earlier in this chapter).

For example, to add $6\frac{3}{4}+2\frac{4}{5}$, begin by adding the whole number parts of the mixed numbers. This is a breeze: $6+2=8$.

Next, add the two fractional parts:

$$\frac{3}{4}+\frac{4}{5}=\frac{15}{20}+\frac{16}{20}=\frac{31}{20}$$

This value is an improper fraction, so change it to a mixed number:

$$=1\frac{11}{20}$$

To complete the problem, add the results from the first two steps:

$$8+1\frac{11}{20}=9\frac{11}{20}$$

Subtracting mixed numbers — without borrowing!

Generally, subtracting mixed numbers is the most difficult calculation with fractions. This is a shame, because it doesn't have to be.

The problem here is usually because of borrowing, which involves a bunch of difficult, tedious steps. The method here allows you to subtract mixed numbers in three steps *without borrowing*:

1. **Subtract the whole number parts.**

2. **Subtract the fractional parts. (The result may be positive or negative.)**

3. **Add the results from Steps 1 and 2.**

For example, suppose you want to subtract $4\frac{2}{3}-1\frac{8}{15}$. To begin, subtract $4-1=3$.

Next, subtract $\frac{2}{3}-\frac{8}{15}$. You can use cross-multiplication here, but to avoid big numbers, try increasing the terms of $\frac{2}{3}$, changing the denominator to 15, before you subtract:

$$\frac{2}{3}-\frac{8}{15}=\frac{10}{15}-\frac{8}{15}=\frac{2}{15}$$

To complete the problem, you need to add the results from the first two steps: $3+\frac{2}{15}=3\frac{2}{15}$.

If those steps made sense to you, you're ready to tackle a hard problem that would normally require you to borrow in order to solve it: $6\frac{1}{3}-2\frac{3}{4}$.

To begin, subtract $6-2=4$. Easy, right?

Next, subtract $\frac{1}{3}-\frac{3}{4}$. This step involves some work:

$$\frac{1}{3}-\frac{3}{4}=\frac{4}{12}-\frac{9}{12}=-\frac{5}{12}$$

The result is a negative number. Don't be afraid, it won't bite you.

Now, to complete the problem, you need to add the results from the first two steps. But because your result from Step 2 is negative, this *addition* looks more like *subtraction*: $4-\frac{5}{12}$.

If this subtraction looks easy to you, please feel free to jump over this explanation and complete the problem. If not, here's a quick way to subtract a whole number from a fraction:

1. Subtract 1 from the whole number: $4 - 1 = 3$.

2. Copy the denominator: 12.

3. Subtract the denominator from the numerator to find the numerator of the answer: $12 - 5 = 7$.

$$4 - \frac{5}{12} = 3\frac{7}{12}$$

This method for subtracting mixed numbers without borrowing always works. And if you've struggled to master the skill of subtracting mixed numbers, or simply given up on it, I think you'll agree that each of these steps is a lot easier than the usual way that mixed number subtraction is taught.

Multiplying mixed numbers

To multiply mixed numbers, first turn them into improper fractions (see "Converting mixed numbers to improper fractions"). Then, multiply as you normally would (see "Multiplying fractions").

For example, to multiply $3\frac{1}{2} \times 1\frac{3}{5}$, begin by changing the mixed numbers to their equivalent improper fractions:

$$3\frac{1}{2} \times 1\frac{3}{5} = \frac{7}{2} \times \frac{8}{5}$$

Now, simplify the problem by canceling a factor of 2, then multiply:

$$= \frac{7}{\cancel{2}^1} \times \frac{\cancel{8}^4}{5} = \frac{28}{5}$$

Complete the problem by converting this improper fraction to a mixed number (see "Converting mixed numbers to improper fractions"). To do so, divide $28 \div 5 = 5r3$, so:

$$= 5\frac{3}{5}$$

Dividing mixed numbers

You can divide mixed numbers by changing them to improper fractions (see "Converting mixed numbers to improper fractions") and performing the division as you normally would (see "Dividing fractions"). For example, to divide $2\frac{1}{4}$ by $1\frac{5}{7}$, begin by converting these two mixed numbers to improper fractions:

$$2\frac{1}{4} \div 1\frac{5}{7} = \frac{9}{4} \div \frac{12}{7}$$

Next, change the problem to multiplication using **Keep-Change-Flip**:

$$= \frac{9}{4} \times \frac{7}{12}$$

You can make this problem easier by canceling a factor of 3 before you multiply:

$$= \frac{\cancel{9}^3}{4} \times \frac{7}{\cancel{12}^4} = \frac{21}{16}$$

To complete the problem, convert the resulting improper fraction to a mixed number as I show you in "Converting smaller improper fractions." To do this, write down the number 1, then copy the denominator of 16, and subtract $21 - 16 = 5$ to find the numerator:

$$= 1\frac{5}{16}$$

Chapter 11

Decimals, Percents, and Word Problems

This chapter continues to build on arithmetic skills discussed in Chapter 10. Starting with decimals and percents, you bring together what you know about fractions. You discover how to solve problems that involve comparing values and converting among fractions, decimals, and percents. And to finish up, you apply your arithmetic skills to word problems.

By the end of this chapter, you'll be well prepared to take on the arithmetic practice questions in Chapter 12.

Decimals

Decimals are the simplest way to represent positive values that are smaller than 1 — that is, parts of a larger whole value. And even if math is not your favorite thing, I promise you that you have no reason to dislike decimals.

For starters, you work with decimals just about every day, whenever you use money. A dollar is divided into 100 cents, each of which can be represented by the decimal $.01. Furthermore, working with decimals is a lot easier than working with fractions.

In the next section, I show you what you need to know about decimals to do well on the ACCUPLACER.

Decimal basics

Like fractions, decimals offer a way to represent small values that are part of a greater whole. And the good news is that you will almost certainly find decimals easier to work with than fractions!

Writing a whole number as a decimal

You can write any whole number as a decimal by ending it with a decimal point and placing one or more zeros after it. For example,

$$7 = 7.0 \qquad 65 = 65.0 \qquad 239 = 239.0$$

In fact, people change whole numbers to decimals all the time when working with money, by adding two extra zeros — that is, by extending the number to two decimal places:

$$\$25 = \$25.00 \qquad \$999 = \$999.00 \qquad \$1,000,000 = \$1,000,000.00$$

Leading and trailing zeros

A *leading zero* in a decimal is a 0 placed just in front of the decimal point — for example, the 0 in 0.999. Leading zeros have no value, but are often included for clarity, so that the decimal cannot be mistaken for a whole number.

A *trailing zero* is a decimal placed after the last significant (non-zero) digit in the decimal. This is most common when working with money — for example, the 0 in $2.50.

Both leading and trailing zeros don't change the value of a decimal, and can be safely dropped or attached whenever you like.

Multiplying and dividing decimals by 10, 100, 1,000, and beyond

Any number that starts with 1 and is followed by all 0s is called a *power of 10*. Decimals are tailor-made for multiplying and dividing decimals by powers of 10. In this section, I show you how to perform this simple but useful calculation.

MULTIPLYING DECIMALS BY POWERS OF 10

The number 10 has *one* zero, so to multiply any decimal by 10, move the decimal point *one* place to the right. For example,

$$8.2 \times 10 = 82 \qquad 6.23 \times 10 = 62.3 \qquad 174.531 \times 10 = 1,745.31$$

Similarly, 100 has *two* zeros, so to multiply a decimal by 100, move the decimal point *two* places to the right. For example,

$$1.54 \times 100 = 154 \qquad 0.888 \times 100 = 88.8 \qquad 234.5 \times 100 = 23,450$$

Generally speaking, to multiply a decimal by a power of 10, you count the zeros in that number and move the decimal point that number of places to the right. For example,

$$3.6 \times 1000 = 3600 \qquad 0.721 \times 10,000 = 7210 \qquad 0.0046 \times 100,000 = 460$$

DIVIDING DECIMALS BY POWERS OF 10

The number 10 has *one* zero, so to divide a decimal by 10, move the decimal point *one* place to the left. For example,

$$42.6 \div 10 = 4.26 \qquad 5.731 \div 10 = 0.5731 \qquad 223.07 \div 10 = 22.307$$

Likewise, the number 100 has *two* zeros, so to divide a decimal by 100, move the decimal point *two* places to the left. For example,

$$77.2 \div 100 = 0.772 \qquad 924.1 \div 100 = 9.241 \qquad 0.4567 \div 100 = 0.004567$$

In general, to divide a decimal by a power of 10, you count the zeros in that number and move the decimal point that number of places to the left. For example,

$$912 \div 1000 = 0.912 \qquad 7.632 \div 10,000 = 0.0007632 \qquad 65,123.9 \div 100,000 = 0.651239$$

Operations on decimals

The good news here is that adding, subtracting, multiplying, and dividing decimals is not much more difficult than performing these operations on whole numbers. In this section, you discover how to set up and solve decimal calculations.

Adding decimals

REMEMBER

To add two or more decimals, line up the decimal points and add.

For example, suppose that you want to add $7.3 + 0.085 + 21.02$. Begin by stacking these three numbers so that the decimal points are aligned, one above the next:

```
    7.3
    0.085
+  21.02
```

Now, add as you normally would, placing the decimal point of the answer just below those above the line:

```
    7.3
    0.085
+  21.02
   28.405
```

So $7.3 + 0.085 + 21.02 = 28.405$.

Subtracting decimals

REMEMBER

Subtracting decimals is similar to adding them: Place the first decimal above the second, with the decimal points lined up, then subtract as you normally would.

For example, to subtract $25.4 - 1.019$, stack these decimals so that the decimal points match up:

```
   25.400
-   1.019
```

Notice that I filled in 25.400 with trailing zeros to make the problem look more like regular subtraction. To begin this problem, I need to borrow:

```
   25.4³10⁹10
-   1.0  1  9
   24.3  8  1
```

Therefore, $25.4 - 1.019 = 24.381$.

Multiplying decimals

REMEMBER

When multiplying decimals — unlike when adding or subtracting them — you don't have to worry about stacking them so that the decimal points line up. In fact, you can just ignore the decimal points until the end of the problem.

For example, to multiply 35.2×0.048, set up and solve the problem as you would with regular multiplication:

```
      35.2
×    0.048
      2816
    14080
    16896
```

When the calculation is complete, place the decimal point in the problem by counting and adding up the decimal places in the two numbers that you're multiplying. The decimal 35.2 has one decimal place — that is, one digit (2) to the right of the decimal point. Similarly, 0.048 has three digits (048) to the right of the decimal point, so it has 3 decimal places. Thus, the result has $3 + 1 = 4$ decimal places:

```
      35.2
×    0.048
      2816
    14080
   1.6896
```

Therefore, $35.2 \times 0.048 = 1.6896$.

Dividing decimals

REMEMBER

When dividing decimals, follow these steps:

1. Change the divisor (the number you're dividing by) to a whole number by moving the decimal point to the right.

2. Move the decimal point in the dividend (the number you're dividing) the *same number of decimal places.*

3. Place the decimal point in the quotient (the answer) *just above* the decimal point in the dividend.

For example, to divide $0.365 \div 0.05$, set up the problem as follows:

$$.05\overline{)\,.365}$$

Note that I dropped the leading zeros in both numbers, but *not* the place-holding zero in .05. Now, to change this decimal to a whole number, move the decimal point in .05 two places to the right, and do the same with the decimal point in .365:

$$5.\overline{)\,36.5}$$

Next, place a decimal point directly above the one in 36.5, and do the long division.

```
       07.3
5.)   36.5
      35
      ──
      15
      15
      ──
       0
```

Therefore, $0.365 \div 0.05 = 7.3$.

Percents

Like fractions and decimals, percents are a third way of representing portions of a whole number value. In this section, I show you how to solve the types of percent problems you are most likely to see on the ACCUPLACER. I also show you how to understand and calculate more advanced percent decrease and increase problems.

Calculating percents

Finding a percent of a number is a common calculation typically performed in a wide variety of business transactions. In this section, you become an expert in calculating percents.

Percents everyone should know by heart

Some percent values are used so commonly, and are so easy to work with, that you can often compute them without a calculator, and in some cases in your head. What follows is a quick guide to making fast, accurate percent calculations for a few common percents.

CALCULATING 50%

TIP

Remember that 50% means *half*. So, to calculate 50% of a number, divide that number by 2. For example,

50% of 80 is 40.   50% of 222 is 111.   50% of 17 is 8.5.

CALCULATING 25%

TIP

The value 25% equals $\frac{1}{4}$. Calculate 25% of a number by dividing that number by 4 — or, by cutting it in half twice. For example,

25% of 88 is 22.   25% of 600 is 150.   25% of 14 is 3.5.

CALCULATING 10%

TIP

You can calculate 10% of any number by moving the decimal point one place to the left.

10% of 28 is 2.8.   10% of 1776 is 177.6.

If the number happens to end with zero, just drop this zero:

10% of 760 is 76.   10% of 45,000 is 4500.

CALCULATING 20%

Next time you receive great service in a restaurant, make your server happy by leaving a 20% tip. The value of 20% is double the value of 10%. So, calculate 20% of a number by finding 10% and then doubling the number. For example,

20% of 40 is 4 doubled, which equals 8.

20% of 310 is 31 doubled, which equals 62.

20% of 23 is 2.3 doubled, which equals 4.6.

CALCULATING 5%

Notice that 5% is half of 10%. This fact helps you to calculate 5% of a number easily by finding 10% and then taking half of this result. For example,

5% of 28 is half of 2.8, which equals 1.4.

5% of 900 is half of 90, which equals 45.

5% of 1250 is half of 125, which is 62.5.

Reverse calculation

In some cases, you can calculate an apparently difficult percent problem easily by reversing it.

For example, consider 48% of 10. You can rewrite this problem by reversing it, as 10% of 48, which equals 4.8.

Here are a few more examples:

18% of 50 = 50% of 18 = 9
120% of 25 = 25% of 120 = 30
64% of 5 = 5% of 64 = 3.2

This trick always works, so keep an eye open for places to try it out!

Calculating with percents by multiplying decimals

When the percentage you want to calculate doesn't include one or more numbers that are easy to work with, calculate it using decimal multiplication as follows:

1. **Change the percent to a decimal by moving the decimal point two places to the left and dropping the percent sign.**

2. **Multiply this value by the number.**

For example, to find 62% of 350, calculate

$0.62 \times 350 = 217$.

Similarly, 48% of 85 can be rewritten and solved as

$0.48 \times 85 = 40.8$.

And calculate 120% of 22 as follows:

$1.2 \times 22 = 26.4$.

Solving more difficult percent problems using word equations

Percent problems usually involve three parts: a percent, a starting number, and an ending number. For example,

75%	of	40	is	30
Percent		Starting number		Ending number

The previous section describes how to find the ending number by direct calculation. For example,

75%	of	40	is	?
Percent		Starting number		Ending number

Percent problems get trickier, though, when they ask you to supply either the percent or the starting number. For example,

?%	of	40	is	30
Percent		Starting number		Ending number

75%	of	?	is	30
Percent		Starting number		Ending number

Problems like these can be confusing. To solve them, set up an equation using the information in Table 11-1 as a guide for how to change each word.

TABLE 11-1 Key to Converting a Word Problem to an Equation

Of	%	Is, equals	What (number)
× (Multiplication sign)	× 0.01	=	n

Here is an example:

What percent of 40 is 30?

To solve this problem, change the words in this question into an equation as follows:

What	percent	of	40	is	30?
n	× 0.01	×	40	=	30

Now, gather all these elements together as an equation:

$$n \times 0.01 \times 40 = 30$$

The parentheses around (0.01) remind you to multiply, and then solve the problem:

$$0.40n = 30$$

$$\frac{0.40n}{0.40} = \frac{30}{0.40}$$

$$n = 75$$

Therefore, 75% of 40 is 30.

As another example, consider the following problem:

24 equals 12.5% of what number?

Again, begin by changing every word in this question to a symbol:

24	equals	12.5	%	of	what number?
24	=	12.5	× 0.01	×	n

Gather these symbols up as an equation, and solve:

$$24 = 12.5 \times 0.01 \times n$$

$$24 = 0.125n$$

$$\frac{24}{0.125} = \frac{0.125n}{0.125}$$

$$192 = n$$

Therefore, 24 is 12.5% of 192.

Percent decrease and increase

Percent decrease and increase problems usually involve money that is added to or subtracted from the dollar amount of an item or commodity. In this section, I show you how to approach and solve percent decrease and increase problems directly with a single calculation.

Percent decrease

Percent decrease problems include the following:

>> The price of an item on sale

>> Loss on an investment

>> Depreciation of an asset

Here's an example:

Rommell bought a TV that normally sells for $2700 on sale for 35% off. What was the sale price of the item?

Notice that when the price of the TV is reduced by 35%, its sale price becomes 65% of the original price, because 100% − 35% = 65%. So to find the sale price, you want to calculate 65% of $2700. To do this, convert 65% to a decimal (65% = 0.65) and multiply:

$$\$2700 \times 0.65 = \$1755$$

Percent increase

Percent increase problems are commonly about the following:

» Tip added to a check at a restaurant

» Tax added to a purchase

» Profit on an investment

For example, consider the following problem:

Kathleen invested $1200 in a stock and made a 15% profit on the investment. How much was her stock worth when she sold it?

Notice here that the question asks not how much money Kathleen made, but rather the entire value of the stock at the end of the transaction. To solve this problem, add 100% to the profit and then turn this value into a decimal:

$$100\% + 15\% = 115\% = 1.15$$

Now, multiply the original price of the stock by this value:

$$\$1200 \times 1.15 = \$1380$$

Converting Fractions, Decimals, and Percents

One important skill when working with fractions, decimals, and percents is converting values from one form to another. This skill is especially important on the ACCUPLACER, and you will probably see more than one question that requires you to demonstrate your knowledge in this area. In this section, you get comfortable with these types of conversions.

Common conversions among fractions, decimals, and percents

Some values of fractions, decimals, and percents are used so frequently that they may look familiar to you — and, if possible, should be committed to memory. Table 11-2 provides a handy reference for equivalent values of fractions, decimals, and percents.

TABLE 11-2 Common Conversions for Fractions, Decimals, and Percents

Fractions	Decimals	Percents
1	1.0	100%
$\frac{1}{2}$	0.5	50%
$\frac{1}{3}$	$0.\overline{3}$	$33\frac{1}{3}\%$
$\frac{2}{3}$	$0.\overline{6}$	$66\frac{2}{3}\%$
$\frac{1}{4}$	0.25	25%

(continued)

Fractions	Decimals	Percents
$\frac{3}{4}$	0.75	75%
$\frac{1}{5}$	0.2	20%
$\frac{2}{5}$	0.4	40%
$\frac{3}{5}$	0.6	60%
$\frac{4}{5}$	0.8	80%
$\frac{1}{8}$	0.125	12.5%
$\frac{3}{8}$	0.375	37.5%
$\frac{5}{8}$	0.625	62.5%
$\frac{7}{8}$	0.875	87.5%
$\frac{1}{10}$	0.1	10%
$\frac{3}{10}$	0.3	30%
$\frac{7}{10}$	0.7	70%
$\frac{9}{10}$	0.9	90%
$\frac{1}{20}$	0.05	5%
$\frac{1}{25}$	0.04	4%
$\frac{1}{50}$	0.02	2%
$\frac{1}{100}$	0.01	1%

Fraction-decimal conversions

ACCUPLACER questions often ask you to change a fraction to a decimal, or vice versa. In this section, I show you how to make both types of conversions.

Changing fractions to decimals

Remember that a fraction is simply another way to express division. To change a fraction to a decimal, divide the numerator (top number) by the denominator (bottom number).

For example, to convert the fraction $\frac{7}{16}$ to a decimal, divide 7 by 16. To do this, append enough trailing zeros to 7 so that you can complete the division.

$$
\begin{array}{r}
0.4375 \\
16\overline{)7.0000} \\
\underline{6\,4} \\
60 \\
\underline{48} \\
120 \\
\underline{112} \\
80 \\
\underline{80} \\
\end{array}
$$

Therefore, $\frac{7}{16} = 0.4375$.

In some cases, this process may not terminate, but instead may result in a repeating decimal. For example, here's what happens when you convert $\frac{2}{11}$ to a decimal:

$$
\begin{array}{r}
0.18 \\
11\overline{)2.00} \\
\underline{11} \\
90 \\
\underline{88} \\
2 \\
\end{array}
$$

In this last line, the 2 that you started with has reappeared, so if you continue dividing the numbers 1 and 8 will repeat forever, so $\frac{2}{11} = 0.\overline{18}$.

Changing decimals to fractions

REMEMBER

To convert a decimal to a fraction, begin by writing that decimal as the numerator of a fraction with 1 in the denominator. Then multiply both the numerator and denominator by a power of 10 large enough to turn the numerator into a whole number.

For example, here's how you change 0.79 to a fraction:

$$
\frac{0.79 \times 100}{1 \times 100} = \frac{79}{100}
$$

In some cases, you may need to simplify the resulting fraction. For example, suppose that you want to change 0.62 to a fraction:

$$
\frac{0.62 \times 100}{1 \times 100} = \frac{62}{100}
$$

To simplify this result, divide the numerator and the denominator by a common factor of 2:

$$
= \frac{62 \div 2}{100 \div 2} = \frac{31}{50}
$$

Decimal-percent conversions

Decimals and percents are similar to each other, and you can convert between them easily by moving the decimal point. In this section, I show you how it's done.

Changing decimals to percents

REMEMBER

To change a decimal to a percent, move the decimal point two places to the right and attach a percent sign. For example,

$$0.59 = 59\% \qquad 0.125 = 12.5\% \qquad 2.09 = 209\%$$

In some cases, you may need to include one or more place-holding zeros:

$$0.3 = 30\% \qquad 1.8 = 180\% \qquad 5.0 = 500\%$$

Changing percents to decimals

REMEMBER

To change a percent to a decimal, move the decimal point two places to the left and remove the percent sign.

For example, to change 37% to a decimal, remember that the whole number 37 has an unwritten decimal point after the 7. Move this decimal point two places to the left and remove the percent sign:

$$37\% = 0.37$$

Similarly, to change 60% to a decimal, recall that 60 has an unseen decimal point after the 0. Move this decimal point two places to the left and remove the percent sign:

$$60\% = 0.6$$

As you can see, after making the conversion, you can safely drop the trailing 0 after the 6.

Fraction-percent conversions

In most cases, converting between fractions and percents in either direction requires an intermediate conversion to decimals. Here, you discover how to leverage the methods from earlier in this section to make these conversions.

Changing fractions to percents

REMEMBER

To change a fraction to a percent, first change the fraction to an equivalent decimal (see "Changing fractions to decimals"), and then change this decimal to a percent (see "Changing decimals to percents").

For example, to change $\frac{4}{9}$ to its equivalent percent, divide 4 by 9:

$$4 \div 9 = 0.\overline{4}$$

The result is a repeating decimal. Now, change this decimal to a percent by moving the decimal point two places to the right:

$$= 44.4...\%$$

When a percent includes a repeating decimal, this repeating portion is most commonly expressed as a fraction:

$$44\frac{4}{9}\%$$

Changing percents to fractions

REMEMBER

To change a percent to a fraction, first convert the percent to a decimal (see "Changing percents to decimals"), and then change this decimal to a fraction (see "Changing decimals to fractions"). For example, to change 55% to a fraction, first change it to a decimal by moving the decimal point two places to the left and dropping the percent sign:

$$55\% = 0.55$$

Now, place this decimal in the numerator of a fraction with 1 in the denominator, multiply both by 100, and then simplify:

$$= \frac{0.55 \times 100}{1 \times 100} = \frac{55 \div 5}{100 \div 5} = \frac{11}{20}$$

Comparing Values

One common type of problem on the ACCUPLACER asks you to compare values.

Comparing fractions with cross-multiplication

When working with whole numbers, you can easily tell when one value is greater than another. Not so with fractions! The ACCUPLACER almost always includes one or more questions that ask you to compare the values of fractions.

For example, consider the fractions $\frac{2}{5}$ and $\frac{4}{11}$. Which of these two fractions is larger?

An easy way to find out is to use cross-multiplication, which is a simple but useful tool with a variety of applications. In Chapter 10, I show you how to use cross-multiplication to add and subtract fractions with different denominators.

REMEMBER

To compare a pair of fractions using cross-multiplication, always begin with the numerator (top number) of the first fraction:

1. **Multiply the numerator of the first fraction by the denominator (bottom number) of the second fraction, and pair this value with the first fraction.**

2. **Multiply the denominator of the first fraction by the numerator of the second fraction, and pair this value with the second fraction.**

3. **Compare the two values that you calculated.**

    The greater value is paired with the greater fraction; if the values are equal, the fractions are equal.

For example, here's how you cross-multiply $\frac{2}{5}$ and $\frac{4}{11}$:

$$\frac{2}{5} \qquad\qquad \frac{4}{5}$$
$$2 \times 11 = 22 \qquad 5 \times 4 = 20$$

As you can see, $\frac{2}{5}$ is paired with the greater value of 22, so $\frac{2}{5} > \frac{4}{11}$.

Comparing decimals

Comparing decimals is relatively easy, provided you take an orderly approach so you don't get confused. Here's what I recommend:

REMEMBER

1. **Add enough trailing zeros so that all the decimals have the same length.**

2. **Drop the decimal points and all leading zeros.**

3. **Compare the numbers.**

Here's an example:

EXAMPLE

Which of the following decimals is greatest?

(A) 0.02

(B) 0.021

(C) 0.003

(D) 0.0029

The longest of these four decimals is 0.0029, which has four decimal places, so extend all four decimals to four places with trailing zeros:

| 0.0200 | 0.0210 | 0.0030 | 0.0029 |

Now, drop the decimal points and all leading zeros:

| 200 | 210 | 30 | 29 |

The greatest number here is 210, so the answer is B.

Comparing fractions, decimals, and percents

Another common type of ACCUPLACER problem asks you to compare values that are expressed as fractions, decimals, or percents. Perhaps the easiest way to handle problems like this is to convert *every* value to a decimal, and then compare the decimals (flip to "Comparing decimals" earlier in this chapter). Here's an example:

EXAMPLE

Which of the following shows the correct ordering of the three values 49%, 0.049, and $\frac{4}{9}$ from least to greatest?

(A) $\frac{4}{9} < 49\% < 0.049$

(B) $\frac{4}{9} < 0.049 < 49\%$

(C) $0.049 < \frac{4}{9} < 49\%$

(D) $49\% < 0.049 < \frac{4}{9}$

To solve this problem, begin by converting 49% to a decimal:

$49\% = 0.49$

Next, change $\frac{4}{9}$ to a decimal:

$$
\begin{array}{r}
0.444 \\
9\overline{)4.000} \\
\underline{3\,6} \\
40 \\
\underline{36} \\
40 \\
\underline{36} \\
4
\end{array}
$$

As you can see, $\frac{4}{9} \approx 0.\overline{4}$.

Now, arrange the three decimal values 0.049, 0.49, and 0.44 from lowest to highest:

$$0.049 < 0.44... < 0.49$$

Thus, the smallest value is 0.049, so the answer is C.

Word Problems

Many students find word problems confusing, because they are unclear exactly what the problem is asking them to do. Word problems on the ACCUPLACER can usually be split into three types:

» Addition, subtraction, multiplication, and division problems in which the operation you need to use isn't given explicitly

» Whole number maximum or minimum — real-world problems in which you need to calculate a whole-number value without going over or under another value

» Parts of a whole — finding a missing portion of a whole item using fractions, decimals, or percents

Whose operation is this, anyway?

On the Arithmetic portion of the ACCUPLACER, a word problem often tests whether you can assess which operation — addition, subtraction, multiplication, or division — you need to apply to solve the problem.

Addition

Word problems that require addition often give you two or three values and then ask you to find the *sum* or *total*. Here's an example:

EXAMPLE

Coming home from work, Kaleb spent $48.85 on groceries at the supermarket, then bought a $5.00 box of candy from a group of local students who were outside the store. After that, he picked up his dry cleaning at a cost of $12.50. What was the total amount that he spent on these three transactions?

To find the answer, add the three values:

$$\$48.85 + \$5.00 + \$12.50 = \$66.35$$

So Kaleb spent $66.35.

Subtraction

Subtraction word problems usually ask you to find a *difference*, how much **more or less**, or in some cases the distance between two places. Here's an example:

EXAMPLE

Travis noticed that at the beginning of a car trip to his grandmother's house, the odometer read 82,745 miles. When his family arrived, the odometer read 83,107 miles. How many miles did the car travel on this trip?

To solve, subtract the greater value from the lesser one:

$$83,107 - 82,745 = 362$$

Thus, the car traveled 362 miles.

Multiplication

Word problems that require you to multiply usually mention that a group of people each receive a set number of items or perform a set number of tasks. Here's an example:

EXAMPLE

As part of a term project for her drama class, Elizabeth wants to give a 39-page copy of a play to each of the 14 students who will participate in it. How many pages will all these copies contain?

To find the solution, multiply 39 by 14:

$$39 \times 14 = 546$$

So Elizabeth's copies will contain 546 pages.

Division

Division word problems often present you with a large number of items, and ask you to split them into equal groups. Here's an example:

EXAMPLE

Ms. Blackboard brought a bag containing 432 candies as a treat for her class and wants to give each of her 27 students an equal number of them. How many candies does each student receive?

To calculate the answer, divide 432 by 27:

$$432 \div 27 = 16$$

In some cases, a division problem asks you to find a remainder. Here's an example:

EXAMPLE

Sean recently found a box of 467 vintage baseball cards in his attic. He gave each of his 7 grandchildren an equal number of them. When this was done, he still had a few cards left over, which he gave to the youngest grandchild. How many extra cards did the youngest child receive?

This problem asks you to divide 467 by 7, and then find the remainder:

$$467 \div 7 = 66r5$$

So, each child received 66 baseball cards, and the youngest child received an additional 5 baseball cards.

To the max (or min)

A whole-number maximum or minimum problem asks you to calculate a whole-number value without going over or under another value. An example or two will make this clear:

EXAMPLE

An old freight elevator can hold no more than 1000 pounds. Tyler wants to load it with boxes that weigh 130 pounds each. What is the maximum number of whole boxes that Tyler can place on the elevator at one time?

A good first step here is to divide 1000 by 130:

$$1000 \div 130 \approx 7.7$$

Seeing this answer, you may be tempted to round 7.7 up to 8. But the problem asks for the maximum number of *whole* boxes that the elevator can hold. So Tyler can place at most 7 boxes on the elevator, because $130 \times 7 = 910$. If he tries to place 8 boxes on the elevator, $130 \times 8 = 1040$, which exceeds the weight allowed on the elevator.

Here's another example of this type of problem:

EXAMPLE

In return for a pair of pricey tickets to a show for her best friend and herself, Veronica promised her aunt and uncle that she would spend at least 20 hours mowing their lawn. The job takes 1.5 hours each time she does it, and she must complete each job after she begins. How many times does Veronica have to mow the lawn to keep her promise?

To solve this problem, begin by calculating $20 \div 1.5 = 13.\overline{3}$. This time, you may be tempted to round this value down to 13. But, if Veronica mows the lawn only 13 times, she has not kept her promise, because $1.5 \times 13 = 19.5$. However, $1.5 \times 14 = 21$, so if she completes the job 14 times, she has fulfilled her obligation.

Parts of the whole

Another common type of word problem on the ACCUPLACER requires you to find a part of a whole. In this section, I show you how to do this type of problem using fractions and percentages.

Fractions

Remember that all the fractional parts of a single item add up to 1. Here's an example:

EXAMPLE

After their grandmother baked a plate of chocolate chip cookies, Hazel ate $\frac{3}{8}$ of the cookies and her sister Edith ate $\frac{2}{5}$ of them. Their cousin, Claire, ate the rest. What fraction of the cookies did Claire eat?

To solve, let c stand for the fraction of the cookies that Claire ate, then set up the following equation:

$$\frac{3}{8} + \frac{2}{5} + c = 1$$

To begin, add $\frac{3}{8} + \frac{2}{5} = \frac{15}{40} + \frac{16}{40} = \frac{31}{40}$. So,

$$\frac{31}{40} + c = 1$$

Now, subtract $\frac{31}{40}$ from each side of the equation and simplify the result:

$$c = 1 - \frac{31}{40} = \frac{9}{40}$$

Therefore, Claire ate $\frac{9}{40}$ of the cookies.

Percents

A whole item divided into parts based on percentage will add up to 100%. Here's an example:

EXAMPLE

In an election for county dogcatcher, Ms. Gibbons received 34.5% of the vote, Mr. Thwack received 21.9%, and Ms. Kalabash received 18.7%. The rest of the votes went to Mr. Fassbinder. What percentage of the vote did Mr. Fassbinder receive?

To solve this problem, let f equal the percentage of votes that Mr. Fassbinder received, and then set up the following equation:

$$34.5\% + 21.9\% + 18.7\% + f = 100\%$$

Begin by adding the three percentages on the left side of the equation:

$$75.1\% + f = 100\%$$

Now, subtract 75.1% from both sides of the equation:

$$f = 100\% - 75.1\% = 24.9\%$$

So, Mr. Fassbinder received 24.9% of the votes.

» Solving problems using fractions, decimals, and percents

» Setting up and solving word problems

Chapter **12**
Arithmetic Practice Questions

Chapters 10 and 11 describe the skills necessary to do well on the ACCUPLACER Arithmetic Test. In this chapter, you put these skills to work with practice problems. All answers appear at the end of the chapter.

Arithmetic Practice Problems

Ready to start testing your arithmetic skills? This section includes practice problems from every topic covered in Chapters 10 and 11.

Whole numbers practice

1. What is 8,765,432 rounded to the nearest thousand?

2. If you multiply 786×2971, which of the following best approximates the answer?

  (A) 140,000

  (B) 240,000

  (C) 1,400,000

  (D) 2,400,000

3. What is the value of $15 - 4^2 \div (32 - 8 \times 3)$?

4. Which of the following inequalities is equivalent to the one shown in the figure?

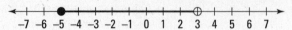

(A) $-5 \leq x < 3$

(B) $-5 < x \leq 3$

(C) $x \leq -5$ or $x > 3$

(D) $x < -5$ or $x \geq 3$

5. Which of the following would be the best estimation for $451,327 \div 8946$?

(A) 5

(B) 50

(C) 500

(D) 5000

6. Evaluate $(17 - 2^3) \times (6 + 10 \div 2)$.

7. Which of these inequalities expresses the solution set for the number line shown in the following figure?

(A) $-4 \leq x < 2$

(B) $-4 < x \leq 2$

(C) $x \leq -4$ or $x > 2$

(D) $x < -4$ or $x \geq 2$

8. What is 939,594 rounded to the nearest hundred?

Fractions practice

1. $\frac{1}{10} + \frac{1}{6} =$

2. $\frac{7}{9} + \frac{5}{9} =$

3. $\frac{6}{7} + \frac{2}{3} =$

4. $\frac{3}{4} + \frac{9}{20} =$

5. $\frac{11}{16} - \frac{5}{16} =$

6. $\frac{5}{8} - \frac{1}{24} =$

7. $\frac{1}{4} - \frac{7}{8} =$

8. $\frac{2}{7} - \frac{1}{13} =$

9. $\frac{1}{5} \times \frac{3}{10} =$

10. $\frac{2}{3} \times \frac{9}{14} =$

11. $\frac{15}{16} \times \frac{12}{25} =$

12. $\frac{9}{10} \times \frac{7}{100} =$

13. $\frac{3}{4} \div \frac{4}{5} =$

14. $\frac{9}{10} \div \frac{3}{5} =$

15. $\frac{8}{15} \div \frac{10}{21} =$

16. $\frac{11}{27} \div \frac{22}{23} =$

17. $2\frac{3}{8} + 4\frac{2}{3} =$

18. $3\frac{4}{5} + 5\frac{3}{4} =$

19. $4\frac{7}{16} - 2\frac{1}{4} =$

20. $5\frac{1}{5} - 3\frac{5}{6} =$

21. $2\frac{1}{5} \times 3\frac{3}{4} =$

22. $6\frac{2}{7} \times 1\frac{3}{11} =$

23. $5\frac{1}{2} \div 3\frac{5}{8} =$

24. $6\frac{2}{3} \div 10\frac{10}{11} =$

Decimals practice

1. $0.835 + 1.09 =$

2. $4.4 + 82.39 + 0.027 =$

3. $42.563 - 8.05 =$

4. $2.8 - 0.074 =$

5. $25.6 \times 0.14 =$

6. $0.0748 \times 3.6 =$

7. $1.74 \div 0.06 =$

8. $32.175 \div 1.8 =$

Percent practice

1. What is 20% of 140?

2. What is 48% of 50?

3. Jacqui took her friend, Bethany, out to lunch for her birthday. When the check arrived, it was for $45.00. If Jacqui included an 18% tip for the server, what did she pay in total for the lunch?

4. What percent of 39 is 117?

5. By changing phone plans, Blair received a 30% discount on a phone that normally sells for $780. How much did she pay for the phone?

6. What is 24% of 408?

7. What is 65% of 135?

8. Drew invested $6,600 in a stock that lost 15% of its value before Drew sold it. How much was the stock worth at the time of sale?

9. What is 5% of 680?

10. What is 132% of 25?

11. What percent of 120 is 72?

12. 28 is 70% of what number?

Fraction, decimal, percent conversion practice

1. Convert the decimal 0.72 to a fraction.

2. Write the decimal 0.375 as a percent.

3. Change $\frac{13}{16}$ to a percent.

4. Write the fraction $\frac{6}{25}$ as a decimal.

5. How do you rewrite 64% as a fraction?

6. Convert 30.2% to a decimal.

7. Change 0.444 to a fraction.

8. How do you rewrite 32.5% as a fraction?

9. What is the value of 188.1% expressed as a decimal?

10. Change $\frac{17}{20}$ to a decimal.

11. Rewrite $\frac{5}{11}$ as a percent.

12. Convert 0.086 to a percent.

Comparing values practice

1. Of $\frac{2}{3}$, $\frac{5}{7}$, and $\frac{7}{10}$, which is the greatest fraction?

2. Which of the following decimals is the greatest?

    (A) 0.1811

    (B) 0.088

    (C) 0.0188

    (D) 0.801

3. Which of the following values is the greatest: $\frac{57}{10,000}$, 5.7%, or 0.0057?

4. Which of these decimals is the greatest?

    (A) 0.3054

    (B) 0.306

    (C) 0.32

    (D) 0.321

5. Of 0.98%, $\frac{9}{800}$, 0.098, which value is the least?

6. Which of the following properly orders the fractions $\frac{4}{7}$, $\frac{6}{11}$, and $\frac{7}{12}$ from smallest to greatest?

    (A) $\frac{4}{7} < \frac{6}{11} < \frac{7}{12}$

    (B) $\frac{6}{11} < \frac{4}{7} < \frac{7}{12}$

    (C) $\frac{6}{11} < \frac{7}{12} < \frac{4}{7}$

    (D) $\frac{7}{12} < \frac{4}{7} < \frac{6}{11}$

Word problem practice

1. Anjelika spent $35.75 on a sweater and $56.99 on a pair of shoes. How much more did the shoes cost than the sweater?

2. Serena baked a batch of brownies before work. When she returned home, she found that her roommate Wendy had eaten $\frac{1}{6}$ of the brownies and Wendy's boyfriend, Carter, had eaten $\frac{1}{4}$ of them. What fraction of the brownies were left?

3. Eric is supervising a group of 57 children at a summer camp. He wants to rent a set of row-boats, each of which will hold a maximum of six children. What is the minimum number of rowboats that Eric will need to rent to make sure that every child is included?

4. As an incentive for her sales staff of 12 people, Shondra allocated $7500 for a quarterly bonus, which she promised to split evenly among all salespeople who made their quotas. If all 12 people succeeded, how much money did each person receive?

5. Norma produced a report on how the non-profit corporation that she works for had raised funds in the preceding year. She found that 36.8% of funding came from a Rockefeller Foundation grant, 22.7% came from the United Way, and 19.4% came from the Bill and Melinda Gates Foundation. If the remaining funding came from local donations, what percentage of the total did this constitute?

6. Randolph weighs himself and his two children. He finds that he weighs 186 pounds, his son weighs 77 pounds, and his daughter weighs 39 pounds. If they all stand on the scale together, what does the scale read?

7. David hires day laborers to help with local construction projects, paying $145 per day. If his daily budget cannot go over $1250, what is the maximum number of people he can hire in a single day?

8. Every year, Roy gives each of his grandchildren $75 for his or her birthday. If he has 17 grandchildren, what is his total expense each year?

9. An art teacher bought a box of 256 pastels and distributed them evenly among 13 art tables in his classroom. He kept the remaining pastels in his desk to use as replacements. How many pastels were among these replacements?

10. Candice read 25% of a novel during the first week it was assigned and 40% of it the following week. At that point, she had 105 pages left to read. How many pages did the novel have altogether?

Arithmetic Practice Problem Answers and Explanations

So, how did you do? Here are the answers to the problems from this chapter, with explanations to help you understand how to solve each problem.

Whole numbers practice answers

1. **8,765,000.** To round 8,765,432 to the nearest thousand, focus on the hundreds digit, which is 4. Change this digit and all digits to its right to 0, and keep the thousands digit the same:

   $$8,765,432 \to 8,765,000$$

2. **D. 2,400,000.** Estimate as follows:

   $$786 \times 2971 \approx 800 \times 3000 = 2,400,000$$

3. **13.** Begin by evaluating the value inside the parentheses (first the multiplication, then the subtraction). Next, evaluate the exponent, then the division, and finally the subtraction.

   $$15 - 4^2 \div (32 - 8 \times 3)$$
   $$= 15 - 4^2 \div (32 - 24)$$
   $$= 15 - 4^2 \div 8$$
   $$= 15 - 16 \div 8$$
   $$= 15 - 2$$
   $$= 13$$

4. **A. $-5 \le x < 3$.** The solution set is the area between the two numbers, so this can be any value from -5 to 3. Additionally, the closed circle on -5 is represented as the inequality $\le$ and the open circle on 3 is represented as the inequality $<$.

5. **B. 50.** Estimate as follows:

   $$451,327 \div 8946 \approx 450,000 \div 9000 = 50$$

6. **99.** Begin by evaluating the values inside the first parentheses (exponent, then subtraction). Then move on to the second parentheses (division, then addition). Finally, multiply these values.

   $$(17 - 2^3) \times (6 + 10 \div 2)$$
   $$= (17 - 8) \times (6 + 10 \div 2)$$
   $$= 9 \times (6 + 10 \div 2)$$
   $$= 9 \times (6 + 5)$$
   $$= 9 \times 11$$
   $$= 99$$

7. **D. $x < -4$ or $x \ge 2$.** The solution set includes two separate sections on the number line, so this must be represented as two separate inequalities with *or*. Additionally, the open circle on -4 is represented as the inequality $<$ and the closed circle on 2 is represented as the inequality $\le$.

8. **939,600.** To round 939,594 to the nearest hundred, focus on the tens digit, which is 9. Change this digit and all digits to its right to 0, and add 1 to the hundreds digit 5:

   $$939,594 \to 939,600$$

Fractions practice answers

1. $\frac{4}{15}$. Both fractions have numerators of 1, so add the denominators to find the numerator, and multiply the denominators to find the denominator, then reduce: $\frac{1}{10} + \frac{1}{6} = \frac{16}{60} = \frac{4}{15}$

2. $1\frac{1}{3}$. Add the numerators and keep the denominator, then convert to a mixed number and simplify: $\frac{7}{9} + \frac{5}{9} = \frac{12}{9} = 1\frac{3}{9} = 1\frac{1}{3}$

3. $1\frac{11}{21}$. Cross-multiply to add, then convert to a mixed number: $\frac{6}{7} + \frac{2}{3} = \frac{18}{21} + \frac{14}{21} = \frac{32}{21} = 1\frac{11}{21}$

4. $1\frac{3}{20}$. Increase the terms of the first fraction to make the denominators equal, then add the fractions and convert to a mixed number, then simplify the fraction: $\frac{3}{4} + \frac{9}{20} = \frac{15}{20} + \frac{9}{20} = \frac{24}{20} = 1\frac{4}{20} = 1\frac{1}{5}$

5. $\frac{3}{8}$. Subtract the first numerator from the second, then simplify the result: $\frac{11}{16} - \frac{5}{16} = \frac{6}{16} = \frac{3}{8}$

6. $\frac{7}{12}$. Increase the terms of the first fraction to make the denominators equal, then subtract and reduce: $\frac{5}{8} - \frac{1}{24} = \frac{15}{24} - \frac{1}{24} = \frac{14}{24} = \frac{7}{12}$

7. $-\frac{5}{8}$. Increase the terms of the first fraction and subtract.

$$\frac{1}{4} - \frac{7}{8} = \frac{2}{8} - \frac{7}{8} = -\frac{5}{8}$$

8. $\frac{19}{91}$. Cross-multiply to subtract: $\frac{2}{7} - \frac{1}{13} = \frac{26}{91} - \frac{7}{91} = \frac{19}{91}$

9. $\frac{3}{50}$. Multiply the numerators and then multiply the denominators: $\frac{1}{5} \times \frac{3}{10} = \frac{3}{50}$

10. $\frac{3}{7}$. Cancel out factors, then multiply across:

$$\frac{2}{3} \times \frac{9}{14} = \frac{2^1}{3^1} \times \frac{9^3}{14^7} = \frac{3}{7}$$

11. $\frac{9}{20}$. Cancel out factors, then multiply across:

$$\frac{15}{16} \times \frac{12}{25} = \frac{15^3}{16^4} \times \frac{12^3}{25^5} = \frac{9}{20}$$

12. $\frac{63}{1000}$. Multiply across: $\frac{9}{10} \times \frac{7}{100} = \frac{63}{1000}$

13. $\frac{15}{16}$. Keep the first fraction, change from division to multiplication, flip the second fraction to its reciprocal, and then multiply across:

$$\frac{3}{4} \div \frac{4}{5} = \frac{3}{4} \times \frac{5}{4} = \frac{15}{16}$$

14. $1\frac{1}{2}$. Keep, change, flip, then cancel out factors and multiply:

$$\frac{9}{10} \div \frac{3}{5} = \frac{9}{10} \times \frac{5}{3} = \frac{9^3}{10^2} \times \frac{5^1}{3^1} = \frac{3}{2} = 1\frac{1}{2}$$

15. $1\frac{3}{25}$. Keep, change, flip, then cancel out factors and multiply:

$$\frac{8}{15} \div \frac{10}{21} = \frac{8}{15} \times \frac{21}{10} = \frac{8^4}{15^5} \times \frac{21^7}{10^5} = \frac{28}{25} = 1\frac{3}{25}$$

16. $\frac{23}{54}$. Keep, change, flip, then cancel out factors and multiply:

$$\frac{11}{27} \div \frac{22}{23} = \frac{11}{27} \times \frac{23}{22} = \frac{\cancel{11}^1}{27} \times \frac{23}{\cancel{22}^2} = \frac{23}{54}$$

17. $7\frac{1}{24}$. Add the whole numbers, then add the fractions by cross-multiplying. Convert the second fraction to a mixed number and add again:

$$2\frac{3}{8} + 4\frac{2}{3} = 6 + \frac{9}{24} + \frac{16}{24} = 6 + \frac{25}{24} = 6 + 1\frac{1}{24} = 7\frac{1}{24}$$

18. $9\frac{11}{20}$. Add the whole numbers, then add the fractions by cross-multiplying. Convert the second fraction to a mixed number and add again:

$$3\frac{4}{5} + 5\frac{3}{4} = 8 + \frac{16}{20} + \frac{15}{20} = 8 + \frac{31}{20} = 8 + 1\frac{11}{20} = 9\frac{11}{20}$$

19. $2\frac{3}{16}$. Increase the terms of the second fraction to make the denominators the same. Then subtract the whole numbers and subtract the fractions: $4\frac{7}{16} - 2\frac{1}{4} = 4\frac{7}{16} - 2\frac{4}{16} = 2\frac{3}{16}$

20. $1\frac{11}{30}$. Subtract the whole numbers, then subtract the fractions by cross-multiplying. This results in a negative fraction, so complete the final subtraction to get the result:

$$5\frac{1}{5} - 3\frac{5}{6} = 2 + \frac{6}{30} - \frac{25}{30} = 2 - \frac{19}{30} = 1\frac{11}{30}$$

21. $8\frac{1}{4}$. Convert both mixed numbers to improper fractions, then cancel factors and multiply across. Convert the result back to a mixed number: $2\frac{1}{5} \times 3\frac{3}{4} = \frac{11}{5} \times \frac{15}{4} = \frac{11}{\cancel{5}^1} \times \frac{\cancel{15}^3}{4} = \frac{33}{4} = 8\frac{1}{4}$

22. 8. Convert both mixed numbers to improper fractions, then cancel factors and multiply across. The result is a fraction with 1 in the denominator, so change this to a whole number: $6\frac{2}{7} \times 1\frac{3}{11} = \frac{44}{7} \times \frac{14}{11} = \frac{\cancel{44}^4}{\cancel{7}^1} \times \frac{\cancel{14}^2}{\cancel{11}^1} = 8$

23. $1\frac{15}{29}$. Convert both mixed numbers to improper fractions, then keep, change, flip. Cancel factors and multiply across. Convert the result back to a mixed number:

$$5\frac{1}{2} \div 3\frac{5}{8} = \frac{11}{2} \div \frac{29}{8} = \frac{11}{2} \times \frac{8}{29} = \frac{11}{\cancel{2}^1} \times \frac{\cancel{8}^4}{29} = \frac{44}{29} = 1\frac{15}{29}$$

24. $\frac{11}{18}$. Convert both mixed numbers to improper fractions, then keep, change, flip. Cancel factors and multiply across: $6\frac{2}{3} \div 10\frac{10}{11} = \frac{20}{3} \div \frac{120}{11} = \frac{20}{3} \times \frac{11}{120} = \frac{\cancel{20}^1}{3} \times \frac{11}{\cancel{120}^6} = \frac{11}{18}$

Decimals practice answers

1. 1.925. Line up the decimal points and add:

$$\begin{array}{r} 0.835 \\ + \quad 1.09 \\ \hline 1.925 \end{array}$$

2. 86.817. Line up the decimal points and add:

$$\begin{array}{r} 4.4 \\ 82.39 \\ + \quad 0.027 \\ \hline 86.817 \end{array}$$

3. 34.513. Line up the decimal points, filling with trailing zeros as needed, and subtract:

$$
\begin{array}{r}
42.563 \\
- \quad 8.050 \\
\hline
34.513
\end{array}
$$

4. 2.726. Line up the decimal points, filling with trailing zeros as needed, and subtract:

$$
\begin{array}{r}
2.800 \\
- \quad 0.074 \\
\hline
2.726
\end{array}
$$

5. 3.584. Multiply as with whole numbers, then add up the decimal places $1 + 2 = 3$ to find the number of decimal places in the answer:

$$
\begin{array}{r}
25.6 \\
\times \quad 0.14 \\
\hline
1024 \\
2560 \\
\hline
3.584
\end{array}
$$

6. 0.26928. Multiply as with whole numbers, then add up the decimal places $4 + 1 = 5$ to find the number of decimal places in the answer:

$$
\begin{array}{r}
0.0748 \\
\times \quad 3.6 \\
\hline
4488 \\
22440 \\
\hline
0.26928
\end{array}
$$

7. 29. To change the divisor 0.06 to a whole number, multiply both numbers by 100, changing the problem to $174 \div 6$, and then divide:

$$
\begin{array}{r}
029. \\
6.\overline{)174.} \\
12 \\
\hline
54 \\
54 \\
\hline
0
\end{array}
$$

8. 17.875. To change the divisor 0.06 to a whole number, multiply both numbers by 10, changing the problem to $321.75 \div 18$. Place the decimal point in the answer immediately above the decimal point inside the division, add a trailing zero, and then divide:

$$
\begin{array}{r}
017.875 \\
18.\overline{)321.750} \\
18 \\
\hline
141 \\
126 \\
\hline
157 \\
144 \\
\hline
135 \\
126 \\
\hline
90 \\
90 \\
\hline
0
\end{array}
$$

Percent practice answers

1. **28.** To calculate 20% of 140, find 10% and double the result: 20% of 140 is 14 doubled, which equals 28.

2. **24.** To find 48% of 50, calculate 50% of 48, which equals 24.

3. **$53.10.** To calculate a percent increase of 18% on top of $45, find 118% of $45:

   $$1.18 \times 45 = 53.1$$

4. **300%.** Rewrite the elements of the question as follows:

What	percent	of	39	is	117
n	(0.01)	×	39	=	117

   Now, gather all these elements together as an equation, and solve:

   $$n(0.01) \times 39 = 117$$
   $$0.39n = 117$$
   $$\frac{0.39n}{0.39} = \frac{117}{0.39}$$
   $$n = 300$$

5. **$546.** To calculate a percent decrease of 30% off of $780, calculate 70% of 780:

   $$.7 \times 780 = 546$$

6. **97.92.** Multiply 0.24 by 408:

   $$0.24 \times 408 = 97.92$$

7. **87.75.** Multiply 0.65 by 135:

   $$0.65 \times 135 = 87.75$$

8. **$5,610.** To calculate a percent decrease of 15% off of $6,600, find 85% of 6,600:

   $$.85 \times 6600 = 5610$$

9. **34.** To find 5% of 680, calculate 10% of 680 and then find half of the result: 5% of 680 is half of 68, which is 34.

10. **33.** 132% of 25 equals 25% of 132. To find this value, take half of 132, and then half of the result: 25% of 132 is half of 66, which is 33.

11. **60%.** Rewrite the elements of the question as follows:

What	percent	of	120	is	72
n	(0.01)	×	120	=	72

    Now, gather all these elements together as an equation, and solve:

    $$n(0.01) \times 120 = 72$$
    $$1.2n = 72$$
    $$\frac{1.2n}{1.2} = \frac{72}{1.2}$$
    $$n = 60$$

12. 40. 28 is 70% of what number? Again, begin by changing every word in this question to a symbol:

28	is	70	%		of	what number
28	=	70	(0.01)	×		n

Gather these symbols up as an equation, and solve:

$$28 = 70(0.01) \times n$$
$$28 = 0.7n$$
$$\frac{28}{0.7} = \frac{0.7n}{0.7}$$
$$40 = n$$

Fraction, decimal, percent conversion practice answers

1. $\frac{18}{25}$. Place the decimal in the numerator of a fraction with a denominator of 1, move the decimal points in both the numerator and denominator until both are whole numbers, then simplify the resulting fraction:

$$\frac{0.72}{1} = \frac{7.2}{10} = \frac{72}{100} = \frac{18}{25}$$

2. 37.5%. Move the decimal point two places to the right and attach a percent sign (%):

$$0.375 = 37.5\%$$

3. 81.25%. Divide 13 by 16 to change the fraction to a decimal, then move the decimal point two places to the right and attach a percent sign (%):

$$13 \div 16 = 0.8125 = 81.25\%$$

4. 0.24. Divide 6 by 25:

$$6 \div 25 = 0.24$$

5. $\frac{16}{25}$. Change 64% to a decimal by moving the decimal point two places to the left and dropping the percent sign (%). Then place this decimal in the numerator of a fraction with a denominator of 1 and move the decimal points of both the numerator and the denominator to the right until both are whole numbers:

$$0.64 = \frac{0.64}{1} = \frac{6.4}{10} = \frac{64}{100} = \frac{16}{25}$$

6. 0.302. Move the decimal point two places to the left and drop the percent sign:

$$30.2\% = 0.302$$

7. $\frac{111}{250}$. Change 0.444 to a fraction. Place the decimal in the numerator of a fraction with a denominator of 1, move the decimal points in both the numerator and denominator until both are whole numbers, then simplify the resulting fraction:

$$\frac{0.444}{1} = \frac{4.44}{10} = \frac{44.4}{100} = \frac{444}{1000} = \frac{111}{250}$$

8. $\frac{13}{40}$. Change 32.5% to a decimal by moving the decimal point two places to the left and dropping the percent sign (%). Then place this decimal in the numerator of a fraction with a denominator of 1 and move the decimal points of both the numerator and the denominator to the right until both are whole numbers:

$$0.325 = \frac{0.325}{1} = \frac{3.25}{10} = \frac{32.5}{100} = \frac{325}{1000} = \frac{13}{40}$$

9. **1.881.** Move the decimal point two places to the left and drop the percent sign:

$$188.1\% = 1.881$$

10. **0.85.** Divide 17 by 20:

$$17 \div 20 = 0.85$$

11. $45\frac{5}{11}\%$. Divide 5 by 11 to change the fraction to a decimal, then move the decimal point two places to the right and attach a percent sign (%):

$$5 \div 11 = 0.\overline{45} = 45.\overline{45}\%$$

Now, change the repeating decimal back to a fraction:

$$= 45\frac{5}{11}\%$$

12. **8.6%.** Convert 0.086 to a percent. Move the decimal point two places to the right and attach a percent sign (%):

$$0.086 = 8.6\%$$

Comparing values answers

1. $\frac{5}{7}$. Begin by cross-multiplying $\frac{2}{3}$ and $\frac{5}{7}$:

$$\frac{2}{3} \qquad \frac{5}{7}$$
$$2 \times 7 = 14 \quad 5 \times 3 = 15$$

As you can see, $\frac{5}{7}$ is paired with the greater value of 15, so you can discard $\frac{2}{3}$. Now, compare $\frac{5}{7}$ and $\frac{7}{10}$:

$$\frac{5}{7} \qquad \frac{7}{10}$$
$$5 \times 10 = 50 \quad 7 \times 7 = 49$$

This time, $\frac{5}{7}$ is paired with the greater value of 50, so $\frac{5}{7}$ is greatest.

2. **D. 0.801.** The longest decimals have four decimal places, so use trailing zeros to extend all the decimals to this length:

0.1811   0.0880   0.0188   0.8010

Now, drop the decimal points and all leading zeros:

1811   880   188   8010

Here, 8010 is greatest, so the answer is D.

3. 5.7%. Convert all values to decimals:

$$\frac{57}{10,000} = 0.0057$$
$$5.7\% = 0.057$$

Thus, the greatest value is 5.7%.

4. D. 0.321. Which of these decimals is the greatest? The longest decimal has four decimal places, so use trailing zeros to extend all the decimals to this length:

0.3054   0.3060   0.3200   0.3210

Now, drop the decimal points and all leading zeros:

3054   3060   3200   3210

Here, 3210 is greatest, so the answer is D.

5. 0.98%. Of 0.98%, $\frac{9}{800}$, 0.098, which value is the least? Convert all values to decimals:

$$0.98\% = 0.0098$$
$$\frac{9}{800} = 0.01125$$

Thus, the least value is 0.98%.

6. B. $\frac{6}{11} < \frac{4}{7} < \frac{7}{12}$. Begin by cross-multiplying $\frac{4}{7}$ and $\frac{6}{11}$:

$$\frac{4}{7} \quad \frac{6}{11}$$
$$4 \times 11 = 44 \quad 6 \times 7 = 42$$

As you can see, $\frac{4}{7}$ is paired with the greater value of 44, so $\frac{4}{7}$ is greater than $\frac{6}{11}$. Now, compare $\frac{4}{7}$ and $\frac{7}{12}$:

$$\frac{4}{7} \quad \frac{7}{12}$$
$$4 \times 12 = 48 \quad 7 \times 7 = 49$$

This time, $\frac{7}{12}$ is paired with the greater value of 49, so $\frac{7}{12}$ is greatest. Therefore, $\frac{6}{11} < \frac{4}{7} < \frac{7}{12}$, so the answer is B.

Word problem practice answers

1. 21.24. $56.99 - $35.75 = $21.24.

2. $\frac{7}{12}$. First, find out what portion of the brownies Wendy and Carter ate:

$$\frac{1}{6} + \frac{1}{4} = \frac{4+6}{24} = \frac{10}{24} = \frac{5}{12}$$

So, the remaining portion of the brownies was $1 - \frac{5}{12} = \frac{12-5}{12} = \frac{7}{12}$.

3. 10 rowboats. Begin by dividing $57 \div 6 = 9.5$. However, Eric can only rent a whole number of boats. If he rents 9 boats, then he only has room for $9 \times 6 = 54$ children. Therefore, he needs to rent a minimum of 10 boats, because $10 \times 6 = 60$.

4. $625. $7500 \div 12 = $625.

5. 21.1%. Begin by adding $36.8\% + 22.7\% + 19.4\% = 78.9\%$, which is the total funding from the three agencies. To find the remaining percentage of funding, subtract this value from 100%: $100\% - 78.9\% = 21.1\%$

6. 302 pounds. $186 + 77 + 39 = 302$.

7. 8 people. Begin by dividing: $\$1250 \div \$145 \approx 8.6$. However, if David hires 9 people, he will be over budget, because $\$145 \times 9 = \$1305$. So David can hire a maximum of 8 people, because $\$145 \times 8 = \$1160$.

8. $1,275. $\$75 \times 17 = \$1275$.

9. 9 pastels. $256 \div 13 = 19r9$ — that is, 19 with a remainder of 9.

10. 300 pages. Candice read 25% of a novel during the first week it was assigned and 40% of it the following week. At that point, she had 105 pages left to read. How many pages did the novel have altogether? In the first two weeks, Candice read $25\% + 40\% = 65\%$ of the novel, so she had $100\% - 65\% = 35\%$ remaining. This 35% represented 105 pages, so set up a proportion as follows:

$$\frac{35}{105} = \frac{100}{x}$$
$$35x = 10,500$$
$$x = \frac{10,500}{35}$$
$$x = 300$$

5

The Quantitative Reasoning, Algebra, and Statistics Test

IN THIS PART . . .

Bolster your pre-algebra skills, such as evaluating and simplifying algebraic expressions.

Shore up the algebra and geometry skills you need most on the ACCUPLACER.

Work with statistics, probability, and sets.

Use the skills explained in this part to answer ACCUPLACER practice questions.

Chapter **13**

Quantitative Reasoning and Pre-Algebra

In this chapter, you get started on the second of the three ACCUPLACER math tests: the Quantitative Reasoning, Algebra, and Statistics Test (QAS for short).

Although this material is technically more difficult than what's covered on the Arithmetic test, you may in fact be more familiar with it — especially if you've taken a math class in the last year or two. This is the math that's generally covered in high school Algebra 1 and Geometry classes. So, as an added bonus, if you're taking a math class right now, some of what I discuss in this part of the book may well be immediately useful in that class!

Quantitative Reasoning

Quantitative reasoning is the ability to think abstractly about numbers, and to use math to solve more sophisticated problems.

In this section, I give you some basic information about sets of numbers and number properties, as well as absolute value. I also describe how to solve problems involving rates, ratios, and proportions, including problems that require you to convert units of measurement. Finally, I focus on some of the trickier aspects of exponents (powers) that you may encounter on the ACCUPLACER.

Understanding the basic sets of numbers

On the ACCUPLACER, your understanding of the following sets of numbers may be tested: natural numbers (also called counting numbers), integers, rational numbers, irrational numbers, and real numbers. In this section, I discuss all five of these important number sets.

Natural numbers (counting numbers)

The *natural numbers* (or *counting numbers*) are the numbers that you learned when you first started to count: {1, 2, 3, 4, 5...}. You can also think of these as the *positive whole numbers*.

Here's a representation of this set of numbers on a number line:

```
1  2  3  4  5  6  7  8  9  10
```

Note that 0 is *not* in this set of numbers.

Integers

The *integers* are an extension of the natural numbers to include 0 and all the negative whole numbers: {... − 3, − 2, − 1, 0, 1, 2, 3...}. On the number line, the integers look like this:

```
−7 −6 −5 −4 −3 −2 −1  0  1  2  3  4  5  6  7
```

Rational numbers

The *rational numbers* are an extension of the integers to include all numbers that can be represented as fractions — that is, as a value with an integer in the numerator and the denominator. For example,

$$\frac{2}{3} \qquad -\frac{9}{5} \qquad \frac{179}{678} \qquad \frac{31}{1,000,000}$$

All decimals that are either terminating or repeating are rational numbers. For example,

$$0.671 = \frac{671}{1000} \qquad -3.7 = \frac{-37}{10} \qquad 0.\overline{2} = \frac{2}{9} \qquad 0.\overline{497} = \frac{497}{999}$$

And all the natural numbers and integers are rational numbers, because you can always rewrite each of these numbers as a fraction with 1 in the denominator. For example,

$$3 = \frac{3}{1} \qquad -917 = \frac{-917}{1} \qquad 55,790 = \frac{55,790}{1} \qquad 0 = \frac{0}{1}$$

Irrational numbers

The *irrational numbers* are all the numbers on the number line that are *not rational numbers*. In practice on the ACCUPLACER, these fall into two basic categories:

>> Pi (π), which is a non-repeating, non-terminating decimal that equals approximately 3.14.

>> Square roots of non-square numbers, such as $\sqrt{2}$, $\sqrt{3}$, $\sqrt{5}$, and so forth. These numbers, too, are non-repeating, non-terminating decimals that can be approximated by rounding:

$$\sqrt{2} \approx 1.414 \quad \sqrt{3} \approx 1.732 \quad \sqrt{5} \approx 2.236$$

Real numbers

The *real numbers* are all the numbers that appear on the number line. They include all the rational and irrational numbers. On the ACCUPLACER, you work only with real numbers.

Absolute value

REMEMBER

The *absolute value* of a number — represented by a pair of bars ($|\ |$) — is its non-negative value.

For example,

$$|-7| = 7 \qquad |5| = 5 \qquad |0| = 0$$

As you can see, when a number is negative (such as –7), its absolute value is its positive value — that is, its distance from zero on the number line. However, when a number is positive (such as 5) or 0, its absolute value remains unchanged.

When evaluating absolute value using the order of operations (PEMDAS), you can think of the absolute value bars as parentheses; that is, you need to evaluate everything inside the parentheses down to a single number before you can evaluate the absolute value of what's inside.

For example, suppose you want to evaluate the following:

$$-4 \div |5 + 2 \times (-3)|$$

Because the absolute value bars are similar to parentheses, you want to begin evaluating inside them, starting with multiplication and then subtraction:

$$= -4 \div |5 - 6|$$
$$= -4 \div |-1|$$

Now that the value inside the absolute value bars is a single number, you can evaluate it ($|-1| = 1$) and complete the problem:

$$= -4 \div 1$$
$$= -4$$

As another example, consider the following:

$$|3 - 5| - |-7 + 10|$$

To begin, find the values inside both sets of absolute value bars:

$$= |-2| - |3|$$

Next, evaluate the absolute values of each number ($|-2| = 2$ and $|3| = 3$), and complete the problem.

$$= 2 - 3 = -1$$

Rates, ratios, and proportional relationships

Some important applications of the rational numbers (see "Rational numbers") to real-world situations are rates, ratios, and proportional relationships. In this section, you get a handle on all of these important ideas.

Rates

REMEMBER

A *rate* is a regular change in a quantity that is usually (but not always) expressed over time using the word *per* (which means *for each*). For example, if you earn *$100 per day*, then you earn $200 in 2 days, $500 in 5 days, $3000 in 30 days, and so on.

Similarly, if you are a teacher who needs to provide *3 slices of pizza per child*, then you need to provide 6 slices for 2 children, 15 slices for 5 children, 90 slices for 30 children, and so on.

One of the most common types of rate problems requires you to divide distance by time to calculate the rate of a moving object. For example, if a car travels 432 miles in 8 hours, you can discover its average rate as follows:

432 miles ÷ 8 hours = 54 miles per hour

So the average rate of the car is 54 miles per hour. Note that the unit *miles per hour* comes from the two units (miles and hours) that were used in the calculation.

This idea can be extended to other types of problems. Here's an example:

Eugene works at a restaurant, and at top speed he once set up 91 place settings in 13 minutes. What was his rate per minute?

Calculate by dividing:

91 place settings ÷ 13 minutes = 7 place settings per minute

Ratios and proportions

REMEMBER

A *ratio* is a relationship between quantities that is similar to the relationship of the numerator and denominator of a fraction. For example,

Suppose the ratio of boys to girls in a club is 4:3 (or 4 to 3). This means that for every 4 boys in the classroom, there are exactly 3 girls. So, if the club has 8 boys, it has 6 girls. Similarly, if the class has 30 <u>girls</u>, it has 40 <u>boys</u>.

Typically, ratios are used to set up *proportions* — equations involving equivalent ratios expressed as fractions — to solve problems on the ACCUPLACER. For example,

A club that has a 4-to-3 ratio of boys to girls includes exactly 21 girls. How many boys are in the club?

To solve this problem, set up a proportion as follows:

$$\frac{\text{boys}}{\text{girls}} = \frac{4}{3}$$

Next, substitute the number 21 for *girls* and a variable for *boys*, then cross-multiply and solve for *b*:

$$\frac{b}{21} = \frac{4}{3}$$
$$3b = 21 \times 4$$
$$3b = 84$$
$$\frac{3b}{3} = \frac{84}{3}$$
$$b = 28$$

Therefore, the club has 28 boys.

Unit conversions

A useful application of ratios and proportions can be found when doing *unit conversions* — that is, changing one unit of measurement to another. For example, consider the following problem:

If 1 inch equals 2.54 centimeters, how many inches does 21 centimeters equal? (Round your answer to the nearest two decimal places.)

To solve this problem, use information about how to convert inches to centimeters to set up the ratio *1 inch to 2.54 centimeters* as a fraction:

$$\frac{1 \text{ in.}}{2.54 \text{ cm}}$$

Next, using this fraction, set up a proportion using the remaining information in the problem:

$$\frac{1 \text{ in.}}{2.54 \text{ cm}} = \frac{x \text{ in.}}{21 \text{ cm}}$$

Note that on both sides of the equation, inches appear in the numerator and centimeters in the denominator. To solve, begin by cross-multiplying to get rid of the fractions (you can also cancel out all units):

$$21 = 2.54x$$
$$\frac{21}{2.54} = \frac{2.54x}{2.54}$$
$$8.27 \approx x$$

So the answer rounded to two decimal places is 8.27.

Exponents (Powers)

At its simplest, an exponent is repeated multiplication written in a compact form. In this section, you expand your knowledge of exponents.

Positive whole-number exponents

REMEMBER

When an exponent is a positive whole number, you can evaluate it by turning it into multiplication. For example,

$$7^3 = 7 \times 7 \times 7 = 343$$
$$10^6 = 10 \times 10 \times 10 \times 10 \times 10 \times 10 = 1,000,000$$
$$2^8 = 2 \times 2 \times 2 \times 2 \times 2 \times 2 \times 2 \times 2 = 256$$

Here, the numbers 7, 10, and 2 are the *bases* (the numbers being multiplied), and 3, 6, and 8 are the *exponents* (the number of times the base is multiplied).

You can apply an exponent to any base, even those that are negative or rational numbers (such as fractions or decimals). For example,

$$(-8)^3 = (-8) \times (-8) \times (-8) = -512$$
$$\left(\frac{2}{5}\right)^2 = \frac{2}{5} \times \frac{2}{5} = \frac{4}{25}$$
$$(-1.2)^5 = (-1.2) \times (-1.2) \times (-1.2) \times (-1.2) \times (-1.2) = -2.48832$$

When a minus sign is placed in front of a number that is raised to an exponent *without parentheses*, the exponent is evaluated first, and then the minus sign is tacked on. An example will make this clear:

$$(-5)^4 = (-5) \times (-5) \times (-5) \times (-5) = 625$$
$$-5^4 = -(5 \times 5 \times 5 \times 5) = -625$$

Here, $(-5)^4$ is evaluated by multiplying -5 by itself 4 times, so the result is 625. In contrast, -5^4 is evaluated by multiplying 5 by itself 4 times and then negating, so the result is -625.

Squares and square roots

REMEMBER

Applying an exponent of 2 is so common that this operation is given a special name: *squaring a number*. Table 13-1 shows the results when you square the numbers from 1 to 10. Each of these resulting numbers (1, 4, 9, 100) is called a *square number*.

TABLE 13-1 ## Squares and Square Roots (Radicals)

Squares	Square roots (radicals)
$1^2 = 1$	$\sqrt{1} = 1$
$2^2 = 4$	$\sqrt{4} = 2$
$3^2 = 9$	$\sqrt{9} = 3$
$4^2 = 16$	$\sqrt{16} = 4$
$5^2 = 25$	$\sqrt{25} = 5$
$6^2 = 36$	$\sqrt{36} = 6$
$7^2 = 49$	$\sqrt{49} = 7$
$8^2 = 64$	$\sqrt{64} = 8$
$9^2 = 81$	$\sqrt{81} = 9$
$10^2 = 100$	$\sqrt{100} = 10$

The inverse of squaring a number is called *taking the square root of a number*. When you take the square root of a square number, the result is the number that you originally squared. Thus, squaring a number and taking the square root of a number are inverse operations.

The symbol $\sqrt{\ }$ can be read as either *the square root of* or *radical*. For example, $\sqrt{9}$ is read as either *the square root of 9* or *radical 9*.

Zero and negative exponents

REMEMBER

When you raise a base to an exponent of 0, the result is 1:

$$2^0 = 1 \qquad 10^0 = 1 \qquad 765^0 = 1$$

The only exception to this rule is 0^0, which is undefined.

When you raise a base to an exponent of -1, the result is the reciprocal of that number. For example,

$$4^{-1} = \frac{1}{4} \qquad \left(\frac{1}{7}\right)^{-1} = 7 \qquad \left(\frac{3}{5}\right)^{-1} = \frac{5}{3}$$

When you raise a base to an exponent of –2, the result is the reciprocal of the square of that number. For example:

$$6^{-2} = \frac{1}{6^2} = \frac{1}{36} \qquad \left(\frac{1}{9}\right)^{-2} = 9^2 = 81 \qquad \left(\frac{7}{8}\right)^{-2} = \left(\frac{8}{7}\right)^2 = \frac{64}{49}$$

Generally speaking, when you raise a base to any negative exponent, the result is the reciprocal of that number raised to the positive version of that exponent. For example,

$$5^{-3} = \frac{1}{5^3} = \frac{1}{125} \qquad \left(\frac{1}{10}\right)^{-4} = 10^4 = 10,000 \qquad \left(\frac{2}{3}\right)^{-5} = \left(\frac{3}{2}\right)^5 = \frac{243}{32}$$

Powers of 10

One of the most useful applications of exponents is *powers of 10* — that is, applying an exponent to a base of 10. Although you can express these results as fractions, exponents of 10 are even easier to express as decimals. Table 13-2 lists a few powers of 10 with exponents that are negative numbers, 0, and positive numbers.

TABLE 13-2 **Positive and Negative Powers of Ten**

Positive exponents	Negative exponents
$10^0 = 1$	
$10^1 = 10$	$10^{-1} = 0.1$
$10^2 = 100$	$10^{-2} = 0.01$
$10^3 = 1,000$	$10^{-3} = 0.001$
$10^4 = 10,000$	$10^{-4} = 0.0001$
$10^5 = 100,000$	$10^{-5} = 0.00001$
$10^6 = 1,000,000$	$10^{-6} = 0.000001$

Scientific notation

Scientific notation is a compact way to express very large and very small positive numbers. Scientific notation uses a combination of decimals and powers of 10 to express numbers in a way that makes them less confusing (when you get the hang of it!).

A number written in scientific notation includes a number greater than 0 and less than 10 multiplied by a power of 10.

To see how this works, consider the number 8,000,000,000,000 (8 trillion, which is the world population rounded to the nearest trillion). You can rewrite this number in an equivalent form by expressing it as a number times a power of 10:

$$8,000,000,000,000 = 8 \times 1,000,000,000,000 = 8 \times 10^{12}$$

As you can see, 8×10^{12} captures the same information as 8,000,000,000,000, but in a way that is easier to read and, therefore, less likely to be copied incorrectly.

REMEMBER

To change a number that is 10 or greater from standard notation (the numbers you are used to) to scientific notation, follow these steps:

1. Move the decimal point to the left so that the resulting number has a single digit to the left of the decimal point.

2. Count the number of spaces that you moved the decimal point.

3. Multiply the result by 10 raised to the power of this number.

In a similar way, you can change the number 602,000,000,000,000,000,000,000 (602 sextillion), which is an approximation of Avogadro's constant (the number of atoms in 12 grams of carbon), to scientific notation as follows:

$$602,000,000,000,000,000,000,000 = 6.02 \times 10^{23}$$

In this case, the number you started with was extremely unwieldy, but the resulting expression of this value is quite easy to read.

REMEMBER

You can also use scientific notation to write small decimals in a way that is more compact and easier to read than standard notation:

1. Move the decimal point to the right so that the resulting number has a single digit to the left of the decimal point.

2. Count the number of spaces that you moved the decimal point.

3. Multiply the result by 10 raised to the *negative* power of this number.

For example, here's how you rewrite the number 0.00000000345 in scientific notation:

$$0.000000003457 = 3.457 \times 10^{-9}$$

As when working with extremely large numbers, changing an extremely small number to scientific notation greatly improves the readability of the number.

Algebraic rules for exponents

REMEMBER

When you know how to calculate exponents numerically (as I discuss earlier in this chapter — flip to "Exponents [Powers]"), you're ready to move on to the algebra of exponents. Here are four rules you need to know so that you can simplify expressions with exponents:

$$x^a \cdot x^b = x^{a+b}$$

$$\frac{x^a}{x^b} = x^{a-b}$$

$$(x^a)^b = x^{ab}$$

$$x^{-a} = \frac{1}{x^a}$$

Don't worry if these don't make sense right now — just keep them in your back pocket for reference. In this section, I discuss these four important rules for simplifying expressions and how to apply them.

MULTIPLYING EXPONENTS WITH THE SAME BASE

The rule $x^a \cdot x^b = x^{a+b}$ tells you that to multiply a pair of exponents with the same base, keep the base and add the exponents. For example,

$$x^4 \cdot x^3 = x^7 \qquad y^{10} \cdot y^9 = y^{19} \qquad 10^{20} \cdot 10^3 = 10^{23}$$

You can use this rule to multiply more complicated expressions:

$$x^2y^3z^5 \cdot xy^8z^5 = x^3y^{11}z^{10}$$

As you can see, to multiply a pair of 1-term expressions, just add the exponents. As another example, here's how you multiply a 1-term expression by a 2-term expression using distribution:

$$x^2y^3(x^5y + x^{10}y^6) = x^7y^4 + x^{12}y^9$$

Here, add the exponents in each respective case.

DIVIDING EXPONENTS WITH THE SAME BASE

The rule $\frac{x^a}{x^b} = x^{a-b}$ tells you that to simplify a fraction with exponents (that is, to divide a pair of exponents with the same base), subtract the exponents from top to bottom and keep the bases the same.

For example,

$$\frac{x^9}{x^5} = x^4 \qquad \frac{2^{18}}{2^8} = 2^{10} \qquad \frac{y^7}{y^{10}} = y^{-3}$$

Notice in the last example that when the exponent above is less than the exponent below, the result is a negative exponent (see "Zero and negative exponents").

Using this rule allows you to simplify more complicated rational expressions:

$$\frac{x^7y^9z^5}{x^2y^{11}z} = x^5y^{-2}z^4$$

You can also combine the rules for multiplication and division to simplify relatively complex rational expressions:

$$\frac{x^3y^4(xy^{12})}{x^9y^3(x^6y^2)} = \frac{x^4y^{16}}{x^{15}y^5} = x^{-11}y^{11}$$

In the first step, I used the multiplication rules to simplify the numerator and denominator. Then, in the second step, I used the division rules to remove the fraction. The result is a much simpler expression with two variables.

RAISING AN EXPONENT TO AN EXPONENT

The rule $(x^a)^b = x^{ab}$ tells you that when raising an exponent to an exponent, you can simplify by multiplying the exponents. For example,

$$(x^3)^5 = x^{15} \qquad (y^{10})^4 = y^{40} \qquad (10^3)^2 = 10^6$$

When applying an exponent to a single-term expression, multiply the outer exponent by every exponent in the expression:

$$(x^2y^7z^{10})^3 = x^6y^{21}z^{30}$$

You can use this rule in conjunction with the other rules that you know. For example, suppose you want to simplify the following:

$$\frac{x^2y^3(xy^4)^3}{x^{10}y(x^4y^2)^2}$$

Begin by simplifying the exponents of exponents by multiplying:

$$= \frac{x^2 y^3 x^3 y^{12}}{x^{10} y x^8 y^4}$$

Next, multiply in both the numerator and denominator by adding exponents of variables with the same base:

$$= \frac{x^5 y^{15}}{x^{18} y^5}$$

Finally, simplify the fraction by subtracting exponents:

$$= x^{-13} y^{10}$$

REWRITING NEGATIVE EXPONENTS AS FRACTIONS

The rule $x^{-a} = \frac{1}{x^a}$ gives you a method to simplify negative exponents by switching them to positive exponents and placing them in the denominator. For example,

$$x^{-7} = \frac{1}{x^7} \qquad \frac{1}{y^{-4}} = y^4 \qquad \frac{x^{-2}}{y^{-3}} = \frac{y^3}{x^2}$$

Notice that in the second example, the negative exponent in the denominator simplifies to a positive exponent in the numerator. And in the third example, the two negative exponents switch places and become positive.

You can use this rule to simplify more complicated expressions that have negative exponents. For example,

$$\frac{(x^{-4} y^7 z^5)^{-2}}{(x^{-5} y z^{-2})^3}$$

To begin, remove the parentheses by multiplying exponents:

$$= \frac{x^8 y^{-14} z^{-10}}{x^{-15} y^3 z^{-6}}$$

Now, change negative exponents to positive ones by moving all negative exponents to the opposite side of the fraction bar:

$$= \frac{x^8 x^{15} z^6}{y^3 y^{14} z^{10}}$$

Next, multiply in both the numerator and denominator by adding exponents:

$$= \frac{x^{23} z^6}{y^{17} z^{10}}$$

Simplify the variable z by subtracting:

$$= \frac{x^{23} z^{-4}}{y^{17}}$$

To finish, change the negative exponent to positive by placing this variable in the denominator:

$$= \frac{x^{23}}{y^{17} z^4}$$

Fractional exponents

REMEMBER

The idea of a fractional exponent, such as $9^{\frac{1}{2}}$, may seem strange, but the good news is that the fractional exponents you will see on the ACCUPLACER are not difficult to calculate. The most common fractional exponent is $\frac{1}{2}$, which can be rewritten as a square root. For example,

$$9^{\frac{1}{2}} = \sqrt{9} = 3 \qquad 36^{\frac{1}{2}} = \sqrt{36} = 6 \qquad 100^{\frac{1}{2}} = \sqrt{100} = 10$$

As you can see, when you apply an exponent of $\frac{1}{2}$ to a square number (flip to "Squares and square roots"), the result is always an integer.

Knowing Some Pre-Algebra Basics

Algebra is the main focus of study in high school math. And, as you may have guessed, algebra looms large on the ACCUPLACER. In this section, I set you up for success with algebra by giving you a review of some important pre-algebra concepts.

Clarifying some basic algebra terminology

As a math student, you've probably heard your teachers toss around words like *expression*, *coefficient*, and *constant* with reckless abandon. And while you may have a sense of what they mean, you might feel weird about asking. In this section, I clarify all these words and put them in perspective, so you have a handy reference at your disposal whenever you need it.

Arithmetic and algebra

Arithmetic is the mathematics of numbers and operations on them. For example, here are a few typical arithmetic problems:

$$7 + 3 = \underline{\quad} \qquad 21 - (3 + 4 \div 2) = \underline{\quad} \qquad \sqrt{\frac{6 - 3 \times 1.5}{2}} + \frac{(10 \div 2)^{10}}{17^2 - 15^2} = \underline{\quad}$$

As you can see, arithmetic problems can be easy, medium, or hard. In all cases, however, your mission is to figure out how to *evaluate* the numbers and operators — that is, use the order of operations (PEMDAS) to condense them down to an answer that's a single number (see "Evaluating").

In contrast, *algebra* introduces the concept of the *variable*, which is a letter that can stand in for a number. Here are some typical equations that can be solved using algebra.

$$x + 9 = 11 \qquad 4(2x + 3) = 11(x - 1) + 8 \qquad x^3 + 2x^2 - 12x - 7 = 0$$

As with arithmetic, algebra problems also run the gamut from simple to complex. Note here, though, that each algebra problem includes at least one occurrence of a variable. Your goal in solving each of these problems is to find the value of the variable.

Equations and inequalities

An *equation* is a mathematical statement that includes an equals sign (=), which tells you that one quantity is equal to another quantity. For example,

$$6x + 5 = 11x - 10$$

This equation tells you that the quantity $6x + 5$ is equal to the quantity $11x - 10$.

An *inequality* is a mathematical statement that typically includes one of four inequality signs: < (less than), > (greater than), ≤ (less than or equal to), or ≥ (greater than or equal to). For example,

$$6x + 5 \leq 11x - 10$$

This inequality tells you that the quantity $6x + 5$ is less than or equal to the quantity $11x - 10$.

Expressions

An *expression* is a mathematical phrase. Typically, an equation includes two expressions, on opposite sides of the equals sign. For example, $7x - 3$ and $19x$ are both expressions, because both can be placed on opposite sides of an equals sign to create the equation $7x - 3 = 19x$.

Similarly, you could use these two expressions to form an inequality by placing an inequality sign, such as >, between them: $7x - 3 > 19x$.

Expressions tend to fall into two main categories:

>> An *arithmetic expression* (also called a *numerical expression*) includes only numbers and operators, but no variable.

For example, $8(5-3)^2 + 10$ is an arithmetic expression. You can use the order of operations (PEMDAS) to evaluate an arithmetic expression, turning it into a number:

$$8(5-3)^2 + 10 = 42$$

>> An *algebraic expression* (or *variable expression*) includes at least one variable. For example, $8(x-3)^2 + 10$.

An algebraic expression can't be evaluated unless you know the value of every variable it includes.

Terms

Every expression includes one or more *terms* separated by plus signs (+) or minus signs (−). For example, consider this expression:

$$5x^3 - 3x^2 + 9x - 7$$

This expression includes four terms: $5x^3$, $-3x^2$, $9x$, and -7. Notice that a minus sign is always included with the term that it immediately precedes.

All terms fall into one of two types:

>> A *variable term* is any term that includes a variable — that is, a letter, such as x, that stands for a number. For example, $5x^3$, $-3x^2$, and $9x$ are all variable terms.

>> A *constant term* (or *constant*, for short) is simply a term that includes no variable — that is, a number. For example, -7 is a constant.

You can rearrange the terms in an expression in any way you like, provided that you keep the sign with the term that it's part of. For example, here's an equivalent way to write the previous expression:

$$-7 + 9x - 3x^2 + 5x^3$$

Coefficients and variables

Every term within an expression can be separated into two parts, the coefficient and the variable part:

>> The *coefficient* is the number (including the sign) that precedes the term.

>> The *variable part* comprises whatever variables are included in that term.

For example, consider the following expression:

$$y^4 - 6xy^3 - x^2y^2 + 4x^3y - 15x^4 - 17$$

This expression has six terms, each of which breaks down as shown in Table 13-3.

TABLE 13-3 **Breaking an Expression into Terms**

Term	Type of term	Coefficient	Variable part
y^4	Variable term	1	y^4
$-6xy^3$	Variable term	-6	xy^3
$-x^2y^2$	Variable term	-1	x^2y^2
$4x^3y$	Variable term	4	x^3y
$-15x^4$	Variable term	-15	x^4
-17	Constant	-17	No variable

Notice that the first term (y^4), which appears to have no coefficient, has a coefficient of 1. Similarly, the third term ($-x^2y^2$) has a coefficient of -1.

Knowing three key tools for working with algebraic expressions

REMEMBER

You have three key tools when working with algebraic expressions:

>> *Evaluating* — Substituting (plugging in) numbers for variables, then using the order of operations (PEMDAS) to change the resulting arithmetic expression to a number

>> *Simplifying* — Removing parentheses and combining like terms to put the expression into its simplest possible form

>> *Factoring* — Introducing parentheses to change the expression into a factored form (that is, a form in which components are combined by multiplication)

In this section, I review evaluating and simplifying in depth. I also give you a brief review of the most common type of factoring, GCF factoring. In Chapter 17, I discuss more complex factoring. Together, these three tools provide you with a great deal of power for solving a wide variety of different ACCUPLACER algebra problems.

Evaluating

The word *evaluating* is related to the word *value*. When you evaluate an expression, you discover its value, changing it to a single number.

To evaluate an algebraic expression (an expression with at least one variable), you need to know the values of every variable in that expression. When this is the case, evaluate the expression by doing the following:

REMEMBER

1. **Plug in the numerical value of every variable to turn the algebraic expression into an arithmetic expression.**

2. **Use the order of operations (PEMDAS) to evaluate this arithmetic expression.**

For example, suppose you know that $x = 5$ and $y = -2$, and you want to evaluate this expression:

$$3x^2 + 2xy - 6y^2$$

Begin by plugging in 5 for x and -2 for y throughout the expression:

$$= 3(5)^2 + 2(5)(-2) - 6(-2)^2$$

Now, you can use the order of operations (PEMDAS) to evaluate this arithmetic expression.

P: Parentheses. Although this expression has parentheses, every set of parentheses contains a single number, so you have no further evaluation to do.

E: Exponents. The expression has two exponents:

$$= 3(25) + 2(5)(-2) - 6(4)$$

M: Multiplication. The expression has numerous opportunities to multiply:

$$= 75 - 20 - 24$$

D: Division. The expression has no division.

A: Addition. The expression has no addition.

S: Subtraction. The expression contains two subtractions:

$$= 31$$

The evaluation problem is complete after you have reduced the expression to a single number.

Simplifying

Simplifying an expression means to put it in an equivalent form with the fewest terms possible and without parentheses. In this section, I show you some of the most basic ways to simplify expressions.

COMBINING LIKE TERMS

The most basic way to simplify an algebraic expression is to combine like terms — that is, add together any terms that have the same variable part (flip to "Coefficients and variables").

To combine like terms, follow these steps:

1. **Add the coefficients of the terms.**
2. **Keep the variable part the same.**

For example, here is an expression that can be simplified by combining like terms:

$$10x^3 + 8xy + 4x^3 - 5x^2 - 3xy - 2x^3 + x^2$$

To begin simplifying, notice that the expression includes three terms with a variable part of x^3: $10x^3$, $4x^3$, and $-2x^3$. You can combine these three terms by adding their coefficients and keeping the variable part the same, as follows:

$$10x^3 + 4x^3 - 2x^3 = 12x^3$$

The expression also includes two x^2 terms: $-5x^2$ and x^2. Combine these two expressions in this way:

$$-5x^2 + x^2 = -4x^2$$

Finally, the expression includes two xy terms: $8xy$ and $-3xy$. Combine them like this:

$$8xy - 3xy = 5xy$$

To complete the simplification, gather the three results together:

$$10x^3 + 8xy + 4x^3 - 5x^2 - 3xy - 2x^3 + x^2 = 12x^3 - 4x^2 + 5xy$$

As you can see, combining like terms simplifies the expression from seven terms to three terms.

DISTRIBUTING TO REMOVE PARENTHESES

In some cases, an algebraic expression is complicated by parentheses, and can be simplified by removing them. To do this, *distribute* the term immediately preceding the parentheses, multiplying it by each of the terms inside the parentheses. (Remember that *terms* are the combinations of numbers and variables that can be separated by addition and subtraction.)

For example,

$$4x^5 + 7xy + 2x^2(3 - 6x^3) - x(3y - x^4)$$

At first glance, this expression appears to have no like terms. However, you can remove each set of parentheses by distributing the term immediately preceding it. Begin by removing the first set of parentheses:

$$2x^2(3 - 6x^3) = 6x^2 - 12x^5$$

Here, I multiplied $2x^2$ by both 3 and $-6x^3$, and then removed the parentheses. Next, distribute $-x$ over $3y$ and $-x^4$ as follows:

$$-x(3y - x^4) = -3xy + x^5$$

Thus, you can rewrite the original expression in this way:

$$4x^5 + 7xy + 6x^2 - 12x^5 - 3xy + x^5$$

The result is a 6-term expression that includes some like terms, which you can simplify as follows:

$$= -7x^5 + 6x^2 + 4xy$$

As you can see, distributing to remove parentheses is often a good step toward significantly simplifying an expression.

FOILING

REMEMBER

Although it's technically a type of distribution, I can almost always count on my students to recognize this next tool by its acronym FOIL. The four capital letters in *FOIL* stand for *First, Outer, Inner, Last.* This acronym helps you to remember all the steps necessary when multiplying a pair of binomials in parentheses.

For example, suppose you want to simplify the following expression:

$$(3x + 5)(4x - 9)$$

Begin by multiplying the two *First* terms:

$$3x \cdot 4x = 12x^2$$

Then, multiply the two *Outer* terms:

$$3x \cdot (-9) = -27x$$

Next, multiply the two *Inner* terms:

$$5 \cdot 4x = 20x$$

Finally, multiply the two *Last* terms:

$$5 \cdot (-9) = -45$$

Gather all four of these terms together as a single expression:

$$12x^2 - 27x + 20x - 45$$

In this case, you can simplify further by combining like terms:

$$= 12x^2 - 7x - 45$$

GCF Factoring

Factoring out a greatest common factor — *GCF factoring* for short — is the most basic type of polynomial factoring. A good way to understand GCF factoring is as reverse distribution. For example, consider the expression $2x(x + 2)$. You can simplify this expression by distributing the $2x$, which involves multiplying it by each of the two terms inside the parentheses:

$$2x(x + 2) = 2x^2 + 4x$$

The result is a *simplified* expression — in this case, a polynomial in its most basic form, without parentheses. However, in some cases, you may need to reverse this process. To see how this works, let's begin with the *result* of the previous problem:

$$2x^2 + 4x$$

This polynomial is fully simplified, but it can be factored because both terms ($2x^2$ and $4x$) are divisible by $2x$, which is their greatest common factor (GCF). As a result, you can begin to factor $2x^2 + 4x$ by factoring out $2x$, placing this value just outside a pair of parentheses:

$2x(\quad)$

Now, examine the two terms to see what's left after you've factored out $2x$:

$2x^2$ becomes x, because $x \cdot 2x = 2x^2$

$4x$ becomes 2, because $2 \cdot 2x = 4x$

So the factored form of $2x^2 + 4x$ is

$2x(x+2)$

As another example, consider the following expression:

$8x^4y^6 + 4x^3y^5 - 20x^2y^3$

Here, each of the three terms has a coefficient that is divisible by 4 (8, 4, and 20), so you can factor out a 4 from each term as follows:

$= 4(2x^4y^6 + x^3y^5 - 5x^2y^3)$

Additionally, each term includes the variable x with an exponent that is at least 2, so you can also factor out an x^2 from each term:

$= 4x^2(2x^2y^6 + xy^5 - 5y^3)$

Finally, each term also includes the variable y, with the lowest exponent being 3, so you can factor out a y^3:

$= 4x^2y^3(2x^2y^3 + xy^2 - 5)$

This result is the final answer. In this process, the GCF of $4x^2y^3$ has been factored out of the original expression, leaving behind the factor $2x^2y^3 + xy^2 - 5$.

To see how this procedure is the reverse of distribution, notice that you can distribute this factored result to produce the original expression:

$4x^2y^3(2x^2y^3 + xy^2 - 5) = 8x^4y^6 + 4x^3y^5 - 20x^2y^3$

As you can see, this equation includes a pair of expressions that are different in form but equivalent.

Chapter **14**

Algebra and Geometry

n this chapter, you apply some of the basic pre-algebra skills discussed in Chapter 13 to more complex algebra problems.

I begin with a focus on solving equations, systems of equations, and inequalities. Next, I show you how to set up algebra word problems in both one and two variables. The chapter continues with a discussion of linear functions — that is, equations of the form $y = mx + b$. I conclude by giving you a few basic geometric formulas and showing you how to apply them.

Using Algebra to Solve Equations, Systems of Equations, and Inequalities

The main purpose of algebra is to solve problems that would be too difficult or time-consuming with just arithmetic. In this chapter, you put the tools you discovered in the previous chapter to work to solve the main types of equations, systems of equations, and inequalities that you will see on the ACCUPLACER.

Solving basic algebraic equations

In the previous chapter, you found a variety of tools and terminology for working with algebraic expressions. Now, get ready to put these tools to use to solve equations using algebra. I start you off with a review of the basics and bring you up to speed at a steady pace. Ready, set, go!

Solving simple equations in your head

Although algebra can be a complex topic, it doesn't always have to be. For example, here is an algebraic equation that I bet you can solve with little to no effort:

$$x + 2 = 5$$

If you solved this equation as $x = 3$, because $3 + 2 = 5$, good job! Here's another problem:

$$3x = 15$$

This time, the answer is $x = 5$, because $3 \times 5 = 15$. Now try this one:

$$4x - 1 = 39$$

Here, you may need to step back for a moment and think: What number, when you multiply it by 4 and then subtract 1, gives you a result of 39? The answer is $x = 10$, because $4 \times 10 - 1 = 40 - 1 = 39$.

One last problem:

$$25x - 17 = 83$$

You *may* be able to solve this in your head. However, if the numbers are just too big and awkward to work with comfortably, you may need a method that enables you to tackle them. This method is called *isolating the variable*.

Isolating the variable to solve basic equations

REMEMBER

Isolating the variable is the most widely used method in algebra; it involves using a variety of techniques to get the variable (such as x) alone on one side of the equals sign with a number on the other side.

How about another look at the last problem in the previous section:

$$25x - 17 = 83$$

Notice here that the variable x is part of a 2-term expression on the left side of the equation. To solve this problem, you first want to move the constant -17 to the opposite side of the equation. And to do this, you need to *undo* this term by adding 17 to *both* sides of the equation:

$$\begin{aligned} 25x - 17 &= 83 \\ +17 \quad & +17 \end{aligned}$$

The result is an equation that you may like much better:

$$25x = 100$$

At this point, you may be able to see the correct answer. But if not, you can complete the problem by dividing both sides by 25:

$$\frac{25x}{25} = \frac{100}{25}$$

Simplifying both sides now leads to the solution:

$$x = 4$$

Consider another example:

$$5x + 19 = -4x - 8$$

This equation has two terms on each side of the equals sign. To isolate the variable, begin by getting all the x's to one side. The simplest way to do this is to add $4x$ to both sides:

$$\begin{aligned} 5x + 19 &= -4x - 8 \\ +4x \quad & +4x \end{aligned}$$

The result is a simplified equation that includes only one variable x:

$$9x + 19 = -8$$

At this point, you're in roughly the same position as you were at the start of the previous problem. Simplify this equation by subtracting 19 from both sides:

$$9x + 19 = -8$$
$$ -19 -19$$

This step makes the equation much simpler:

$$9x = -27$$

To finish up, divide both sides by 9:

$$\frac{9x}{9} = \frac{-27}{9}$$
$$x = -3$$

Combining like terms to solve more complicated equations

Some equations look complicated at first, but when you simplify both sides by combining like terms, they start to look much easier. For example,

$$6x + 4 + 7x = 21 + 8x - 7$$

A good first step here is to combine like terms on both sides of the equation:

$$13x + 4 = 14 + 8x$$

Now, you've simplified the equation down to a basic equation that you know how to solve:

$$5x + 4 = 14$$
$$5x = 10$$
$$x = 2$$

Distributing to solve equations with parentheses

Equations become even more complicated when they include parentheses that need to be removed. For example,

$$5x - (4x - 3) = 2(-3x - 5) + 14$$

Your only move here is to distribute to remove parentheses:

$$5x - 4x + 3 = -6x - 10 + 14$$

This problem can now be further simplified by combining like terms:

$$x + 3 = -6x + 4$$

And again, the problem yields to the methods you already know for isolating the variable:

$$7x + 3 = 4$$
$$7x = 1$$
$$x = \frac{1}{7}$$

Multiplying by the denominator to solve equations with a single fraction

TIP

Equations begin to look tougher when fractions are involved. Your first task is to get rid of the fractions as quickly as possible. Consider the following equation:

$$\frac{5x}{9} = 2x - 1$$

Fortunately, this equation has only one fraction. To remove it, multiply both sides of the equation by the denominator of the fraction, which is 9:

$$9\left(\frac{5x}{9}\right) = 9(2x - 1)$$

This multiplication cancels out the denominator on the left side of the equation, and you can distribute to remove the parentheses on the right side:

$$5x = 18x - 9$$

Now, throw all your usual tricks at the problem:

$$-13x = -9$$
$$x = \frac{9}{13}$$

Cross-multiplying to solve equations with fractions on both sides of the equals sign

When both sides of an equation are fractions, you can remove both of them by cross-multiplying:

REMEMBER

1. **Multiply the numerator (top number) of the left side by the denominator (bottom number) of the right side.**

2. **Multiply the numerator of the right side by the denominator of the left side.**

3. **Set these two results equal.**

For example, suppose you are faced with the following equation:

$$\frac{3x - 2}{5} = \frac{2x + 6}{7}$$

The difficulty here is that both sides of the equation are fractions. Cross-multiply to take care of the problem:

$$7(3x - 2) = 5(2x + 6)$$

Now, the equation looks like those that you have handled before. Distribute and solve:

$$21x - 14 = 10x + 30$$
$$11x - 14 = 30$$
$$11x = 44$$
$$x = 4$$

Multiplying by the lowest common denominator to solve equations with multiple fractions

More difficult problems include multiple fractions. For example,

$$\frac{2}{5}(x - 3) = \frac{3}{4}x - \frac{1}{3}(x + 7)$$

TIP

Although you may be tempted to distribute first, the real problem here is that the three fractions all have different denominators. Instead, multiply both sides of the equation — that is, every term — by the lowest common denominator of the three fractions, which is 60:

$$60 \cdot \frac{2}{5}(x - 3) = 60 \cdot \frac{3}{4}x - 60 \cdot \frac{1}{3}(x + 7)$$

At first glance, this may look more complicated. However, you can now cancel factors:

$$\cancel{60}^{12} \cdot \frac{2}{\cancel{5}^1}(x-3) = \cancel{60}^{15} \cdot \frac{3}{\cancel{4}^1}x - \cancel{60}^{20} \cdot \frac{1}{\cancel{3}^1}(x+7)$$

By changing all the denominators to 1, you've made the fractions go away:

$$12 \cdot 2(x-3) = 15 \cdot 3x - 20 \cdot 1(x+7)$$

Now, multiply, then distribute and solve:

$$24(x-3) = 45x - 20(x+7)$$
$$24x - 72 = 45x - 20x - 140$$
$$24x - 72 = 25x - 140$$
$$-72 = x - 140$$
$$68 = x$$

Multiplying by a power of 10 to solve decimal equations

My personal favorite way to solve any equation with decimals is to get rid of the decimals in the first step. To do this, multiply by a power of 10 (10, 100, 1000, and so forth) that is just large enough to get rid of the decimals. Use this chart to figure out exactly what to multiply by.

Maximum number of decimal places	Multiply by
1	10
2	100
3	1000
4	10,000

For example, consider this hairy equation:

$$3.5(2x-1) = 1.25(x-8) + 2.4(4x+2)$$

Although you may be inclined to distribute first, this step would only compound the trouble by spreading decimals to more terms.

Instead, multiply by a power of 10 to get rid of the decimals before you distribute. Here, the decimal with the most decimal places is 1.25, which has 2 of them, so multiply *every term* by 100:

$$100 \cdot 3.5(2x-1) = 100 \cdot 1.25(x-8) + 100 \cdot 2.4(4x+2)$$

This may seem to make the problem worse rather than better. However, remember that when you multiply a decimal by 100, you move the decimal point two places to the right:

$$350(2x-1) = 125(x-8) + 240(4x+2)$$

Now it's safe to distribute:

$$700x - 350 = 125x - 1000 + 960x + 480$$

Sure, the numbers are a tad larger than you'd like, but now you can proceed to the solution without worrying about those pesky decimals:

$$700x - 350 = 1085x - 520$$
$$520 - 350 = 385x$$
$$170 = 385x$$
$$\frac{170}{385} = x$$

You're almost done, but both the numerator and denominator are divisible by 5, so reduce this fraction by a factor of 5 to complete the problem:

$$\frac{34}{77} = x$$

Solving equations in terms of other variables

Throughout this section, all the equations you've worked with have included just one variable, x. When an equation on the ACCUPLACER has a single variable, you can usually count on being able to solve it.

In contrast, when an equation has more than one variable, it is usually not solvable. However, you can solve it for one variable *in terms of* another variable — that is, you can isolate one variable, with the remaining variables on the other side of the equation.

For example, consider this equation:

$$3ax + 4b = 5$$

You can't find the value of x as a number, because the equation includes more than one variable. However, you can solve this equation for the variable x in terms of the other variables, a and b. To do this, isolate x just as you would normally do. To begin, subtract $4b$ from both sides of the equation:

$$3ax = 5 - 4b$$

Notice here that the term $4b$ moves to the other side of the equals sign, and is negated. Now, the only term on the left side of the equals sign is the term that includes x. To isolate x, divide both sides of the equation by $3a$:

$$\frac{3ax}{3a} = \frac{5 - 4b}{3a}$$

When you simplify, the problem is complete:

$$x = \frac{5 - 4b}{3a}$$

Notice that the variables a and b appear in the solution. This is typical when solving for one variable in terms of other variables.

By the way, in the original equation, there is nothing particularly special about the variable x: You can just as easily solve for the variable b in terms of a and x. To do this, isolate b:

$$3ax + 4b = 5$$
$$4b = 5 - 3ax$$
$$\frac{4b}{4} = \frac{5 - 3ax}{4}$$
$$b = \frac{5 - 3ax}{4}$$

This time, the variables a and x appear in the solution, with b alone on the left side of the equation.

Solving systems of two equations

A system of two equations is a pair of equations, each of which includes two variables (such as x and y). When a system is relatively simple, you can solve it using substitution; otherwise, you need to solve it by combining equations. In this section, I show you when and how to use both methods.

Using substitution to solve simple systems of equations

The simplest method for solving a system of equations is *substitution* — that is, plugging the value of a variable directly into one of the equations. This works best when one equation in the system already has a variable isolated on one side. For example,

$$y = 2x - 5$$
$$3x + 4y = 46$$

In this system, the first equation gives you the value of y in terms of x (flip to "Solving equations in terms of other variables"). Plug this value of y into the second equation:

$$3x + 4(2x - 5) = 46$$

This step turns the system into a single equation in one variable. You know what to do:

$$3x + 8x - 20 = 46$$
$$11x - 20 = 46$$
$$11x = 66$$
$$x = 6$$

To complete the solution, plug this value of x into either equation — whichever seems easiest:

$$y = 2(6) - 5$$
$$y = 12 - 5$$
$$y = 7$$

Therefore, the solution to this system of equations is $x = 6$ and $y = 7$. In some questions on the ACCUPLACER, this solution may be placed in coordinate form (x, y) — that is, $(6, 7)$.

When working with some systems of equations, you may need to do a little algebraic sleight-of-hand to solve one equation in terms of a variable before you substitute. For example,

$$7x + y = -13$$
$$2x + 9y = 5$$

In this system, neither equation tells you the value of one variable in terms of the other. However, notice that in the first equation, the value of y has no coefficient. This fact makes solving for y in terms of x relatively simple — just subtract $7x$ from both sides:

$$y = -7x - 13$$

Now, substitute $-7x - 13$ for y into the second equation:

$$2x + 9(-7x - 13) = 5$$

Again, you've turned a system in two variables into a single equation with one variable, which you can solve:

$$2x - 63x - 117 = 5$$
$$-61x - 117 = 5$$
$$-61x = 122$$
$$x = \frac{122}{-61}$$
$$x = -2$$

Now, plug in –2 for x into *any version* of either equation, whichever looks simplest to work with:

$$y = -7x - 13$$
$$y = -7(-2) - 13$$
$$y = 14 - 13$$
$$y = 1$$

So the answer this time is $(-2, 1)$.

Combining equations to solve more difficult systems

Consider the following system of equations:

$$5x - 7y = 26$$
$$6x + 7y = -15$$

This system is not solved easily with substitution, because neither variable can be solved easily in terms of the other. For example, if you try to solve for y in terms of x in the first equation, here's what you get:

$$5x - 7y = 26$$
$$-7y = 26 - 5x$$
$$y = \frac{26 - 5x}{-7}$$

Not a pretty picture.

For a problem like this, a better method is to combine equations, either by adding them or subtracting them. To show you how this works, look what happens when you add the left sides of both equations and right sides of both equations, keeping every result on its respective side of the equals sign:

$$5x - 7y = 26$$
$$6x + 7y = -15$$
$$11x \qquad = 11$$

As you can see, the two x-terms add up to $11x$, while the two y-terms cancel each other out. The result is a single equation in one variable that you can solve easily:

$$x = 1$$

As usual with systems of equations, when you know the value of one equation, you can plug this back into any of the original equations and solve for the second variable:

$$5x - 7y = 26$$
$$5(1) - 7y = 26$$
$$5 - 7y = 26$$
$$-7y = 21$$
$$y = -3$$

So the answer this time is $(1, -3)$.

As you may have guessed, not every system of equations will be solved so easily. Consider the following example:

$$2x + 5y = 19$$
$$5x - 3y = 1$$

This time, if you were to add or subtract the two equations, neither variable would cancel out, so you wouldn't be able to solve the problem. In cases like these, you need to multiply one or both equations by numbers that will cause one variable to drop out of the equation when you combine them. Here, multiply the first equation by 3 and the second equation by 5:

$$6x + 15y = 57$$
$$25x - 15y = 5$$

Now, the coefficients of y in the two equations are 15 and -15, so when you add the two equations, this variable drops out:

$$31x = 62$$

Now, you can easily solve for x:

$$x = 2$$

Plugging in 2 for x into one of the two equations allows you to complete the solution:

$$5(2) - 3y = 1$$
$$10 - 3y = 1$$
$$-3y = -9$$
$$y = 3$$

So the answer is (2, 3).

Solving inequalities

REMEMBER

An inequality is a mathematical statement that doesn't include an equals sign ($=$), but instead uses an inequality sign ($<$, $>$, $\leq$, or $\geq$). (For more, flip to Chapter 13 and read "Equations and inequalities.")

Solving inequalities is very similar to solving equations, with one important difference: When you multiply or divide by a negative number, you flip the inequality sign as shown in the following table.

Inequality sign	Flipped
$<$	$>$
$>$	$<$
$\leq$	$\geq$
$\geq$	$\leq$

To see how this works, consider the following inequality:

$$5x + 4 < 7x + 8$$

You can work with this inequality in much the same way as you work with an equation. To begin, isolate the variable x by subtracting $7x$ from both sides:

$$-2x + 4 < 8$$

Next, subtract 4 from both sides:

$$-2x < 4$$

To finish the problem, you want to divide by –2. However, when you divide by a negative number, you need to flip the inequality <, changing it to >:

$$\frac{-2x}{-2} > \frac{4}{-2}$$

Now, simplify to complete the solution:

$$x > -2$$

On the ACCUPLACER, the solution to an inequality is often displayed on a number line. Here's how to show this solution:

As you can see, the number line includes an open circle at –2, to indicate that this point is not part of the solution set.

As another example, consider this problem:

$$4(2x+1) - 3 \geq 7x + 5(x - 3)$$

This problem is similar to those you have solved before. Begin by distributing and simplifying, then isolate the variable as with an equation:

$$8x + 4 - 3 \geq 7x + 5x - 15$$
$$8x + 1 \geq 12x - 15$$
$$-4x + 1 \geq -15$$
$$-4x \geq -16$$

All these steps are exactly the same as they would be if the problem were an equation. However, to finish the problem, you now want to divide both sides by –4, so you need to flip the inequality sign:

$$\frac{-4x}{-4} \leq \frac{-16}{-4}$$
$$x \leq 4$$

Now, place the solution set on a number line as follows:

Setting Up Algebra Word Problems

Word problems are popular with the creators of tests like the ACCUPLACER, but maybe not so much with students like you who take the test.

The first hurdle with word problems is figuring out how to rewrite the information in the problem as an equation that enables you to solve the problem. Another hurdle is knowing which algebra skills you need to use to solve the equation.

In this section, I show you how to set up and solve the most common types of ACCUPLACER word problems.

Changing common words and phrases to equations

One of the toughest tasks in tackling word problems is figuring out how to frame them as numbers and operations, so that you can use algebra to solve them.

Fortunately, a lot of word problems on the ACCUPLACER test your understanding of a small set of common words. When you feel comfortable handling sentences and phrases that include these words, you're more than halfway to solving the problem.

Table 14-1 shows a list of typical English sentences that you may find on the ACCUPLACER, along with how to rewrite them as equations.

TABLE 14-1 ## Changing Words to Equations

English sentence	Equation
Five more than a number equals 17.	$x + 5 = 17$
x is three less than y.	$x = y - 3$
Seven times a number is equal to 56.	$7x = 56$
36 divided by a number is equal to 4.	$\dfrac{36}{x} = 4$
When you divide a number by 11, the result is 4.	$\dfrac{n}{11} = 4$
The sum of a number and 12 equals 100.	$x + 12 = 100$
The difference of 45 and a number is 23.	$45 - x = 23$
The product of a number and 8 is 96.	$8x = 96$
The quotient of 99 and a number equals 9.	$\dfrac{99}{x} = 9$
Two times the quantity of a number plus 1 equals 22.	$2(x + 1) = 22$
The quantity 10 minus a number divided by 3 equals 2.	$\dfrac{10 - x}{3} = 2$

Word problems with rates, ratios, and proportions

A common type of word problem on the ACCUPLACER involves rates, ratios, and proportions (see Chapter 13, "Rates, ratios, and proportional relationships"). For example,

Jonas works as a lifeguard at a swimming pool that has a capacity of 19,440 gallons. The pool can be drained at a rate of 45 gallons of water per minute. To the nearest tenth of an hour, how many hours does the pool take to drain completely?

To solve this problem, begin by setting up a proportion that represents 45 gallons of water in 1 minute:

$$\frac{\text{gallons}}{\text{minute}} = \frac{45}{1}$$

Next, fill in the information that you know and declare a variable m that represents the number of *minutes* the pool takes to drain:

$$\frac{19,440}{m} = \frac{45}{1}$$

Cross-multiply to remove the fractions, then isolate the variable:

$$19,440 = 45m$$
$$\frac{19,440}{45} = \frac{45m}{45}$$
$$432 = m$$

So the pool takes 432 minutes to drain completely. To find out how many hours this is, divide by 60:

$$432 \div 60 = 7.2$$

Therefore, the pool takes 7.2 hours to drain.

Creating equations in 1 variable

You can solve many ACCUPLACER problems by creating an equation in 1 variable. For example,

Aaron and his younger brother, Brett, measure their heights. They find that Aaron's height is 9 inches shorter than twice Brett's height, and that together their total height is 84 inches. What is Brett's height in inches?

Begin by letting b stand for Brett's height. Then you can represent Aaron's height in terms of b as $2b - 9$. And together, their height is $2b - 9 + b$. Set this value equal to 84 inches and solve:

$$2b - 9 + b = 84$$
$$3b = 93$$
$$b = 31$$

So Brett is 31 inches tall. Here's another example:

Wanda, Daria, and Fiona were all selling tickets to support their school play. Wanda sold twice as many tickets as Daria, and Daria sold 5 fewer tickets than Fiona. Together, the three girls sold 77 tickets. How many tickets did Wanda sell?

Although you're looking for the number of tickets Wanda sold, if you let w equal this number, then you will need to use the fraction $\frac{w}{2}$ to represent Daria's tickets. To avoid using fractions, try this instead:

$$\text{Daria} = d$$
$$\text{Wanda} = 2d$$
$$\text{Fiona} = d + 5$$

With this setup, you can build and solve the following equation:

$$d + 2d + d + 5 = 77$$
$$4d + 5 = 77$$
$$4d = 72$$
$$d = 18$$

Daria sold 18 tickets, so Wanda sold twice as many, which is 36.

Creating 2-variable systems of equations

One classic type of problem is easiest to solve by setting up a system of two equations and solving it. Here is an example:

Aaron worked at a booth during a fair, where he sold hot dogs for $2.25 each and drinks for $1.50. Altogether, he sold a total of 235 hot dogs and drinks, and brought in a total of $420.75 in revenue. How many hot dogs did Aaron sell?

A good way to begin thinking about this problem is to make a chart and fill in whatever information you can:

	Price per Item	Number Sold	Total Revenue
Hot Dogs	2.25	h	$2.25h$
Drinks	1.50	d	$1.50d$
Total	_____	235	420.75

Here, I declare the variable h to equal the number of hot dogs that Aaron sold, and d to equal the number of drinks. So, when you add these two numbers together, the result is 235, which gives you the following equation:

$$h + d = 235$$

You can find the total amount of revenue (money that Aaron made) from hot dogs by multiplying $2.25 \times h = 2.25h$. Similarly, multiply $1.50 \times d = 1.50d$ to find the total revenue from drinks. When you add these two values together, the result is $420.75, which gives you another equation:

$$2.25h + 1.50d = 420.75$$

These two equations form a system of equations that you can solve by either of the methods I outline in (see "Solving systems of two equations"). As I mention there, solving by substitution is usually easier when you can isolate one of the variables:

$$h + d = 235$$
$$d = 235 - h$$

Now, you can plug $235 - h$ into the second equation in place of d:

$$2.25h + 1.50(235 - h) = 420.75$$

This method reduces the system to a single equation in one variable, which you can begin solving by distributing:

$$2.25h + 352.50 - 1.50h = 420.75$$

If you're like me, you may want to get rid of decimals by multiplying every term by 100 (that is, by moving every decimal point two places to the right):

$$225h + 35,250 - 150h = 42,075$$

Now, combine like terms and then isolate h to solve the problem:

$$75h + 35,250 = 42,075$$
$$75h = 6825$$
$$\frac{75h}{75} = \frac{6825}{75}$$
$$h = 91$$

Therefore, Aaron sold 91 hot dogs.

Linear Functions

REMEMBER

A *linear function* is an equation of the form $y = mx + b$. Typically, m and b are replaced by numbers. For example,

$$y = 3x + 2 \qquad y = \frac{1}{4}x - 5 \qquad y = 125x + 1250$$

Linear functions are a big part of any standard Algebra 1 course. For this reason, the ACCUPLACER places great emphasis on linear functions as a subject that you are expected to know in detail. In this section, you drill down on what you need to know about linear functions.

Understanding linear functions in the real world

A good way to understand how linear functions work is with a real-world example. In this section, I introduce you to the basics of linear functions with an example that shows you how time really is money.

Seeing four ways to represent the same idea

You can encapsulate the information in a linear function in four different but related ways: a story, a chart, a graph, or an equation.

For example, consider the following scenario:

At the beginning of the year, Peg's grandfather started a college fund for her by placing $1250 in a bank account. Every month since then, he's added $125 to the account.

You can represent this word problem in a variety of other ways. For example, here is a chart that captures this same information:

Month	0	1	2	3	4	5	6	7	8
Balance	1250	1375	1500	1625	1750	1875	2000	2125	2250

Still a third way to organize this information is in a graph.

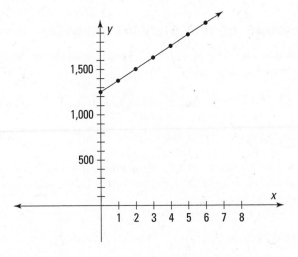

Finally, a fourth method is to use the following equation:

$$y = 125x + 1250$$

Understanding the input variable (x) and the output variable (y)

In the equation $y = 125x + 1250$, the variable x represents the number of months after Peg's grandfather opened the bank account, and the variable y represents the dollar value in the account. These variables are called, respectively, the *input variable* and the *output variable*.

While the value of this equation may not be immediately evident, imagine that this word problem ends with the following question:

How much money will be in the account after 60 months?

Although you can extend the chart or the graph out to 60 months, this will probably take a long time, and a very large piece of paper.

Or, instead, you can just plug in 60 for x as the input variable to this equation, and solve for the output variable, y:

$$y = 125(60) + 1250$$
$$= 7500 + 1250$$
$$= 8750$$

Therefore, in 5 years, the bank account will have grown to $8750.

As another example, suppose the question were as follows:

In how many months will the money in the bank account reach $10,000?

This time, the question gives you the value of the output variable, y, and asks you to solve for the input variable, x:

$$10,000 = 125x + 1250$$
$$8750 = 125x$$
$$\frac{8750}{125} = \frac{125x}{125}$$
$$70 = x$$

Thus, the account will reach $10,000 in 70 months.

Understanding the slope (*m*) and the y-intercept (*b*)

The equation $y = 125x + 1250$ is an example of a general class of equations called *linear functions*, which all have the form $y = mx + b$.

In this example, $m = 125$ and $b = 1250$. Each of these two values has a specific part to play.

FOCUSING ON THE Y-INTERCEPT

The value *b* is the *y-intercept* of this function. It is the point where the function crosses the *y*-axis. The y-intercept tells you the *starting value* of the function — that is, the value that was originally placed into the bank account.

UNDERSTANDING THE SLOPE

The value *m* is the *slope* of this function. It tells you the *increment value* of the function — that is, the amount that is regularly added to the bank account.

REMEMBER

The *slope* of a line is defined in the following way:

$$\text{slope} = \frac{\text{rise}}{\text{run}} = \frac{y_2 - y_1}{x_2 - x_1}$$

The *rise* is the distance up or down, and the *run* is the distance from left to right (always in that direction). You can measure these distances using two points on a line.

Working with linear functions slope-intercept form ($y = mx + b$)

In the previous section, you worked with a real-world linear function $y = mx + b$ to see how all four parts of this function (*x*, *y*, *m*, and *b*) work together.

In this section, you continue to build on this understanding.

Graphing linear functions in 2 variables

You can graph any linear function $y = mx + b$ by using the *m* and *b* values, as follows:

1. Plot the *y*-intercept *b* as (0, *b*).

2. Use the slope *m* to plot a second point.

3. Draw a line through these two points.

To see how this works, here's how you plot the equation $y = 3x + 1$:

1. Plot the *y*-intercept as (0, 1).

2. Use the slope 3 (up 3, over 1) to plot the point (1, 4).

3. Draw a line through these two points.

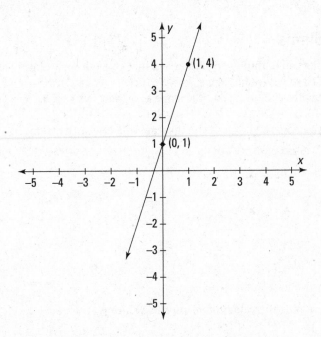

As another example, here's how you plot the equation $y = -\dfrac{3}{4}x - 2$:

1. Plot the *y*-intercept as (0, − 2).

2. Use the slope $-\dfrac{3}{4}$ (down 3, over 4) to plot the point (4, − 5).

3. Draw a line through these two points.

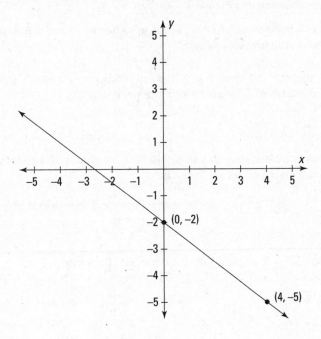

Point-slope problems

A *point-slope problem* asks you to find the equation of a line given the coordinates of one point on that line plus its slope. To solve a point-slope problem, use the equation $y = mx + b$, plugging the values of the two coordinates as x and y, and the slope as m, and solve for b.

For example, suppose you are asked to find the equation of a line with a slope of $-\frac{1}{2}$ that passes through the point $(6, -1)$. Start with $y = mx + b$ and plug in the three values that you know:

$$-1 = -\frac{1}{2}(6) + b$$

Now, solve for b:

$$-1 = -\frac{1}{2}(6) + b$$
$$-1 = -3 + b$$
$$2 = b$$

To complete the problem, plug the values of m and b back into $y = mx + b$:

$$y = -\frac{1}{2}x + 2$$

Two-point problems

In a *two-point problem*, you are given two points on a line and asked to provide the equation for that line. You solve two-point problems in two steps:

1. Find the slope of the equation using slope $= \frac{y_2 - y_1}{x_2 - x_1}$.

2. Solve the problem as a point-slope problem, by plugging one point and the slope into $y = mx + b$, as shown in the previous section.

To see how this works, suppose you know that a line intersects the two points $(1, 5)$ and $(4, -7)$. Begin by plugging the coordinates of the points into the slope formula:

$$\text{slope} = \frac{5 - (-7)}{1 - 4} = \frac{12}{-3} = -4$$

Therefore, the slope is $m = -4$. Now, plug this slope plus the coordinates for *either* point (whichever looks like it has easier numbers) into $y = mx + b$:

$$5 = -4(1) + b$$
$$5 = -4 + b$$
$$9 = b$$

Therefore, the y-intercept is $b = 9$, so plug the m and b values back into $y = mx + b$, and you have the equation for the line:

$$y = -4x + 9$$

Parallel lines

A pair of *parallel lines* on the xy-graph always have the same slope. You can use this information to solve problems that involve parallel lines. For example,

What is the equation of a line that passes through the point $(-9, 1)$ and is parallel to the line whose equation is $y = 3x + 17$?

The equation of the line that you are looking for has the same slope as $y = 3x + 17$, so it has a slope of 3. Thus, its equation is as follows:

$$y = 3x + b$$

You can find the value of b by plugging the coordinates of the point $(-9, 1)$ into this equation and solving for b:

$$1 = 3(-9) + b$$
$$1 = -27 + b$$
$$28 = b$$

Now, plug this b value back into the equation $y = 3x + b$:

$$y = 3x + 28$$

Perpendicular lines

A pair of *perpendicular lines* on the *xy*-graph (lines that meet at a 90-degree angle) always have slopes that are the negative reciprocal of each other. Table 14-2 shows some examples of negative reciprocals.

TABLE 14-2 Negative Reciprocals

Slope	Perpendicular (negative reciprocal)
$\frac{3}{4}$	$-\frac{4}{3}$
$-\frac{1}{7}$	$\frac{7}{1} = 7$
2	$-\frac{1}{2}$
-5	$\frac{1}{5}$

You can use this information to solve problems that involve perpendicular lines. For example,

What is the equation of a line that passes through the point $(-2, 3)$ and is perpendicular to the line whose equation is $y = \frac{2}{5}x + 9$?

The equation of the line that you are looking for has a slope that is perpendicular to $y = \frac{2}{5}x + 9$, so it has a slope that is the negative reciprocal of $\frac{2}{5}$, which is $-\frac{5}{2}$. Thus, its equation is as follows:

$$y = -\frac{5}{2}x + b$$

You can find the value of b by plugging the coordinates of the point $(-2, 3)$ into this equation and solving for b:

$$3 = -\frac{5}{2}(-2) + b$$
$$3 = 5 + b$$
$$-2 = b$$

Now, plug this b value back into the equation $y = -\frac{5}{2}x + b$:

$$y = -\frac{5}{2}x - 2$$

Graphing linear systems and linear inequalities

A system of linear equations consists of two equations in two variables. Similarly, a system of linear inequalities consists of two inequalities in two variables. Systems of linear equations and inequalities can be graphed on the xy-plane to provide a visual representation of these systems.

Graphing systems of two linear equations

Earlier in this chapter (see "Solving systems of two equations"), you worked with systems of linear equations in two variables, x and y. When you graph such a system, the point where the lines intersect is the solution of the system.

For example, consider the following system:

$$y = -3x + 10$$
$$y = x - 2$$

Notice that when you graph these two lines, as shown in the following figure, the point on the graph where the lines cross is (3, 1). This intersection point is the solution of the system.

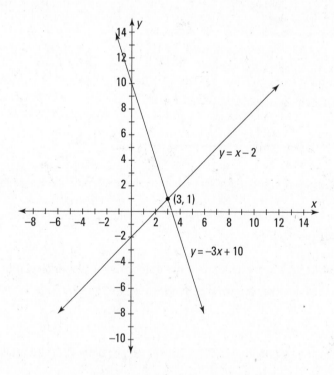

To solve a system of linear equations of the form $y = mx + b$ without graphing, set the right sides of both equations equal to each other, and then solve for x.

Consider the system of equations from the previous example. Because both equations are in slope-intercept form, you can set the right sides of the equations equal to each other:

$$-3x + 10 = x - 2$$

The result is a single equation with only one variable, x, which you can solve:

$$+10 = 4x - 2$$
$$12 = 4x$$
$$\frac{12}{4} = \frac{4x}{4}$$
$$3 = x$$

So $x = 3$. Now, substitute 3 for x in either of the original equations (I always choose the one with easier numbers):

$$y = x - 2 = 3 - 2 = 1$$

So $y = 1$. Therefore, the solution of the system is (3, 1).

Linear inequalities

In a *linear inequality*, the equals sign (=) in $y = mx + b$ is replaced by one of the four inequality signs — less than (<), greater than (>), less than or equal to (≤), or greater than or equal to (≥).

The following figure shows graphs of four linear inequalities: $y < 2x + 1$, $y > -3x + 4$, $y \le \frac{1}{4}x + 2$, and $y \ge x - 1$.

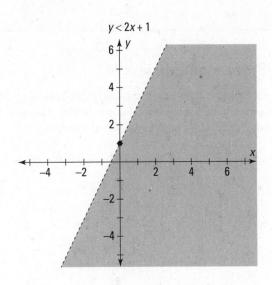

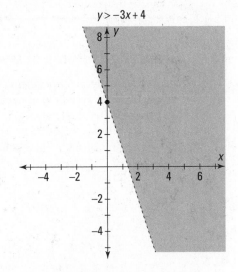

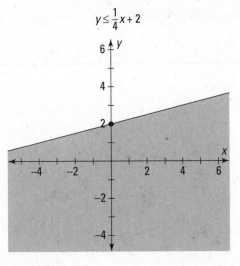

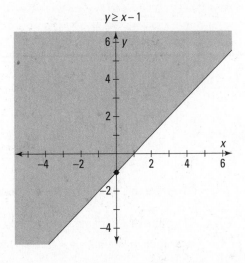

Note that a linear inequality indicates not just a line but a *region* on the graph. When a linear inequality is stated in slope-intercept form, this region is always either *below* the line (for inequalities with < and ≤) or *above* it (for inequalities with > and ≥).

Additionally, recall that when you represent equations with < and > on the number line, you use an *open circle* to express the idea that a given number is *excluded* from the solution set (see "Solving inequalities"). In a similar way, when you graph a linear inequality using < or >, you draw a *dashed line* to represent that these values are excluded from the solution set.

This information is summarized in Table 14-3.

TABLE 14-3 ## Graphing Linear Inequalities

Inequality sign	Shaded region	Type of line	Example
< (less than)	Below the line	Dashed	$y < 2x + 1$
> (greater than)	Above the line	Dashed	$y > -3x + 4$
≤ (less than or equal to)	Below the line	Solid	$y \le \frac{1}{4}x + 2$
≥ (greater than or equal to)	Above the line	Solid	$y \ge x - 1$

Graphing systems of two linear inequalities

A *system of two linear inequalities* is a pair of inequalities, each with two variables (such as x and y). To graph such a system, you need to graph both inequalities independently. The solution set is the region on the graph where the shaded areas from both inequalities overlap.

For example, consider this: What system of linear inequalities does the following graph represent?

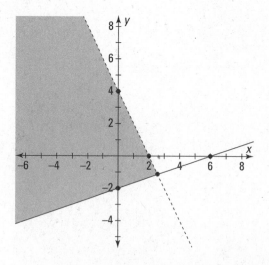

To solve this problem, consider both lines independently:

The line that is rising from left to right has a slope of $\frac{1}{3}$ and a y-intercept of −2, so if it were an equation, it would be $y = \frac{1}{3}x - 2$. However, the region above this line is shaded, and the line itself is solid, so its inequality sign is ≥. Therefore, this inequality is $y \ge \frac{1}{3}x - 2$.

In a similar way, the line that is falling from left to right has a slope of –2 and a y-intercept of 4, so if it were an equation, it would be $y = -2x + 4$. In this case, though, the region below this line is shaded, and the line is dashed, so its inequality sign is <. Therefore, this inequality is $y < -2x + 4$.

Working with linear functions in standard form ($ax + by = c$)

So far, you have worked with linear functions in slope-intercept form ($y = mx + b$). Another common form in which you may encounter linear functions is *standard form*:

$$ax + by = c$$

For example,

$$2x + 3y = 6 \qquad -5x + 13y = -4 \qquad 40x - 15y = 12$$

In this section, you discover how to convert an equation in standard form to slope-intercept form, as well as how to work directly with equations in standard form.

In many cases, the best way to handle a linear equation in standard form ($ax + by = c$) is to change it to slope-intercept form ($y = mx + b$). To do this, solve the equation for y in terms of x (see "Solving systems of two equations"), as follows:

REMEMBER

1. **Subtract over the x-term.**

2. **Divide every term by the coefficient of the y-term.**

For example, to change the equation $5x + 2y = 6$ to slope-intercept form, begin by subtracting over the x-term, which is $5x$:

$$5x + 2y = 6$$
$$2y = -5x + 6$$

Notice that in this step, I place the term $-5x$ in front of the term 6, which is the usual order of terms in slope-intercept form. Now, divide *every* term by the coefficient of the y-term, which is 2, and simplify:

$$\frac{2y}{2} = \frac{-5x}{2} + \frac{6}{2}$$
$$y = -\frac{5}{2}x + 3$$

The result is the same equation in slope-intercept form. You can now tell just by looking at the equation that the slope of this line is $-\frac{5}{2}$ and its y-intercept is 3.

Distance formula

The *distance formula* enables you to find the distance between any two points (x_1, y_1) and (x_2, y_2):

REMEMBER

$$d = \sqrt{(x_2 - x_1)^2 + (y_2 - y_1)^2}$$

For example, suppose you want to find the distance between the points (–3, 2) and (1, 5). Plug these four points into the distance formula as follows:

$$d = \sqrt{(1 - (-3))^2 + (5 - 2)^2}$$

Now, simplify:

$$= \sqrt{4^2 + 3^2}$$
$$= \sqrt{16 + 9}$$
$$= \sqrt{25}$$
$$= 5$$

Thus, the distance between these two points is 5.

Geometric transformations

Geometric transformations are mathematically precise ways of moving shapes on the *xy*-graph. Moving a shape is accomplished by moving every point on that shape, so the key to understanding geometric transformations is knowing how to transform a single point. This operation varies depending on the particular type of transformation.

In this section, you discover how to perform three types of transformations: translations, reflections, and rotations.

Translations

A *translation* is a geometric transformation in which a shape is moved from one place on the *xy*-graph to another. When a shape is translated, every point on that shape moves exactly the same distance both horizontally and vertically. Thus, the changes to *x*- and *y*-values of each point are the same in all cases.

To illustrate how this works, the following figure shows a line segment with endpoints A and B at $(2, 3)$ and $(-2, 5)$, respectively. The position of this line segment is translated so that endpoint A moves to the point $A' = (-1, 4)$. When you compare the points A and A', you can see that the *x*-value decreases by 3 and the *y*-value increases by 1. Thus, when translated, point B moves to $B' = (-5, 6)$.

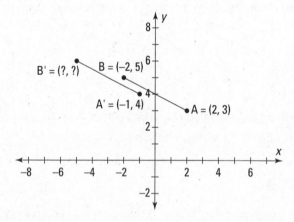

Reflections

A *reflection* is a geometric transformation that resembles a mirror. The two most common reflections on the *xy*-graph are reflections across the *x*-axis and *y*-axis:

>> **Reflection across the *x*-axis:** To reflect a point across the *x*-axis, *negate* the *y*-coordinate (that is, change the sign of this coordinate either from positive to negative or from negative to positive).

>> **Reflection across the *y*-axis:** To reflect a point across the *y*-axis, *negate* the *x*-coordinate.

The following figure shows the results when you reflect the three vertices of triangle *ABC* across both the *x*–axis and *y*–axis.

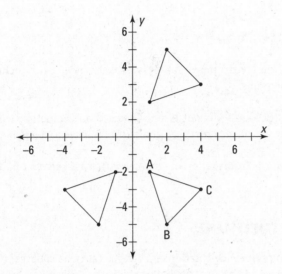

Rotations

A *rotation* is a transformation in a circle around a single central point. Rotations on the *xy*-graph are typically around the origin. The following figure shows the three most common rotations.

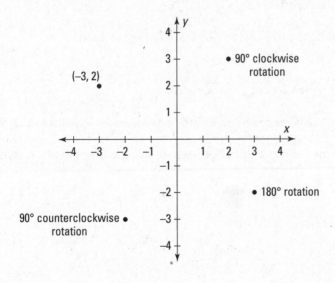

180-degree rotation: Negate both the *x*-value and the *y*-value. In this figure, the 180-degree rotation of the point $(-3, 2)$ changes it to $(3, -2)$.

90-degree clockwise rotation: Switch the x-value and the y-value; then, negate *one* of these two values (depending on the specific rotation). In this figure, the 90-degree clockwise rotation moves the point $(-3, 2)$ to $(2, 3)$.

90-degree counterclockwise rotation: Switch the x-value and the y-value; then, negate *one* of these two values (depending on the specific rotation). In this figure, the 90-degree counterclockwise rotation moves the point $(-3, 2)$ to $(-2, -3)$.

Geometry

In high school, geometry is often a year-long course that takes a deep dive, including a strong focus on proving geometric theorems.

Some good news is that the ACCUPLACER is not heavily focused on geometry. Geometric proofs are not tested, and only a short list of very practical geometry topics is touched upon.

In this section, I give you the basic information on geometry that will set you up for success on the ACCUPLACER.

Basic geometric formulas

Geometric formulas allow you to perform calculations on shapes (such as squares, rectangles, and circles) and solids (such as prisms). In this section, you review the basic geometric formulas that you need to know for the ACCUPLACER.

Squares

Measurements with squares are based on the length of a single side of the square. The area of a square that has a side of length s is $A = s^2$.

So, if the side of a square is 7 inches, you calculate its area as follows:

$$A = 7^2 = 49$$

Therefore, the square has an area of 49 square inches.

The perimeter of a square is the combined length of its four sides, calculated as $P = 4s$.

For example, to find the perimeter of a square whose side has a length of 7 inches, calculate as follows:

$$P = 4 \times 7 = 28$$

So this square has a perimeter of 28 inches.

Rectangles

Rectangle measurements are based on its *length* (longer side) and *width* (shorter side). The area of a rectangle with length l and width w is $A = l \times w$.

So, if the length of a rectangle is 9 centimeters and its width is 4 centimeters, you can determine its area as follows:

$$A = 9 \times 4 = 36$$

Thus, this rectangle has an area of 36 square centimeters.

The formula for the perimeter of a rectangle, measuring the total distance around it, is $P = 2(l + w)$.

So, the perimeter of a rectangle of length 9 centimeters and width 4 centimeters is

$$P = 2(9 + 4) = 2 \times 13 = 26$$

Therefore, the perimeter of this rectangle is 26 centimeters.

Circles

Circle measurement is based on the length of the *radius*: the distance from its center to any point on the circle.

The *diameter* of a circle is the distance from any point on the circle, through the center, to that circle's opposite point. You calculate the diameter of a circle with a radius of length r as $D = 2r$.

So, calculate the diameter of a circle with a radius of 9 kilometers as follows:

$$D = 2 \times 9 = 18$$

Therefore, this circle has a diameter of 18 kilometers.

You can calculate the area of a circle with radius r using the formula $A = \pi r^2$. (The value π is an irrational number that equals approximately 3.14.)

Thus, the area of a circle whose radius equals 9 kilometers is

$$A = \pi \times 9^2 = 81\pi \approx 254.34$$

So the area of this circle is exactly 81π square kilometers, which is approximately equal to 254.34 square kilometers.

Circumference is essentially the same thing as perimeter — the length around a shape — but the word *circumference* only refers to circles.

You calculate the circumference of a circle with radius r as $C = 2\pi r$. So you can calculate the circumference of a circle whose radius equals 9 kilometers as

$$C = 2 \times \pi \times 9 = 18\pi \approx 56.52$$

Therefore, this circle's circumference is exactly 18π kilometers, or approximately 56.52 kilometers.

Boxes

A box is any three-dimensional rectangular shape. You can measure the *volume* of a box — that is, the space inside it — using the formula $V = l \times w \times h$, where these values are the length, width, and height of the box.

For example, consider a room with a length of 15 feet, a width of 10 feet, and a height of 8 feet. You can calculate the volume as

$$V = l \times w \times h = 15 \times 10 \times 8 = 1200$$

So the volume of the room is 1200 cubic feet.

Prisms

A *prism* is the projection of a shape into space. For example, a cylinder is the projection of a circle. You can measure the *volume* of a prism — that is, the space inside it — using the formula $V = A_b h$, where A_b is the area of the base.

For example, suppose that a cylindrical storage tank has a height of 7 feet, and the radius of its circular base is 4 feet. To find the volume of the tank, begin by using the formula for the area of a circle to find the area of the base:

$$A = \pi r^2 = \pi(4)^2 = 16\pi$$

Now, plug this value into the formula for the volume of a cylinder:

$$V = A_b h = 16\pi(7) = 112\pi$$

If needed, you can use 3.14 as an approximation for the value of π:

$$= 112(3.14) \approx 351.68$$

So the volume of the tank is approximately 351.68 cubic feet.

Working with geometric formulas algebraically

When you know the basic area and perimeter formulas that I discuss in the previous section, you can use them to solve slightly trickier geometry problems. In this section, you use these formulas to tease out more subtle information about a variety of shapes.

Applying formulas

When you feel comfortable working with geometric formulas (see "Basic geometric formulas"), you're ready to take them to the next level. For example, consider this problem:

If a rectangle has a length of 8 and an area of 40, what is its width?

You can solve this problem by plugging the length and area into the area formula, and solving for the width:

$$A = l \times w$$
$$40 = 8w$$
$$5 = w$$

Therefore, the width of this rectangle equals 5.

Here's another example:

What is the circumference of a circle whose area equals 64π?

To begin, use the area formula for a circle to find the radius:

$$A = \pi r^2$$
$$64\pi = \pi r^2$$
$$64 = r^2$$
$$\sqrt{64} = \sqrt{r^2}$$
$$8 = r$$

Now, plug this radius into the formula for the circumference:

$$C = 2\pi 8 = 16\pi \approx 50.24$$

Pythagorean theorem

The *Pythagorean theorem* $a^2 + b^2 = c^2$ allows you to find the hypotenuse c (longest side) of a right triangle given the lengths of the legs a and b (shorter sides).

For example, suppose you know that a right triangle has legs of lengths 5 and 12. You can plug these values of a and b (in any order) into the Pythagorean theorem, and solve for c.

$$a^2 + b^2 = c^2$$
$$5^2 + 12^2 = c^2$$
$$25 + 144 = c^2$$
$$169 = c^2$$
$$\sqrt{169} = \sqrt{c^2}$$
$$13 = c$$

So this right triangle has a hypotenuse of 13.

You can also find the length of one leg of a right triangle, given the length of its hypotenuse and one leg. For example, suppose a right triangle has a hypotenuse of 15 inches and a leg that is 12 inches. Plug these numbers into the Pythagorean theorem (make sure to substitute 12 for c!):

$$12^2 + b^2 = 15^2$$

Now, solve this equation for b:

$$144 + b^2 = 225$$
$$b^2 = 81$$
$$\sqrt{b^2} = \sqrt{81}$$
$$b = 9$$

So the length of the remaining leg is 9 inches.

Chapter **15**

Statistics, Probability, and Sets

In this chapter, you take a break from algebra (yay!) and explore a few different areas of math that you'll need for the ACCUPLACER.

To begin, the focus is on statistics, which is math applied to numerical information gathered from the real world, called data sets. You discover how to order and interpret data sets using a variety of mathematical tools such as the minimum and maximum, the range, the mean and the median, and the first and third quartiles.

From there, I show you how to calculate probability, and then apply this technique to three different types of problems: simple probability, compound probability, and conditional probability. To finish up, you dip briefly into basic set theory, solving problems involving both the union and the intersection of sets.

Descriptive Statistics

Statistics is the collection and interpretation of numerical information to form conclusions about the world. When the *collection method* (the way you get the information) and the *interpretation* (the way you crunch the numbers) are both sound, statistical conclusions can be a powerful way to form and prove scientific theories, and even predict what will happen next, or in a similar set of circumstances.

I will admit that there is plenty of math on the ACCUPLACER that you may never use, even in college, after you take the test. Not so with statistics! Most areas of study in both the natural sciences (physics, chemistry, biology, and so on) and the social sciences (psychology, sociology, anthropology, economics, and so on) require at least a passing understanding of statistics — and often much more.

In this section, I give you a basic understanding of statistics that you'll need in order to do well on the ACCUPLACER, and also as you proceed to college.

Calculating statistics

Statistical information usually comes in the form of a *data set* or *sample set,* which is a set of numbers that have been gathered through observation of events. For example,

On Professor Keller's first midterm exam, her 7 students received the following grades (on a scale from 1 to 100):

87, 92, 84, 75, 77, 93, 84

For the remainder of this section, I use this data set as an example to discuss the statistical concepts that you need to know for the ACCUPLACER.

Ordering a data set

REMEMBER

An essential first step when working with a numerical data set is to *order* that set — that is, arrange its values in order from lowest to highest. Here is the *ordered* data set from Professor Keller's exam:

75, 77, 84, 84, 87, 92, 93

For the remainder of this example, I refer only to the ordered version of this data set.

Minimum and maximum

REMEMBER

The *minimum* (*min*) and *maximum* (*max*) of a data set are the lowest and highest values in that set, respectively. When working with an ordered data set, the min is the first value in the set, and the max is the last value.

So the min and max for the exam grades are 75 and 93, respectively.

Spread, or range, of a data set

REMEMBER

The *spread* (or *range*) of a data set is the difference between its maximum and minimum:

Range = Max − Min

To find the spread of grades for Professor Keller's exam, plug in the max and min:

Range = 93 − 75 = 18

Therefore, the range for this data set is 18.

Measures of center

In most statistical studies, finding the center of the data set is important. The two most common measures of center are the *mean* and the *median.*

REMEMBER

The *mean* (or *mean average*) is the center of a data set as calculated by the following formula:

$$\text{Mean} = \frac{\text{Sum of values}}{\text{Number of values}}$$

To find the mean of the exam grades, add up all the grades and divide by 7 (the number of grades):

$$\text{Mean} = \frac{75 + 77 + 84 + 84 + 87 + 92 + 93}{7} = \frac{592}{7} \approx 84.6$$

Thus, the mean is approximately 84.6.

REMEMBER

The *median* is simply the middle value in a data set. (When the data set has an even number of values, the median is the mean of the two middle values.) Thus, the median of the exam grades is 84.

Lower half and upper half of a data set

REMEMBER

The *lower half* of an ordered data set is the set of values that fall to the left of the median. Similarly, the *upper half* is the set of values that fall to the right of the median. Thus, the lower and upper halves of the set of exam grades are as follows:

75	77	84	84	87	92	93
Lower half			Median	Upper half		

Note that the median value (84) is excluded from both the lower half and the upper half. However, because this value occurs twice in the data set, it's included as one of the three values in the lower half. Generally speaking:

When a data set has an *odd* number of values, the median is *excluded* from both the lower half and the upper half.

When a data set has an *even* number of values, all values are *included* in either the lower half or the upper half.

First and third quartiles

REMEMBER

The *first quartile* of a data set is the median of the lower half. Similarly, the *third quartile* is the median of the upper half.

In the set of exam data, the first quartile is 77 and the third quartile is 92. (By the way, in case you were wondering what happened to the *second quartile*, this is another name for the median.)

Displaying statistics

In the previous section, you calculated statistics for a data set culled from Professor Keller's first midterm exam. Here's a recap of some of these statistics:

Lower half				Upper half		
75	77	84	84	87	92	93
Minimum	First quartile		Median		Third quartile	Maximum

Range = 18   Mean approx. = 84.6

Note that five key values — the minimum, the first quartile, the median, the third quartile, and the maximum — are captured here. In this section, you discover how to use a set of standard tools to display the statistics that you've calculated.

Boxplots, or box-and-whisker plots

REMEMBER

A *boxplot* (or *box-and-whisker plot*) is a diagram that displays the minimum, the first quartile, the median, the third quartile, and the maximum. Here is a boxplot for the midterm exam data:

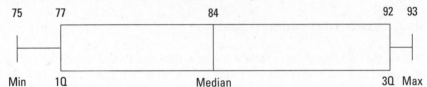

As you can see, this boxplot includes the five values you've been working with. The "box" portion of the boxplot includes two sides that represent the first and third quartiles, with the median drawn as a vertical line between them. Additionally, the minimum and maximum are at their respective ends of the "whiskers."

The shape of a data set

A key feature of a boxplot is that the horizontal dimensions of the two parts of the box and the two whiskers are drawn proportionally to the numbers that are displayed.

So, in this boxplot, the left whisker has a length of 2 units, representing the difference between the minimum and the first quartile. Similarly, the left part of the box measures 7 units, the right part of the box 8 units, and the right whisker 1 unit.

Thus, a boxplot quickly gives you a visual sense of the *shape* of the data set you're working with. In this case, the data is very close to *non-skewed* (or *symmetrical*) — that is, the two sides of the box are roughly the same width, as are the two whiskers.

Recall that the statistics also included a range of 18 and a mean of approximately 84.6. Note that this mean is only slightly higher than the median of 84.

Introducing an outlier

Now, suppose I add some additional data to the set:

The day after the midterm, a student who had been absent from class most of the semester showed up at Professor Keller's office and begged for an opportunity to take the exam that he had missed. She agreed, but only if he would sit down immediately and take it. He did so and, unfortunately, got a grade of 29.

I'll add this extra grade to the data set to see how it affects the statistics:

29, 75, 77, 84, 84, 87, 92, 93

REMEMBER

This low grade — which sticks out like a sore thumb when placed among the rest of the grades — is called an *outlier*: a data point that creates an imbalance with the remaining data in the set. In this case, the *low outlier* has created a *left skew* to the data — that is, an imbalance that tends toward the lower numbers in the set.

Notice, however, that despite this outlier, the median exam grade is still 84. The mean, however, has changed significantly:

$$\text{Mean} = \frac{29 + 75 + 77 + 84 + 84 + 87 + 92 + 93}{8} = \frac{621}{8} \approx 77.6$$

With the inclusion of the low outlier, the mean exam grade has decreased by about 7 points.

The effect on the range, which was previously 18, is even more striking:

Range = 93 – 29 = 64

The range of grades for the exam has increased by 64 points! The introduction of an outlier has also affected the lower half and upper half:

29, 75, 77, 84, 84, 87, 92, 93

These changes also affect the calculations of the first quartile (76 — the median of the lower half) and the third quartile (89.5 — the median of the upper half). The following table sums up this new statistical information.

Lower half				Upper half			
29	75	77	84	84	87	92	93
Minimum	First quartile = 76			Median		Third quartile = 89.5	Maximum

The following boxplot summarizes this information, making it a bit more intuitive visually.

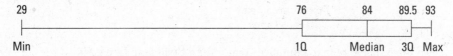

Table 15-1 summarizes information about the effect of outliers on a data set.

TABLE 15-1 The Effect of Outliers on a Data Set

Low outlier(s)	No outliers	High outliers
Left-skewed data	Non-skewed data	Right-skewed data
Median relatively unchanged. Mean decreases.	Median and mean roughly the same.	Median relatively unchanged. Mean increases.
Range increases greatly.		Range increases greatly.

Probability

Probability is the mathematical measurement of how likely an event (or outcome) is to occur. In this section, I discuss the kinds of probability questions that are most likely to occur on the ACCUPLACER. First, I explain how to use the formula for probability to calculate simple, compound, and conditional probability. After that, you apply these skills to a common type of probability problem that includes a table.

Calculating probability

REMEMBER

Probability is calculated using the following formula:

$$\text{Probability} = \frac{\text{Target outcomes}}{\text{Total outcomes}}$$

In this formula, *target outcomes* indicates the number of possible ways the event can occur. *Total outcomes* is the total number of outcomes that could possibly occur.

Because the number of target outcomes is always less than or equal to the number of total outcomes, probability is always a number from 0 (impossibility) to 1 (certainty) — in most cases, falling between these two values.

Simple probability

REMEMBER

Simple probability is the probability that a single event will occur. For example, consider the following question:

What is the probability that when you roll a 6-sided die, a number greater than 3 will land face up?

When you roll a 6-sided die, the number of *total outcomes* equals 6: Each number from 1 to 6 could land face up. In this case, the number of *target outcomes* equals 3: the outcomes in which the numbers 4, 5, or 6 land face up. To find the probability of this event, plug these numbers into the probability formula:

$$\text{Probability} = \frac{\text{Target outcomes}}{\text{Total outcomes}} = \frac{3}{6} = \frac{1}{2}$$

So the probability that a number greater than 3 will land face up is $\frac{1}{2}$.

Compound probability

REMEMBER

Compound probability is the probability that at least two criteria are both met. This makes more sense with an example:

What is the probability that when you roll a 6-sided die, the number that lands face up will be *both* an even number *and* greater than 3?

As in the previous problem, here the number of total outcomes is still 6. However, this time, the number of target outcomes that satisfy both criteria is 2: Either the number 4 **or** the number 6 lands face up. Again, calculate the probability by plugging these numbers into the formula:

$$\text{Probability} = \frac{\text{Target outcomes}}{\text{Total outcomes}} = \frac{2}{6} = \frac{1}{3}$$

So the probability of rolling an even number that's greater than 3 is $\frac{1}{3}$.

Conditional probability

REMEMBER

Conditional probability is the probability that an event will occur *given that* another event also occurs, limiting the possible number of total outcomes. To see how this works, consider the following question:

What is the probability that when you roll a 6-sided die, an even number will land face up *given that* the number rolled is greater than 3?

In this example, the problem states a condition that a number greater than 3 is rolled. This condition limits the number of total outcomes to 3 possibilities: Either a 4, 5, or 6 lands face up. Inside of this condition, there are 2 target outcomes: Either a 4 or a 6 is rolled.

Here's how this lands when calculating probability:

$$\text{Probability} = \frac{\text{Target outcomes}}{\text{Total outcomes}} = \frac{2}{3}$$

So, the probability of rolling an even number *given that* a number greater than 3 is rolled equals $\frac{2}{3}$.

Table-based probability problems

REMEMBER

On the ACCUPLACER, *table-based probability problems* are a potential source of confusion. In this section, I help you sort through a problem of this type, applying your skills for calculating simple, compound, and conditional probability from the previous section.

Throughout this section, I use the following example:

Sister Veronica asked her class of 28 fifth graders to tell her whether they owned at least one dog or at least one cat. She gathered the results into the following table:

	Cat owner	Not a cat owner	Total
Dog owner	7	5	12
Not a dog owner	10	6	16
Total	17	11	28

Before moving on, notice that in this table, all rows and columns are added up in the two totals, and then these totals are added together, with all 28 children accounted for. This is necessary with table problems of this type.

Table-based simple probability

Here's a question that asks you to calculate simple probability based on the table provided:

What is the probability that a child selected at random from the class will be a dog owner?

In this case, the number of total outcomes is 28: the total number of children in the class. The number of target outcomes is 12: the number of dog owners. Calculate the probability as follows:

$$\text{Probability} = \frac{\text{Target outcomes}}{\text{Total outcomes}} = \frac{12}{28} = \frac{3}{7}$$

Thus, the probability of selecting a dog owner at random from the entire class is $\frac{3}{7}$.

Table-based compound probability

A typical compound probability problem looks like this:

What is the probability that a child selected at random from the class will be *both* a dog owner and a cat owner?

Again, the number of total outcomes is 28. However, according to the table, only 7 of the children in the class own *both* a dog *and* a cat. Thus, the probability of this event is as follows:

$$\text{Probability} = \frac{\text{Target outcomes}}{\text{Total outcomes}} = \frac{7}{28} = \frac{1}{4}$$

Thus, $\frac{1}{4}$ of the children own both a dog and a cat.

Here's another example:

What is the probability that a child selected at random from the class will be a cat owner but *not* a dog owner?

Again, the number of total outcomes remains 28. But this time, the table indicates that 10 children own a cat but *not* a dog. Thus, the probability of this event is as follows:

$$\text{Probability} = \frac{\text{Target outcomes}}{\text{Total outcomes}} = \frac{10}{28} = \frac{5}{14}$$

So, if you choose a child from the class at random, the probability is $\frac{5}{14}$ that they will own a cat but not a dog.

Table-based conditional probability

Conditional probability problems based on a table can feel a little tricky, because they can be stated in a few different ways. Here's a typical example:

What is the probability that if a child owns a dog, they also own a cat?

To begin solving this problem, remember that the condition always affects the total outcomes. Here, the condition is preceded by the word *if*: The words *if a child owns a dog* limits the total outcomes to 12. Of these 12 children, 7 own cats, so this is the number of target outcomes. Thus,

$$\text{Probability} = \frac{\text{Target outcomes}}{\text{Total outcomes}} = \frac{7}{12}$$

So, if you randomly select a child from the class who owns a dog, the probability is $\frac{7}{12}$ that they also own a cat.

In contrast, consider the following question:

What is the probability that if a child owns a cat, they also own a dog?

This question may seem identical to the last one, but this time, the condition is *if a child owns a cat*. This condition limits the number of total outcomes to 17, and among these 17 children, 7 own a dog. So calculate as follows:

$$\text{Probability} = \frac{\text{Target outcomes}}{\text{Total outcomes}} = \frac{7}{17}$$

So, $\frac{7}{17}$ of children in the class who own a cat also own a dog.

Here's one final example, in which conditional probability is stated slightly differently:

What's the probability that a child owns a cat *given that* they don't own a dog?

This time, the condition follows the words *given that,* so the limit to the total outcomes is that a child *doesn't own a dog*. So, 16 children in the class fulfill this condition, and of these, 10 are in the target outcome group of cat owners:

$$\text{Probability} = \frac{\text{Target outcomes}}{\text{Total outcomes}} = \frac{10}{16} = \frac{5}{8}$$

Therefore, $\frac{5}{8}$ of children in the class who are *not* dog owners are cat owners.

Set Theory

Set theory is the mathematics of formal collections of items known as *sets*. A set is denoted by a pair of braces ({}). For example,

J = {car, bus, train, airplane}

K = {George Washington, John Adams, Thomas Jefferson}

L = {New York, Seattle, San Francisco, Honolulu, Tokyo, Budapest}

Here, I define three sets J, K, and L. Although these sets include elements that are modes of transportation, U.S. presidents, and cities, questions on the ACCUPLACER usually include sets that are exclusively numbers. For example,

P = {1,2,3,4,5,6,7}   Q = {2,4,6,8,10}   R = {5,6,7,8,9,10}

Throughout this section, I will refer to these three sets.

Elements of a set

REMEMBER

Sets P, Q, and R include, respectively, 7, 5, and 6 *elements*, which are simply things belonging to that set. The notation $\in$ means *is an element of*, and the notation $\notin$ means *is not an element of*. Thus, the following statements are all true:

$1 \in P$ (1 is an element of *P*)

$8 \notin P$ (8 is not an element of *P*)

$8 \in Q$ (8 is an element of *Q*)

$3 \notin R$ (3 is not an element of *R*)

Union of two sets

REMEMBER

The *union* of two sets (denoted by the symbol $\cup$) is the set formed by elements that are in *either* set.

For example,

$P \cup Q = \{1,2,3,4,5,6,7,8,10\}$

$P \cup R = \{1,2,3,4,5,6,7,8,9,10\}$

$Q \cup R = \{2,4,5,6,7,8,9,10\}$

Note that when *both* sets contain an element, this element is included only once in the union set.

Intersection of two sets

REMEMBER

The *intersection* of two sets (denoted by the symbol $\cap$) is the set formed by elements that are in *both* sets.

For example,

$P \cap Q = \{2,4,6\}$

$P \cap R = \{5,6,7\}$

$Q \cap R = \{6,8,10\}$

Combining union and intersection

An ACCUPLACER problem in set theory may contain both union and intersection. When solving this type of problem, always evaluate values that are inside parentheses first.

For example, consider the expression $P \cap (Q \cup R)$. To evaluate this expression, begin with the value of $Q \cup R$, which equals $\{2,4,5,6,7,8,9,10\}$. So,

$$P \cap (Q \cup R) = \{1,2,3,4,5,6,7\} \cap \{2,4,5,6,7,8,9,10\}$$

Now, evaluate this intersection:

$$= \{2,4,5,6,7\}$$

As another example, consider the expression $(P \cap Q) \cup R$. This time, begin evaluating with $P \cap Q$, which equals $\{2,4,6\}$:

$$(P \cap Q) \cup R = \{2,4,6\} \cup \{5,6,7,8,9,10\}$$

To complete the problem, evaluate this union:

$$= \{2,4,5,6,7,8,9,10\}$$

IN THIS CHAPTER

» Practicing your skills working with rates and ratios

» Solving problems involving algebra expressions and equations

» Working with linear functions

» Tackling geometry questions

» Putting your statistics skills to the test

Chapter **16**

Quantitative Reasoning, Algebra, and Statistics Practice Questions

This chapter gives you a chance to test your knowledge in quantitative reasoning, algebra, geometry, and statistics. When you've answered the questions, check your work against the answers and explanations at the end.

Quantitative Reasoning Practice Questions

1. Indicate whether each of the following is a rational number.

(A) 15

(B) −9

(C) $-\frac{17}{35}$

(D) 0.333

(E) $\sqrt{3}$

2. $|6+(-7)|-|5\times(-2)|=$

3. If a car traveled an average of 62 miles per hour over the course of a 713-mile trip, how many hours was it on the road? (Round your answer to the nearest tenth of a mile.)

4. A beach club has a 4-to-7 ratio of men to women. If 203 women belong to the club, how many men are members?

5. On a trip to Europe, Kyle weighed himself and found that his weight was 51.5 kilograms. How much did he weigh in pounds? (One kilogram equals approximately 2.2 pounds.)

6. If one kilometer equals approximately 0.62 miles, how do you represent a distance of 5 miles as kilometers? (Round your answer to the nearest two decimal places.)

7. Scientists estimate that there are approximately 8.7×10^6 different living species on Earth. How do you write that number in standard notation?

8. Mercury is the smallest planet in our solar system, and Jupiter is the largest. The ratio of the mass of Mercury to Jupiter is approximately 0.000174. What is that value in scientific notation?

9. $\dfrac{x^5 \cdot x^6}{x^7} =$

10. $\dfrac{(x^2 y^4)^3}{x^4 y^5} =$

11. What is the simplest equivalent of $\dfrac{a^{-2} b^3}{a^3 b^{10} c}$ written without negative exponents?

12. $64^{\frac{1}{2}} =$

Algebra Practice Questions

In this section, you have a chance to practice your algebra skills: working with algebra expressions, solving algebra equations, making sense of algebra word problems, and solving problems that involve linear functions. If you get stuck, or just want to check your answers, flip to end of the chapter.

Algebra basics

1. Consider the algebraic expression $4x^3 - x^2 + 7x - 9$.

(A) How many terms does this expression have?

(B) What is the constant term in this expression?

(C) What is the coefficient of the x^2 term?

(D) What is the sum of all the coefficients?

2. What is the value of $4x^2 - 8x + 50$ when $x = -3$?

3. What is the value of $7x^2 + xy - 2y^2$ when $x = 2$ and $y = -5$?

4. What do you get when you simplify $3a^2 + 5ab - 4b^3 - 6ab + 10b^3 + ab$?

5. What is the result when you simplify $4(3x + 5) - (3 - 6x) - 2x$?

6. Simplify $(x + 3)(4x - 5)$.

7. If you simplify $(2x - 7)(3x - 1) + 2(x - 4)$, what's the result?

8. Factor $3a^5 - 18a^3 + 6a^2$.

9. What do you get when you factor $10m^7n + 45m^2n^5 - 25m^4n^4 - 30m^9n^3$?

10. When you factor $8x^3yz^2 - 56x^8y^4z^3$, what is the result?

Algebra equations

1. What is the value of x when $5x - 7 = 23$?

2. If $6x + 8 = -3x - 10$, what is the value of x?

3. Solve the following equation for x: $8x + 4 - x = 11 + 5x - 1$

4. If $2(x - 9) - 5x = 7 - (4x + 9)$, what is the value of x?

5. If $\frac{3x - 8}{5} = -2x + 1$, what is the value of x?

6. If $\frac{4x + 5}{3} = \frac{6 - 2x}{5}$, what is the value of x?

7. If $\frac{2}{3}x + \frac{3}{4} = \frac{1}{2}(5x - 7)$, what is the value of x?

8. If $0.3x + 1.2 = 5.3 - 1.75x$, what is the value of x?

9. In the equation $\frac{2a - 3b}{5} = 2b + 7c$, what is the value of a in terms of b and c?

10. Solve this system of equations for x and y:

$$3x + 4y = -1$$
$$x + 5y = 7$$

11. What are the values of x and y in the following system of equations?

$$6x - 4y = 10$$
$$7x - 3y = 25$$

12. Solve the following inequality for x: $4(7 - 2x) + 3x \geq 8$

Algebra word problems

1. Three times a number is equivalent to that number plus 14. What is the number?

2. If 4 times the quantity of a number minus 7 equals that number plus 2, what is the number?

3. Over the summer, Claudine earned $110 more than Sylvie. Altogether, they earned $1,390. How much did Sylvie earn?

4. A supermarket is open for 16 hours every day. On one particular day, a total of 425 customers came through the store. To the nearest hour, how many hours did it take for 100 customers to make purchases? (Assume a constant rate of customers per hour.)

5. Jacob has 19 more comic books than Benjamin. If Benjamin doubles the number of comic books that he has and Jacob buys 9 more, together they will have 79 comic books. How many comic books does Benjamin have right now?

6. A window washing team in a large office building can wash 360 windows in 8 hours. How many windows can they wash in 12 minutes?

7. When you add 7 to a number, then multiply the result by 8, and then subtract 6, the number you end up with is 10. What is the original number?

8. To raise money for their school play, Beatrice sold brownies for $1.25 each, and Chuck sold cookies for $0.75 each. Together, they sold 87 items and made $80.75. How many cookies did Chuck sell?

Linear functions

1. Caren opened her savings account with a deposit of $1200, and every month after that deposited an additional $350. How much money will Caren have in her account 36 months after opening it?

2. Seth owns a vacation timeshare plan that costs $3500 per year plus $65 for each night that he uses the timeshare. Last year, he spent $5710 on the time share. How many nights did he use the timeshare?

3. What is the equation for the line shown in the following graph?

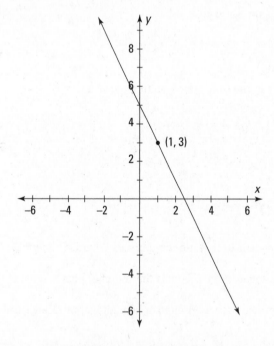

4. What is the equation of a line with a slope of –3 and passing through the point (4,–5)?

5. What is the equation of a line that passes through the points (3, 7) and (–2,–8)?

6. What is the equation of a line that passes through the point (–6, 5) and is parallel to the line whose equation is $y = \frac{1}{3}x - 10$?

7. What is the equation of a line that passes through the point (2, 3) and is perpendicular to the line whose equation is $y = 4x + 9$?

8. At what point on the xy-graph do the two lines $y = 5x + 4$ and $y = -3x + 12$ intersect?

9. What system of linear inequalities is produced in the following graph?

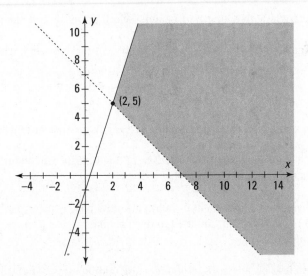

10. Rewrite the equation $3x + 5y = 2$ in slope–intercept form.

11. If the points $(a, 0)$ and $(0, b)$ lie on the graph of the equation $3x - 4y = 24$, what is the value of $a + b$?

12. On an xy-graph, what is the distance between the points $(-1, 5)$ and $(-7, -3)$?

13. If the point $(4, -3)$ is rotated 180 degrees and then reflected across the x-axis, what are the coordinates of the resulting point?

14. The following figure shows Triangle ABC. You want to translate this triangle to a new position as Triangle $A'B'C'$. If A' is located at $(-3, 8)$, what will be the coordinates of the point B'?

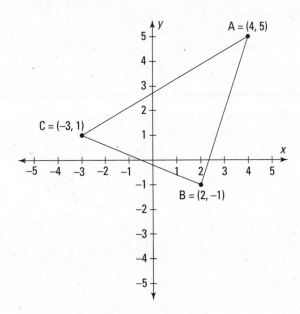

Geometry Practice

1. If a square has a perimeter of 48 inches, what is the length of its side?

2. What is the perimeter of a square that has an area of 25 square inches?

3. What is the area of a rectangle with a length of 15 and a width that is $\frac{1}{3}$ of its length?

4. If a rectangle whose longest side is 8 meters has an area of 24 square meters, what is its perimeter?

5. What is the diameter of a circle that has a circumference of 24π?

6. What is the circumference of a circle that has an area of 25π? (Your answer may include π.)

7. A rectangular room has a length of 15 feet and a height of 8 feet. If its total volume is 1200 square feet, what is the width of the room?

8. A cylindrical can has a height of 10 inches, and the radius of its base equals 3 inches. What is the volume of the can in cubic inches, rounded to two decimal places? (Use 3.14 as an approximation for the value of π.)

9. If a right triangle has two legs of lengths 8 kilometers and 6 kilometers, what is the length of its hypotenuse?

10. Suppose a triangle has a hypotenuse of 17 millimeters and one leg of length 15 millimeters. What is the length of its other leg?

Statistics, Probability, and Sets Practice

Questions 1 and 2 use the following information:

A third-grade class of 17 students had the following grades on a math test:

80, 75, 85, 90, 80, 100, 75, 80, 90, 90, 95, 80, 100, 90, 45, 70, 95

1. Answer the following:

   (A) Place the grades in order from lowest to highest.
   (B) What is the minimum value of the data set?
   (C) What is the maximum value?
   (D) What is the range (spread) of the values?
   (E) What is the mean of the values?
   (F) What is the median?

2. Answer the following:

  (A) Which number is an outlier in the set?

  (B) Does the outlier skew the set right or left?

  (C) Given that there is an outlier, which better represents the center of the data — the mean or the median?

  (D) What is the first quartile (1Q)?

  (E) What is the third quartile (3Q)?

  (F) Draw a boxplot of the data.

Use this information for Questions 3 through 5:

A bag contains 10 tiles, each tile containing a different number from 1 to 10.

3. If you pull a tile at random from the bag, what's the probability that this number will be greater than 6?

4. If you pull a tile at random from the bag, what's the probability that this number will be both even and less than 7?

5. If you pull an odd-numbered tile from the bag, what's the probability that this number will be greater than 4?

Use the following table to answer Questions 6 through 8:

A group of 75 people were asked whether they drink soft drinks and/or coffee. The results were gathered and placed in the following table.

	Coffee	No coffee	Total
Soft drinks	38	19	57
No soft drinks	11	7	18
Total	47	28	75

6. What is the probability that a person picked at random drinks coffee?

7. If a person is picked at random from the group, what is the probability that this person drinks soft drinks but not coffee?

8. If a person who doesn't drink coffee is picked at random, what is the probability that they also don't drink soft drinks?

Use the following information to solve Questions 9 and 10:

$$V = \{1,3,5,7,9\} \quad W = \{4,5,6,7\} \quad X = \{2,3,6,8,9,10\}$$

9. What is the set $V \cup (W \cap X)$?

10. What is the set $(V \cup X) \cap W$?

Answers and Explanations

So how did you do? In this section, you'll find not only answers but also explanations of all the practice questions in the previous section.

Quantitative reasoning answers

1. See below.

   (A) **Rational.** Positive integers are rational.

   (B) **Rational.** Negative integers are rational.

   (C) **Rational.** All fractions are rational.

   (D) **Rational.** Repeating decimals are rational.

   (E) **Irrational.** Square roots of non-square numbers are irrational.

2. **–9.**

$$|6+(-7)|-|5\times(-2)|$$
$$=|-1|-|-10|$$
$$=1-10$$
$$=-9$$

3. **11.5 hours.** Divide 713 by 62: $713 \div 62 = 11.5$

4. **116 men.** Set up the following proportion:

$$\frac{4\,\text{men}}{7\,\text{women}} = \frac{x\,\text{men}}{203\,\text{women}}$$

Next, cross-multiply to remove the fractions, drop the units of measure, and then solve for x:

$$4 \cdot 203 = 7x$$
$$812 = 7x$$
$$116 = x$$

5. **113.3 lbs.** Set up a proportion as follows:

$$\frac{1\,\text{kg}}{2.2\,\text{lbs.}} = \frac{51.5\,\text{kg}}{x\,\text{lbs.}}$$

Cross-multiply and drop the units of measure:

$$x = 51.5 \cdot 2.2 = 113.3$$

6. **8.06 km.** Set up a proportion as follows:

$$\frac{1\,\text{km}}{0.62\,\text{mi.}} = \frac{x\,\text{km}}{5\,\text{mi.}}$$

Cross-multiply and drop the units of measure. Then solve for x:

$$5 = 0.62x$$
$$\frac{5}{0.62} = \frac{0.62x}{0.62}$$
$$8.06 \approx x$$

7. 8,700,000. To multiply 8.7 by 10^6, move the decimal point 6 places to the right.

8. 1.74×10^{-4}. Starting with 0.000174, move the decimal point 4 places to the right and multiply the result by 10^{-4}.

9. x^4.

$$\frac{x^5 \cdot x^6}{x^7} = \frac{x^{11}}{x^7} = x^4$$

10. $x^2 y^7$.

$$\frac{(x^2 y^4)^3}{x^4 y^5} = \frac{x^6 y^{12}}{x^4 y^5} = x^2 y^7$$

11. $\dfrac{1}{a^5 b^7 c}$.

$$\frac{a^{-2} b^3}{a^3 b^{10} c} = a^{-2-3} b^{3-10} c^{-1} = a^{-5} b^{-7} c^{-1} = \frac{1}{a^5 b^7 c}$$

12. 8.

$$64^{\frac{1}{2}} = \sqrt{64} = 8$$

Algebra basics answers

1. See below:

   (A) 4 terms. $4x^3, -x^2, 7x$, and -9

   (B) −9. The constant term is the term without a variable.

   (C) −1.

   (D) 1. $4 - 1 + 7 - 9 = 1$

2. 110.

$$4x^2 - 8x + 50$$
$$= 4(-3)^2 - 8(-3) + 50$$
$$= 4(9) - 8(-3) + 50$$
$$= 36 + 24 + 50$$
$$= 110$$

3. −32.

$$7x^2 + xy - 2y^2$$
$$= 7(2)^2 + 2(-5) - 2(-5)^2$$
$$= 7(4) + 2(-5) - 2(25)$$
$$= 28 - 10 - 50$$
$$= -32$$

4. $3a^2 + 6b^3$.

$$3a^2 + 5ab - 4b^3 - 6ab + 10b^3 + ab$$
$$= 3a^2 - 4b^3 + 10b^3$$
$$= 3a^2 + 6b^3$$

5. $16x + 17$.

$$4(3x + 5) - (3 - 6x) - 2x$$
$$= 12x + 20 - 3 + 6x - 2x$$
$$= 16x + 17$$

6. $4x^2 + 7x - 15$.

$$(x + 3)(4x - 5)$$
$$= 4x^2 - 5x + 12x - 15$$
$$= 4x^2 + 7x - 15$$

7. $6x^2 - 21x - 1$.

$$(2x - 7)(3x - 1) + 2(x - 4)$$
$$= 6x^2 - 2x - 21x + 7 + 2x - 8$$
$$= 6x^2 - 21x - 1$$

8. $3a^2(a^3 - 6a + 2)$.

$$3a^5 - 18a^3 + 6a^2$$
$$= 3a^2(a^3 - 6a + 2)$$

9. $5m^2n(2m^5 + 9n^4 - 5m^2n^3 - 6m^7n^2)$.

$$10m^7n + 45m^2n^5 - 25m^4n^4 - 30m^9n^3$$
$$= 5m^2n(2m^5 + 9n^4 - 5m^2n^3 - 6m^7n^2)$$

10. $8x^3yz^2(1 - 7x^5y^3z)$.

$$8x^3yz^2 - 56x^8y^4z^3$$
$$= 8x^3yz^2(1 - 7x^5y^3z)$$

Algebra equations answers

1. $x = 6$.

$$5x - 7 = 23$$
$$5x = 30$$
$$x = 6$$

2. $x = -2$.

$$6x + 8 = -3x - 10$$
$$9x + 8 = -10$$
$$9x = -18$$
$$x = -2$$

3. $x = 3$.

$$8x + 4 - x = 11 + 5x - 1$$
$$7x + 4 = 10 + 5x$$
$$2x + 4 = 10$$
$$2x = 6$$
$$x = 3$$

4. $x = 16$.

$$2(x - 9) - 5x = 7 - (4x + 9)$$
$$2x - 18 - 5x = 7 - 4x - 9$$
$$-3x - 18 = -2 - 4x$$
$$x - 18 = -2$$
$$x = 16$$

5. $x = 1$.

$$\frac{3x - 8}{5} = -2x + 1$$
$$5(-2x + 1) = 3x - 8$$
$$-10x + 5 = 3x - 8$$
$$13 = 13x$$
$$1 = x$$

6. $x = -\dfrac{7}{26}$.

$$\frac{4x + 5}{3} = \frac{6 - 2x}{5}$$
$$5(4x + 5) = 3(6 - 2x)$$
$$20x + 25 = 18 - 6x$$
$$26x + 25 = 18$$
$$26x = -7$$
$$x = -\frac{7}{26}$$

7. $x = \dfrac{51}{22} = 2\dfrac{7}{22}$.

$$\frac{2}{3}x + \frac{3}{4} = \frac{1}{2}(5x - 7)$$
$$12\left(\frac{2}{3}x + \frac{3}{4}\right) = 12\left(\frac{1}{2}(5x - 7)\right)$$
$$\frac{24}{3}x + \frac{36}{4} = \frac{12}{2}(5x - 7)$$
$$8x + 9 = 6(5x - 7)$$
$$8x + 9 = 30x - 42$$
$$-22x + 9 = -42$$
$$-22x = -51$$
$$x = \frac{51}{22} = 2\frac{7}{22}$$

8. $x = 2.$

$$0.3x + 1.2 = 5.3 - 1.75x$$
$$30x + 120 = 530 - 175x$$
$$205x = 410$$
$$x = 2$$

9. $a = \dfrac{13b + 35c}{2}.$

$$\frac{2a - 3b}{5} = 2b + 7c$$
$$2a - 3b = 5(2b + 7c)$$
$$2a - 3b = 10b + 35c$$
$$2a = 13b + 35c$$
$$a = \frac{13b + 35c}{2}$$

10. $x = -3$
$y = 2.$

$$3x + 4y = -1$$
$$x + 5y = 7$$

Begin by solving the second equation for x in terms of y:

$$x + 5y = 7$$
$$x = 7 - 5y$$

Next, plug in $7 - 5y$ for x in the first equation:

$$3x + 4y = -1$$
$$3(7 - 5y) + 4y = -1$$
$$21 - 15y + 4y = -1$$
$$21 - 11y = -1$$
$$-11y = -22$$
$$y = 2$$

So $y = 2$. Now, plug in 2 for y in either of the two equations, whichever looks simplest:

$$x + 5y = 7$$
$$x + 5(2) = 7$$
$$x + 10 = 7$$
$$x = -3$$

11. $x = 7$
$y = 8.$

Multiply the first equation by –3 and the second by 4:

$$-3(6x - 4y = 10)$$
$$4(7x - 3y = 25)$$

Now, simplify:

$$-18x + 12y = -30$$
$$28x - 12y = 100$$

Add the two equations together, and solve for x:

$$10x = 70$$
$$x = 7$$

Now, plug in 7 for x in either equation — the first equation looks simpler — and solve for y:

$$6(7) - 4y = 10$$
$$42 - 4y = 10$$
$$-4y = -32$$
$$y = 8$$

12. $x \le 4$.

$$4(7 - 2x) + 3x \ge 8$$
$$28 - 8x + 3x \ge 8$$
$$28 - 5x \ge 8$$
$$-5x \ge -20$$

Now, divide both sides by –5, and flip the inequality from $\ge$ to $\le$:

$$\frac{-5x}{-5} \le \frac{-20}{-5}$$
$$x \le 4$$

Algebra word problem answers

1. **7.** Set up and solve the following equation:

$$3x = x + 14$$
$$2x = 14$$
$$x = 7$$

2. **10.** Set up and solve the following equation:

$$4(x - 7) = x + 2$$
$$4x - 28 = x + 2$$
$$3x - 28 = 2$$
$$3x = 30$$
$$x = 10$$

3. **$640.** Let s = Sylvie's earnings. Then Claudine earned $s + 110$. Now, set up and solve the following equation:

$$s + s + 110 = 1390$$
$$2s + 110 = 1390$$
$$2s = 1280$$
$$s = 640$$

4. 4 hours. Begin by setting up the following proportion:

$$\frac{\text{customers}}{\text{hour}} = \frac{425}{16}$$

Now, fill in the information that you know about the customers, substitute h for hours, and solve for h:

$$\frac{100}{h} = \frac{425}{16}$$
$$1600 = 425h$$
$$3.8 \approx h$$

So, approximately 4 hours are needed to serve 100 customers.

5. 17 comic books. If you let b = Benjamin's comic books, then Jacob has $b + 19$ comic books. So, according to the second sentence, you can set up and solve the following equation:

$$2b + b + 19 + 9 = 79$$
$$3b + 28 = 79$$
$$3b = 51$$
$$b = 17$$

6. 9 windows. Begin by setting up the following proportion:

$$\frac{\text{windows}}{\text{hours}} = \frac{360}{8}$$

Next, convert 8 hours into minutes ($8 \times 60 = 480$), and substitute this value into the proportion:

$$\frac{\text{windows}}{\text{minutes}} = \frac{360}{480}$$

Now, fill in the information you know about minutes, substitute w for windows, and solve for w:

$$\frac{w}{12} = \frac{360}{480}$$
$$480w = 4320$$
$$w = 9$$

7. −5. Set up and solve the following equation:

$$8(x + 7) - 6 = 10$$
$$8x + 56 - 6 = 10$$
$$8x + 50 = 10$$
$$8x = -40$$
$$x = -5$$

8. 56 cookies. Let b = Beatrice's brownies and c = Chuck's cookies. So you can set up the following system of equations:

$$b + c = 87$$
$$1.25b + 0.75c = 80.75$$

Working with the first equation, isolate b:

$$b = 87 - c$$

Now, plug $87 - c$ for b into the second equation, and solve for c:

$$1.25(87 - c) + 0.75c = 80.75$$
$$108.75 - 1.25c + 0.75c = 80.75$$
$$10{,}875 - 125c + 75c = 8075$$
$$10875 - 50c = 8075$$
$$-50c = -2800$$
$$c = 56$$

Linear functions answers

1. **$13,800.** Begin by setting up a linear function with a slope of 350 and a y-intercept of 1200:

$$y = 350x + 1200$$

Now, substitute 36 for x and solve:

$$y = 350(36) + 1200$$
$$= 12{,}600 + 1200$$
$$= 13{,}800$$

2. **34 nights.** Begin by setting up a linear function with a slope of 65 and a y-intercept of 3500:

$$y = 65x + 3500$$

Now, substitute 5710 for y and solve:

$$5710 = 65x + 3500$$
$$2210 = 65x$$
$$34 = x$$

3. $y = -2x + 5$. The line crosses the y-axis at $(0, 5)$, so this is the y-intercept. It also crosses the point $(1, 3)$, so you can calculate its slope using the slope formula:

$$m = \frac{y_2 - y_1}{x_2 - x_1} = \frac{3 - 5}{1 - 0} = \frac{-2}{1} = -2$$

Therefore, the slope is -2 and the y-intercept is 5, so the equation is $y = -2x + 5$.

4. $y = -3x + 7$. The equation is a line with a slope of -3, so it takes the form $y = -3x + b$. It includes the point $(4, -5)$, so plug these values for x and y into this equation and solve for b:

$$y = -3x + b$$
$$-5 = -3(4) + b$$
$$-5 = -12 + b$$
$$7 = b$$

Therefore, the y-intercept is 7, so the equation is $y = -3x + 7$.

5. $y = 3x - 2$. To begin, use the slope formula to find the slope of the line:

$$m = \frac{y_1 - y_2}{x_1 - x_2} = \frac{7 - (-8)}{3 - (-2)} = \frac{7 + 8}{3 + 2} = \frac{15}{5} = 3$$

The slope is 3, so the equation is of the form $y = 3x + b$. Plug in the x- and y-values for either point into this equation and solve for b:

$$7 = 3(3) + b$$
$$7 = 9 + b$$
$$-2 = b$$

Thus, the y-intercept is -2, so the equation is $y = 3x - 2$.

6. $y = \frac{1}{3}x + 7$. The line $y = \frac{1}{3}x - 10$ has a slope of $\frac{1}{3}$, so any parallel line has the same slope. So, the equation of the line you are looking for is of the form $y = \frac{1}{3}x + b$. Plug the x- and y-values for the point $(-6, 5)$ into this equation and solve for b:

$$y = \frac{1}{3}x + b$$
$$5 = \frac{1}{3}(-6) + b$$
$$5 = -2 + b$$
$$7 = b$$

Thus, the y-intercept is 7, so the equation is $y = \frac{1}{3}x + 7$.

7. $y = -\frac{1}{4}x + \frac{7}{2}$. The line $y = 4x + 9$ has a slope of 4, so any perpendicular line has a slope that is the negative reciprocal of 4, which is $-\frac{1}{4}$. So the equation is of the form $y = -\frac{1}{4}x + b$. Plug in the x- and y-values for the point $(2, 3)$ and solve for b:

$$y = -\frac{1}{4}x + b$$
$$3 = -\frac{1}{4}(2) + b$$
$$3 = -\frac{1}{2} + b$$
$$\frac{7}{2} = b$$

Therefore, the equation has a y-intercept of $\frac{7}{2}$, so the equation is $y = -\frac{1}{4}x + \frac{7}{2}$.

8. **(1, 9).** Set the right sides of the two equations equal to each other:

$$5x + 4 = -3x + 12$$
$$8x + 4 = 12$$
$$8x = 8$$
$$x = 1$$

So $x = 1$. Now, plug in 1 for x (in whichever equation looks simpler) and solve for y:

$$y = 5x + 4$$
$$y = 5(1) + 4$$
$$y = 9$$

Therefore, $y = 9$, so the point where the two lines intersect is (1, 9).

9. $y \le 3x - 1$ **and** $y > -x + 7$. The graph shows a solid line with a y-intercept of -1 and a slope of 3, with the region below this line shaded, so this inequality is $y \le 3x - 1$. It also shows a dashed line with a y-intercept of 7 and a slope of -1, with the region above this line shaded, so this inequality is $y > -x + 7$.

10. $y = -\dfrac{3}{5}x + \dfrac{2}{5}$. Solve the equation for y in terms of x:

$$3x + 5y = 2$$
$$5y = -3x + 2$$
$$\frac{5y}{5} = \frac{-3x}{5} + \frac{2}{5}$$
$$y = -\frac{3}{5}x + \frac{2}{5}$$

11. **2.** To find the value of a, plug in 0 for y and solve for x:

$$3x - 4(0) = 24$$
$$3x = 24$$
$$x = 8$$

To find the value of b, plug in 0 for x and solve for y:

$$3(0) - 4y = 24$$
$$-4y = 24$$
$$y = -6$$

Therefore, $a + b = 8 + (-6) = 2$.

12. **10.** Plug these values into the distance formula:

$$d = \sqrt{(x_1 - x_2)^2 + (y_1 - y_2)^2}$$
$$= \sqrt{(-1 - (-7))^2 + (5 - (-3))^2}$$
$$= \sqrt{(-1 + 7)^2 + (5 + 3)^2}$$
$$= \sqrt{6^2 + 8^2}$$
$$= \sqrt{36 + 64}$$
$$= \sqrt{100}$$
$$= 10$$

13. $(-4, -3)$. Begin by rotating the point $(4, -3)$ by 180 degrees, which negates both coordinates of the point:

$$(4, -3) \rightarrow (-4, 3)$$

Next, reflecting this new point across the x-axis negates the y-coordinate:

$$(-4, 3) \rightarrow (-4, -3)$$

14. $(-5, 2)$. The translation from A to A' changes the x-coordinate by -7 and the y-coordinate by 3 — that is, a translation of $(x - 7, y + 3)$. Therefore, this translation will have the same effect from B to B':

$$(2, -1) \rightarrow (2 - 7, -1 + 3) = (-5, 2)$$

Geometry answers

1. 12 centimeters. Use the formula for the perimeter of a square:

$$P = 4s$$
$$48 = 4s$$
$$12 = s$$

2. 20 inches. Begin by using the area formula for a square to find the length of the side:

$$A = s^2$$
$$25 = s^2$$
$$\sqrt{25} = \sqrt{s^2}$$
$$5 = s$$

Now, plug this value into the formula for the perimeter of a square:

$$P = 4s = 4(5) = 20$$

3. 75 square units. The rectangle has a length of 15 and a width that's $\frac{1}{3}$ of this length, so its width is 5. Plug these two numbers into the formula for the area of a rectangle:

$$A = L \times W = 15 \times 5 = 75$$

4. 22 meters. Use the formula for the area of a rectangle to find the width:

$$A = L \times W$$
$$24 = 8 \times W$$
$$3 = W$$

Now, plug this value into the formula for the perimeter of a rectangle:

$$P = 2(L + W) = 2(8 + 3) = 2(11) = 22$$

5. 24 units. Use the formula for the circumference of a circle to find the radius:

$$C = 2\pi r$$
$$24\pi = 2\pi r$$
$$\frac{24\pi}{2\pi} = \frac{2\pi r}{2\pi}$$
$$12 = r$$

Now, plug this value into the formula for the diameter of a circle:

$$D = 2r = 2(12) = 24$$

6. 10π. Begin by using the area formula for a circle to find the radius:

$$A = \pi r^2$$
$$25\pi = \pi r^2$$
$$25 = r^2$$
$$5 = r$$

Now, plug this value into the formula for the circumference:

$$C = 2\pi r = 2\pi(5) = 10\pi$$

7. **10 feet.** Use the formula for the volume of a box:

$$V = l \times w \times h$$
$$1200 = 15 \times w \times 8$$
$$1200 = 120w$$
$$10 = w$$

8. **282.6 cubic inches.** Begin by using the formula for the area of a circle to find the area of the base:

$$A = \pi r^2 = \pi(3)^2 = 9\pi$$

Now, plug this value into the formula for the volume of a cylinder:

$$V = A_b h = 9\pi(10) = 90\pi$$

Use 3.14 as an approximation for π:

$$\approx 90(3.14) \approx 282.6$$

9. **10 kilometers.** Use the Pythagorean theorem:

$$a^2 + b^2 = c^2$$
$$8^2 + 6^2 = c^2$$
$$64 + 36 = c^2$$
$$100 = c^2$$
$$\sqrt{100} = \sqrt{c^2}$$
$$10 = c$$

10. **8 millimeters.** Use the Pythagorean theorem, making sure to plug in the length of the hypotenuse as c:

$$a^2 + b^2 = c^2$$
$$15^2 + b^2 = 17^2$$
$$225 + b^2 = 289$$
$$b^2 = 64$$
$$\sqrt{b^2} = \sqrt{64}$$
$$b = 8$$

Statistics, probability, and sets answers

1. See below.

    (A) 45, 70, 75, 75, 80, 80, 80, 80, 85, 90, 90, 90, 90, 95, 95, 100, 100.

    (B) Minimum is 45.

    (C) Maximum is 100.

    (D) Range is $100 - 45 = 55.$

    (E) Mean is approximately 83.53.

$$\frac{45 + 70 + 75 + 75 + 80 + 80 + 80 + 80 + 85 + 90 + 90 + 90 + 90 + 95 + 95 + 100 + 100}{17}$$

$$= \frac{1420}{17}$$

$$\approx 83.53$$

    (F) Median is 85. The median is the central value in the data set:

        45, 70, 75, 75, 80, 80, 80, 80, **85**, 90, 90, 90, 90, 95, 95, 100, 100

2. See below.

    (A) 45.

    (B) Skew left.

    (C) Median.

    (D) 77.5. The first quartile is the median of the lower half.

    (E) 92.5. The third quartile is the median of the upper half.

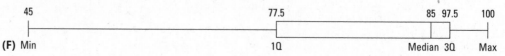

    (F)

3. $\frac{2}{5}$. The total number of possible outcomes is 10, and the target numbers are 7, 8, 9, and 10 — that is, there are 4 target outcomes. Plug these numbers into the formula for probability and simplify:

$$\text{Probability} = \frac{\text{Target outcomes}}{\text{Total outcomes}} = \frac{4}{10} = \frac{2}{5}$$

4. $\frac{3}{10}$. The total number of possible outcomes is 10, and the target numbers are 2, 4, and 6 — that is, there are 3 target outcomes. Plug these numbers into the formula for probability and simplify:

$$\text{Probability} = \frac{\text{Target outcomes}}{\text{Total outcomes}} = \frac{3}{10}$$

5. $\frac{3}{5}$. The total number of possible outcomes is 5 — that is, the five odd numbers 1, 3, 5, 7, and 9. The target numbers are 5, 7, and 9 — that is, there are 3 target outcomes. Plug these numbers into the formula for probability and simplify:

$$\text{Probability} = \frac{\text{Target outcomes}}{\text{Total outcomes}} = \frac{3}{5}$$

6. $\frac{47}{75}$. The number of total outcomes is 75: the total number of people surveyed. The number of target outcomes is 47: the number of people who drink coffee. Calculate the probability as follows:

$$\text{Probability} = \frac{\text{Target outcomes}}{\text{Total outcomes}} = \frac{47}{75}$$

7. $\frac{19}{75}$. Again, the number of total outcomes is 75. However, according to the table, only 19 of the people surveyed drink soft drinks but not coffee. Thus, the probability of this event is:

$$\text{Probability} = \frac{\text{Target outcomes}}{\text{Total outcomes}} = \frac{19}{75}$$

8. $\frac{1}{4}$. To begin solving this problem, remember that the condition always affects the total outcomes. Here, the condition is preceded by the word *if*: The words *if a person who doesn't drink coffee* limits the total outcomes to 28. Of these 28 people, 7 also don't drink soft drinks, so this is the number of target outcomes. Thus:

$$\text{Probability} = \frac{\text{Target outcomes}}{\text{Total outcomes}} = \frac{7}{28} = \frac{1}{4}$$

9. {1, 3, 5, 6, 7, 9}. Begin by finding the intersection of W and X — that is, the set of all elements that are in *both* sets:

$$W \cap X = \{4,5,6,7\} \cap \{2,3,6,8,9,10\} = \{6\}$$

Now, find the union of V and this set — that is, the set of all elements that are in *either* set:

$$V \cup \{6\} = \{1,3,5,7,9\} \cup \{6\} = \{1,3,5,6,7,9\}$$

10. {5, 6, 7}. What is the set $(V \cup X) \cap W$? Begin by finding the union of V and X — that is, the set of all elements that are in *either* set:

$$V \cup X = \{1,3,5,7,9\} \cup \{2,3,6,8,9,10\} = \{1,2,3,5,6,7,8,9,10\}$$

Now, find the intersection of this set and W — that is, the set of all elements that are in *both* sets:

$$\{1,2,3,5,6,7,8,9,10\} \cap W = \{1,2,3,5,6,7,8,9,10\} \cap \{4,5,6,7\} = \{5,6,7\}$$

6

The Advanced Algebra and Functions Test

IN THIS PART . . .

Focus on factoring and functions.

Understand a variety of functions, including polynomials.

Get up to speed with geometry and trigonometry.

Answer practice questions designed to hone the skills you've attained in this part of the book.

Chapter **17**

Factoring and Functions

The Advanced Algebra and Functions Placement Test includes the most sophisticated math found on the ACCUPLACER. In this chapter, you get acquainted with a few key algebra concepts, to set you up for success in Chapters 18 and 19.

To begin, you get a solid review of factoring polynomials, including GCF factoring, quadratic factoring, the difference of squares, and factorable cubic polynomials. Next, you tackle function notation $f(x)$ in a way that will, I hope, steer you around some of the confusion.

After that, I give you an overview of a few basic parent functions, which you see more of in Chapters 18 and 19. Finally, you combine your understanding of function notation and parent functions, working with some of the common transformations of functions that you're likely to see on the ACCUPLACER.

Factoring Polynomials

In Chapter 13, I introduce you to three key tools for working with algebra expressions: evaluating, simplifying, and factoring. I discuss evaluating and simplifying in detail, and also touch upon GCF factoring, then give you a rain check for some of the more advanced forms of factoring.

You can think of factoring as a reversal of some of the basic simplifying tools — such as distribution and FOILing. In this section, I clarify the remaining types of factoring that you will need to master for the ACCUPLACER.

Quadratic factoring

A quadratic polynomial is an expression of the form $ax^2 + bx + c$ — for example,

$$x^2 + 7x + 10 \qquad x^2 - 41x - 42 \qquad 4x^2 + 5x - 6$$

In many cases — and especially on the ACCUPLACER! — quadratic polynomials can be factored into two binomials. In this section, I show you how to do this in the two possible cases: when the leading coefficient is either 1 or not 1.

Factoring when the leading coefficient is 1

The simplest case is when the leading coefficient of a quadratic expression is 1 — that is, the expression has the form $x^2 + bx + c$. To factor an expression of this form, find a pair of numbers that do the following:

Add up to the b value

Multiply to the c value

For example, consider the quadratic expression $x^2 + 7x + 10$. Here, you're looking for a pair of numbers that add up to 7 and multiply to 10. These numbers are 2 and 5, so you can factor the expression as follows:

$$x^2 + 7x + 10 = (x+2)(x+5)$$

Table 17-1 gives eight examples of quadratic expressions that can be factored using this method.

TABLE 17-1 Factoring Quadratic Polynomials

Expression in standard form	b-value when added	c-value when multiplied	Expression in factored form
$x^2 + 11x + 24$	$3 + 8 = 11$	$3 \times 8 = 24$	$(x+3)(x+8)$
$x^2 + 31x + 30$	$1 + 30 = 31$	$1 \times 30 = 30$	$(x+1)(x+30)$
$x^2 - 5x + 6$	$-2 + (-3) = -5$	$-2 \times (-3) = 6$	$(x-2)(x-3)$
$x^2 - 9x + 20$	$-4 + (-5) = -9$	$-4 \times (-5) = 20$	$(x-4)(x-5)$
$x^2 - 28x + 27$	$-1 + (-27) = -28$	$-1 \times (-27) = 27$	$(x-1)(x-27)$
$x^2 - 6x - 16$	$2 + (-8) = -6$	$2 \times (-8) = -16$	$(x+2)(x-8)$
$x^2 + 8x - 33$	$11 + (-3) = 8$	$11 \times (-3) = -33$	$(x+11)(x-3)$
$x^2 - 41x - 42$	$-42 + 1 = -41$	$-42 \times 1 = -42$	$(x-42)(x+1)$

Factoring when the leading coefficient isn't 1

Factoring a quadratic expression when the leading coefficient isn't 1 can be confusing. There are several ways to do this. Here, I give what I believe to be the most reliable method, using what I call an *analogous case* that has 1 as the leading coefficient.

To see how this method works, consider the following quadratic polynomial:

$$4x^2 + 5x - 6$$

To begin, I'm going to do something that, strictly speaking, isn't allowed in math: Pick up the troublesome coefficient 4 and move it next to the constant –6, and then simplify:

$$x^2 + 5x - 6(4)$$
Analogous case
$$= x^2 + 5x - 24$$

This *analogous case* is now a quadratic that you know how to factor, as I show you in "Factoring when the leading coefficient is 1":

$$= (x+8)(x-3)$$

Careful, because we're not done yet. Now, I'm going to take three steps that are easy but may seem weird: First, remember that our original expression began with the term $4x^2$. So, in honor of that memory, I'm now going to replace both the x's inside the parentheses with $4x$:

$(4x+8)(4x-3)$ Weird step #1: Replace x with $4x$.

Second, I notice that I can factor out a 4 from the first set of parentheses:

$= 4(x+2)(4x-3)$ Weird step #2: Factor out the GCF.

Third, I just drop the 4 that I just factored out:

$(x+2)(4x-3)$ Weird step #3: Drop the factor.

This answer is now correct. To check this out, FOIL it:

$$(x+2)(4x-3)$$
$$= 4x^2 + 8x - 3x - 6$$
$$= 4x^2 + 5x - 6$$

Indeed, this is the quadratic expression that I started with. Weird but true, this method always works to factor a quadratic expression.

WARNING

One warning here is that if the expression you're working with has a GCF other than 1, you should factor this GCF out *first*. For example,

$15x^2 + 24x - 12$

Notice here that this quadratic expression has a GCF of 3 that can be factored out:

$$= 3(5x^2 + 8x - 4)$$

This factor of 3 needs to remain part of the problem, and cannot be dropped. I'll quickly work through the next few steps in this problem. First, I change the expression to its analogous case:

$3(x^2 + 8x - (5)4)$
$= 3(x^2 + 8x - 20)$ Analogous case

Now you're ready to do basic quadratic factoring:

$$= 3(x+10)(x-2)$$

Now, replace x with $5x$, then factor the results:

$3(5x+10)(5x-2)$
$= 3[5(x+2)(5x-2)]$

For the final step, drop the factor of 5 but *not* that important factor of 3 from the first step of the problem:

$3(x+2)(5x-2)$

Again, this apparently unsound method yields the correct answer. Crazy, right?

Factoring the difference of squares

When a 2-term expression is of the form $a^2 - b^2$, it's called a *difference of squares*. Note here that a square (a^2) and another square (b^2) are being subtracted. When this is the case, you can factor it using the following pattern:

$$a^2 - b^2 = (a+b)(a-b)$$

To see how this works, consider the expression $x^2 - 25$. Here, notice that the a-value is x, and the b-value is 5, so you can factor this expression as follows:

$$x^2 - 25 = (x+5)(x-5)$$

As another example, factor $4x^2 - 49y^2$. This time, the a-value is $2x$, and the b-value is $7y$, so the expression factors as follows:

$$4x^2 - 49y^2 = (2x+7y)(2x-7y)$$

As one more example, consider the expression $64x^6 - 81y^4$. Here, the a-value is $8x^3$ and the b-value is $9y^2$, so you can factor as follows:

$$64x^6 - 81y^4 = (8x^3 + 9y^2)(8x^3 - 9y^2)$$

All the examples so far have involved coefficients that are square numbers and variables with even exponents, but these requirements aren't strictly necessary. For example,

$$7x - 3 = (\sqrt{7x} + \sqrt{3})(\sqrt{7x} - \sqrt{3})$$

Note that this is not technically factoring, because the resulting coefficients aren't whole numbers. In any case, keep an eye on this example, as problems like this are getting more popular among test makers!

Factoring the sum and difference of cubes

Expressions of the form $a^3 + b^3$ and $a^3 - b^3$ are called, respectively, a *sum of cubes* and a *difference of cubes*. As with a difference of squares, you can factor expressions like these using the following patterns:

Sum of cubes: $a^3 + b^3 = (a+b)(a^2 - ab + b^2)$

Difference of cubes: $a^3 - b^3 = (a-b)(a^2 + ab + b^2)$

These patterns can be a little tricky to follow, so I walk you through them in the next two examples. To begin, suppose you want to factor $8x^3 + 27y^3$. The presence of the plus sign (+) tells you that this is a *sum* of cubes. Begin by setting up the pattern with the plus and minus signs in the correct places:

$$8x^3 + 27y^3 = (\quad + \quad)(\quad - \quad + \quad)$$

Now, take the cube root of both terms and place these values inside the first pair of parentheses:

$$= (2x + 3y)(\quad - \quad + \quad)$$

Next, multiply these two values together and place them in the middle slot:

$$= (2x + 3y)(\quad -6xy + \quad)$$

To finish up, square each of the values inside the first pair of parentheses and place them in order in the remaining two slots:

$$= (2x + 3y)(4x^2 - 6xy + 9y^2)$$

As another example, to factor $125x^3 - 64y^3$, notice first that this is a difference of cubes, so set up the pattern in this way:

$$125x^3 - 64y^3 = (\quad - \quad)(\quad + \quad + \quad)$$

Now, just repeat the steps from the first example. Begin by taking the cube root of both terms:

$$= (5x - 4y)(\quad + \quad + \quad)$$

And again, multiply these two values and place the result in the middle slot (dropping the minus sign when you multiply):

$$= (5x - 4y)(\quad + 20xy + \quad)$$

To complete the problem, square each of the values inside the first pair of parentheses:

$$= (5x - 4y)(25x^2 + 20xy + 16y^2)$$

Cubic factoring

A cubic polynomial has the following form:

$$ax^3 + bx^2 + cx + d$$

While not all cubic polynomials are factorable, the ones you encounter on the ACCUPLACER are likely to be. These expressions can be factored using the following method:

1. **Separate the equation into two parts.**

2. **Apply GCF factoring to both parts (see Chapter 13).**

3. **Split the resulting expression into a pair of factors using reverse distribution.**

4. **If possible, factor one of these resulting factors.**

For example, suppose you want to factor the following expression:

$$4x^3 - 8x^2 - 9x + 18$$

Notice that this is a cubic polynomial that cannot be factored initially using any other method I've discussed so far. Your first step here is simply to separate the first two terms from the last two terms:

$$= 4x^3 - 8x^2 \quad -9x + 18$$

Now, apply GCF factoring to both pairs of terms:

$$= 4x^2(x-2) - 9(x-2)$$

As you can see, here I factored $4x^2$ out of the first two terms, and -9 from the last two, leaving behind a factor of $x-2$ in both cases. The fact that this factor is the same is no accident, and allows for the next step:

$$= (4x^2 - 9)(x-2)$$

Notice here that I took the two terms that were previously *outside* the parentheses and combined them, keeping the factor that was already *inside* the parentheses intact.

At this point, the cubic factoring portion of the problem is complete. However, notice now that you can still do more factoring: The factor $4x^2 - 9$ can be factored further as the difference of squares:

$$= (2x+3)(2x-3)(x-2)$$

This completes the problem, breaking the cubic four-term polynomial into three linear binomials.

When to factor polynomials

Sometimes, you have to take one step back in order to take two steps forward. In a similar way, in some math problems, you need to factor — that is, make an expression *more complicated* — before you can simplify. For example, suppose you want to simplify the following expression:

$$\frac{6x^3 + 2x^2}{3x + 1}$$

As it stands, you can't simplify this expression directly. However, you can do some GCF factoring in the numerator:

$$= \frac{2x^2(3x+1)}{3x+1}$$

The result is a rational expression that looks more complex than the one you started with. But now, with a factor of $3x+1$ in both the numerator and denominator, you can cancel out this factor:

$$2x^2$$

This final expression is equivalent to the first expression, but in a much simpler form. As another example, consider this expression:

$$\frac{x^2 - 4}{x^2 + 4x - 12}$$

WARNING

Again, this rational expression cannot be simplified directly. But before I move forward with this example, allow me to address a common question among my students: "Why can't I cancel an x^2 in the numerator and denominator?" (If you already understand why, feel free to jump over this explanation.)

As the expression stands now, the x^2 in both the numerator and denominator are *terms* — that is, parts of an expression that are joined by addition or subtraction. *Canceling*, however, is a short way of saying *canceling out common factors* — that is, parts of an expression that are joined by multiplication. For example, whenever you simplify a fraction (refer to Chapter 13), you cancel out a common factor in the numerator and denominator.

But you can't cancel out factors if you don't *have* any factors. That's why factoring is a necessary first step. In this case, you can do difference of squares factoring in the numerator and quadratic factoring in the denominator:

$$= \frac{(x+2)(x-2)}{(x-2)(x+6)}$$

Now, you can cancel out a common factor of $x-2$:

$$= \frac{x+2}{x+6}$$

One more example:

$$\frac{x^3 + y^3}{4x^2 y - 4y^3}$$

Here, you can do sum of cubes factoring in the numerator, and GCF factoring in the denominator:

$$= \frac{(x+y)(x^2 - xy + y^2)}{4y(x^2 - y^2)}$$

You still don't have a common factor to cancel out, but you can do difference of squares factoring in the denominator:

$$= \frac{(x+y)(x^2 - xy + y^2)}{4y(x+y)(x-y)}$$

At last! Now, you can cancel a common factor of $x + y$:

$$= \frac{x^2 - xy + y^2}{4y(x-y)}$$

Function Notation

Many students find function notation — that troublesome $f(x)$ — to be confusing and dread problems that include it. However, as I discuss in this section, many problems that include function notation just ask you to plug in a number and then evaluate the result using basic algebra. And even tough-looking problems are often easy when you know how to approach them.

In this section, I give you a good start toward solving a variety of different types of problems.

Evaluating function notation

Working with function notation is often no more complicated than plugging in numbers and simplifying. In this section, I use the following two functions for all the examples:

$$f(x) = x^2 - 7 \qquad g(x) = 3x + 1$$

Evaluating basic functions

To begin thinking about functions, consider this problem:

What is the value of $x^2 - 7$ when $x = 3$?

To solve this problem, plug in 3 for x as follows:

$$3^2 - 7 = 9 - 7 = 2$$

Now, here's the same problem using function notation:

If $f(x) = x^2 - 7$, what is the value of $f(3)$?

To find $f(3)$, plug in 3 for x:

$$f(3) = 3^2 - 7 = 9 - 7 = 2$$

Therefore, $f(3) = 2$. As you can see, function notation is simply a compact way to express this relationship between an input value (x) and an output value $f(x)$ — which is commonly equated with y when graphing on the xy-plane.

As another example, consider the value $g(-2)$ when $g(x) = 3x + 1$. Again, you can find this value simply:

$$g(-2) = 3(-2) + 1 = -6 + 1 = -5$$

Thus, $g(-2) = -5$. Pretty simple, right?

Working with minor complications

When you know how to evaluate a function, a few twists in the road shouldn't be a problem. For example, consider $f(5) + 3$:

$$f(5) + 3 = 5^2 - 7 + 3 = 25 - 7 + 3 = 21$$

Thus, $f(5) + 3 = 21$. Or, consider $2g(6)$, which means multiplying $g(6)$ by 2:

$$2g(6) = 2[3(6) + 1]$$

Now, simply evaluate the result using your trusty PEMDAS rules (refer to Chapter 13):

$$2[18 + 1] = 2(19) = 38$$

So $2g(6) = 38$.

Playing with more than one function

Provided that you follow the rules, you can make sense of a variety of more complicated-looking function problems that include more than one function.

For example, consider the function $\dfrac{-3f(2)}{g(4)-9}$. This looks formidable, but there's nothing here you can't solve. Begin by plugging in the proper values:

$$\frac{-3f(2)}{g(4)-9} = \frac{-3(2^2-7)}{3(4)+1-9}$$

Now, evaluate using the order of operations (PEMDAS — again, refer to Chapter 13 if necessary):

$$= \frac{-3(4-7)}{3(4)+1-9} = \frac{-3(-3)}{12+1-9} = \frac{9}{4}$$

Thus, $\dfrac{-3f(2)}{g(4)-9} = \dfrac{9}{4}$.

Evaluating function notation with variable inputs

Inputting a variable is essentially the same as inputting a numerical value, except that the result typically includes the inputted variable. In this section, again, I use these two functions:

$$f(x) = x^2 - 7 \qquad g(x) = 3x + 1$$

Inputting a single variable

Inputting a single variable is so straightforward that you may not believe it's so easy. Here's how you evaluate $f(a)$:

$$f(a) = a^2 - 7$$

As you can see, you simply plug in a for x into the function. As another example, here's how you evaluate $g(n)$:

$$g(n) = 3n + 1$$

Inputting a single variable with a variation or two

When you know how to evaluate a variable input, the variations you know should look rather simple. For example, here's how you evaluate $f(a) - 5$:

$$f(a) - 5 = a^2 - 7 - 5$$

Again, plug in a for x into the function, but this time you can simplify the resulting expression:

$$= a^2 - 12$$

So $f(a) = a^2 - 12$. As another example, here's how you evaluate $-4g(b) - 3$:

$$-4g(b) - 3 = -4(3b + 1) - 3$$

In this step, I'm just substituting the expression $3b + 1$ for $g(b)$, and then filling in all the other information around this core. I use an extra set of parentheses to clarify that the -4 is only to be multiplied by $3b + 1$. If you perform this substitution step carefully, you'll be set up for success as you simplify:

$$= -12b - 4 - 3 = -12b - 7$$

Therefore, $-4g(b) - 3 = -12b - 7$.

Inputting a variable expression

You can also input a variable expression, substituting it for the input variable and then simplifying. For example, here, to find $g(b - 2)$, substitute $b - 2$ for x into $3x + 1$:

$$g(b - 2) = 3(b - 2) + 1$$

Now, simplify:

$$= 3b - 6 + 1 = 3b - 5$$

Therefore, $g(b - 2) = 3b - 5$. As another example, here's how you find $f(a + 3)$:

$$f(a + 3) = (a + 3)^2 - 7$$

At this point, be sure to simplify the squared part of this expression by expanding:

$$= (a + 3)(a + 3) - 7$$

When you see the expression expanded like this, remember to simplify by FOILing, and then combining like terms.

$$= a^2 + 6a + 9 - 7 = a^2 + 6a + 2$$

Thus, $f(a + 3) = a^2 + 6a + 2$.

Combined functions

So far in this section, all the variations you have worked with have used a single function. You can also combine functions using a variety of operations. Once again, throughout this section, I use the following function definitions:

$$f(x) = x^2 - 7 \qquad g(x) = 3x + 1$$

Understanding combined functions

Evaluate combined functions individually and then simplify. For example, consider $f(3) + g(-1)$:

$$f(3) + g(-1) = (3^2 - 7) + [3(-1) + 1]$$

In this first step, I simply evaluate $f(3)$ and $g(-1)$ individually and then add them together. Now I'm ready to simplify:

$$= 9 - 7 - 3 + 1 = 0$$

As another example, here's how you evaluate $\frac{-g(3)}{2f(5)}$:

$$\frac{-g(3)}{2f(5)} = \frac{-[3(3)+1]}{2(5^2-7)}$$

As usual, simplify to get the answer:

$$= \frac{-[9+1]}{2(25-7)} = \frac{-10}{2(18)} = -\frac{5}{18}$$

Working with notation for combined functions

To make your life even more fun, mathematicians have provided special notation for combining functions:

$$(f+g)(x) = f(x) + g(x) \qquad (f-g)(x) = f(x) - g(x)$$

$$(fg)(x) = f(x)g(x) \qquad \left(\frac{f}{g}\right)(x) = \frac{f(x)}{g(x)}$$

Notice that when you use this notation, both functions $f(x)$ and $g(x)$ take the same input x. For example, consider $(f-g)(6)$:

$$(f-g)(6) = f(6) - g(6)$$

As you can see, the input 6 is in play for both functions. Evaluate $f(6)$ and $g(6)$ just as you normally would:

$$= (6^2 - 7) - [3(6) + 1]$$

Now, you're ready to simplify:

$$= (36-7) - [18+1] = 29 - 19 = 10$$

Here's another example, taking the variable expression $2x$ as an input:

$$\left(\frac{f}{g}\right)(2x) = \frac{f(2x)}{g(2x)}$$

Again, when you've changed the obscure notation to a more workable form, do the substitution:

$$= \frac{(2x)^2 - 7}{3(2x) + 1}$$

Now, simplify:

$$= \frac{4x^2 - 7}{6x + 1}$$

Composite functions

The final twist of the knife when working with function notation is the *composite function*, where you take a function of a function. Although this is where many students throw up their hands in despair, I promise you that this final concept is not as bad as it looks. Continuing our little tradition, I use the following functions:

$$f(x) = x^2 - 7 \qquad g(x) = 3x + 1$$

Understanding composite functions

Here's an example of what a composite function looks like:

$$f(g(3))$$

If you look at this and feel like you've just fallen down the rabbit hole, that's normal. And yet, I assure you that it isn't so terrible. To begin unraveling this mess, remember that the order of operations (PEMDAS) usually requires you to start working inside the parentheses — in this case, with $g(3)$. So begin with a little side work to uncover this value:

$$g(3) = 3(3) + 1 = 10$$

This tells you that $g(3) = 10$, so you can now substitute 10 for $g(3)$ into the original expression:

$$f(g(3)) = f(10)$$

If you are following this process step by step and feel like this is making sense, you're now home free. The last step is simply to find the value of $f(10)$, which you should find pretty simple:

$$= 10^2 - 7 = 100 - 7 = 93$$

Therefore, $f(g(3)) = 93$.

Working with notation for composite functions

Naturally, the great minds of math were not content to just leave composite functions well enough alone, but instead concocted a special notation to assure your ongoing confusion:

$$(f \circ g)(x) = f(g(x))$$

This requires an additional step of unraveling, which I believe you can manage. For example, what is the value of $(f \circ g)(-2)$? As a first step, simply translate the notation back into a more normal-looking composite function:

$$(f \circ g)(-2) = (f(g(-2))$$

Next, focus on the nubby center of this function, $g(-2)$, and evaluate it separately:

$$g(-2) = 3(-2) + 1 = -6 + 1 = -5$$

Now that you know that $g(-2) = -5$, substitute –5 for $g(-2)$ into the original problem:

$$f(g(x)) = f(-5)$$

This should look like a problem you know how to solve:

$$= (-5)^2 - 7 = 25 - 7 = 18$$

Therefore, $(f \circ g)(-2) = 18$

Parent Functions

Parent functions are the simplest forms of a variety of functions. Each parent function is the starting point for an infinite variety of related functions of a given type, sometimes called the set of *child functions* for that parent function.

Table 17-2 gives a list of 11 parent functions that are good to know for the ACCUPLACER. Notice that each of these functions includes a *signature point* that the function contains. This point is a useful place to begin plotting the function on the graph.

TABLE 17-2 **Parent Functions**

Name	Function	Signature point
Linear (1st-degree polynomial)	$f(x) = x$	(0,0)
Quadratic (2nd-degree polynomial)	$f(x) = x^2$	(0,0)
Cubic (3rd-degree polynomial)	$f(x) = x^3$	(0,0)
Quartic (4th-degree polynomial)	$f(x) = x^4$	(0,0)
Exponential	$f(x) = 2^x$	(0,1)
Logarithmic	$f(x) = \log_2 x$	(1,0)
Sine	$f(x) = \sin x$	(0,0)
Cosine	$f(x) = \cos x$	(0,1)
Absolute value	$f(x) = |x|$	(0,0)
Rational	$f(x) = \dfrac{1}{x}$	(1,1)
Radical	$f(x) = \sqrt{x}$	(0,0)

In this section, you get an overview of the most important parent functions on the ACCUPLACER.

Parent functions of polynomials

Polynomials are a large class of functions that are generally of the following form:

$$f(x) = a_n x^n + a_{n-1} x^{n-1} + \ldots + a_2 x^2 + a_1 x + a_0$$

This begins to look a lot simpler when you focus on the parent functions for the four most common polynomial functions: the linear, quadratic, cubic, and quartic functions.

$$f(x) = x \qquad f(x) = x^2 \qquad f(x) = x^3 \qquad f(x) = x^4$$

In this section, I provide an introduction to these four important parent functions (shown in Figure 17-1).

Linear parent function (1st degree polynomial)

The linear parent function is

$$f(x) = x$$

When graphed on the xy-plane as $y = x$, this parent function is a diagonal line, as shown in Figure 17-1a. The set of child functions for this parent function is the set of all functions of the form $y = mx + b$ for all constants m and b see Chapter 14 for more on linear equations.

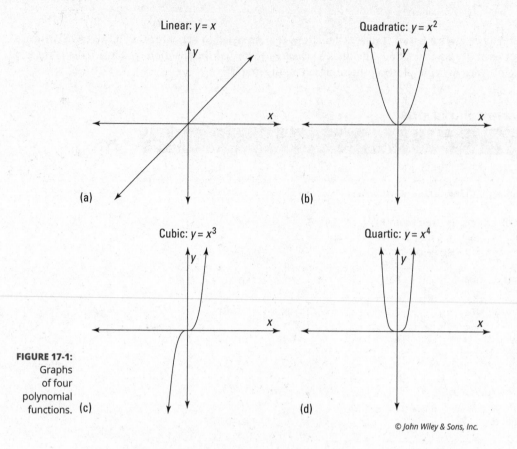

Linear: $y = x$

Quadratic: $y = x^2$

(a)

(b)

Cubic: $y = x^3$

Quartic: $y = x^4$

FIGURE 17-1:
Graphs
of four
polynomial
functions.

(c)

(d)

© John Wiley & Sons, Inc.

Quadratic parent function (2nd degree polynomial)

The quadratic parent function is

$$f(x) = x^2$$

When graphed on the xy-plane as $y = x^2$, this parent function is a parabola line, as shown in Figure 17-1b. The set of child functions for this parent function is the set of all functions of the form $f(x) = ax^2 + bx + c$ for all constants a, b, and c. (You explore quadratic functions in greater detail in Chapter 18.)

Cubic parent function (3rd degree polynomial)

The cubic parent function is

$$f(x) = x^3$$

When graphed on the xy-plane as $y = x^3$, this parent function shows up as variations of the curve shown in Figure 17-1c. The set of child functions for this parent function is the set of all functions of the form $f(x) = ax^3 + bx^2 + cx + d$ for all constants a, b, c, and d. On the ACCUPLACER, cubic functions tend to be factorable, as I discuss earlier in this chapter, in the section, "Factoring Polynomials."

Quartic parent function (4th degree polynomial)

The quartic parent function is

$$f(x) = x^4$$

When graphed on the xy-plane as $y = x^4$, this parent function shows up as variations of the curve shown in Figure 17-1d. The set of child functions for this parent function is the set of all functions of the form $f(x) = ax^4 + bx^3 + cx^2 + dx + e$ for all constants a, b, c, d, and e. You probably won't see many quartic functions on the ACCUPLACER, but it's good to know the general shape of this function.

Exponential and logarithmic parent functions

The parent functions for the exponential and logarithmic curves are inverses of each other. You can see this by looking at Figure 17-2. Notice that these two curves are reflections of each other across the line $y = x$.

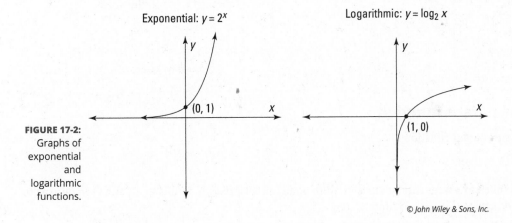

Exponential: $y = 2^x$

Logarithmic: $y = \log_2 x$

$(0, 1)$

$(1, 0)$

© John Wiley & Sons, Inc.

FIGURE 17-2: Graphs of exponential and logarithmic functions.

Exponential parent function

Technically speaking, the exponential parent function is $f(x) = e^x$ (where e is an irrational number approximately equivalent to 2.718). The good news is that the ACCUPLACER doesn't test your knowledge of e, so for the exponential parent function, I switch the base to 2:

$f(x) = 2^x$

The exponential parent function explodes to large numbers very quickly in the positive direction. In the negative direction, it approaches the x-axis as it approaches negative infinity. It's also important to note that this function crosses the y-axis at $(0, 1)$. When plotting this graph, be sure to include this point.

You discover more about exponential functions in Chapter 18.

Logarithmic parent function

As with the exponential parent function, the logarithmic parent function is also based on the value e: $f(x) = \ln x$, where "ln" stands for the natural log, which is a log with a base of e ($\log_e$). On the ACCUPLACER, you can use a base of 2 as a stand-in for e:

$f(x) = \log_2 x$

The domain of this function is $(0, \infty)$ — that is, only positive input values are acceptable, and both 0 and negative inputs are excluded. The range, however, is $(-\infty, \infty)$, so all output values are possible.

You discover more about log functions in Chapter 18.

Trigonometric parent functions

The six trigonometric functions arise from basic facts about right triangles. The two most important of these are the sine and cosine parent functions. Both of these functions, shown in Figure 17-3, resemble wave patterns, and for this reason they are used for scientific modeling of wave phenomena, from sound waves to quantum waves.

In this section, I give you an overview of these two parent functions.

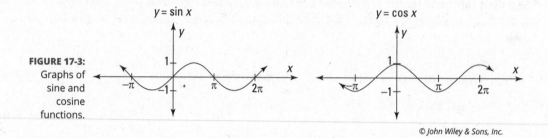

FIGURE 17-3: Graphs of sine and cosine functions.

© John Wiley & Sons, Inc.

Sine parent function

The sine parent function is

$$f(x) = \sin x$$

This function passes through its signature point of $(0,0)$, rises to 1, falls to −1, and then returns to 0 and repeats this cycle infinitely.

You discover more about the sine function in Chapter 19.

Cosine parent function

The cosine parent function is

$$f(x) = \cos x$$

This function passes through its signature point of $(0,1)$, falls to 0 and then to −1, and then returns to $(0,1)$ and repeats this cycle infinitely.

You discover more about the cosine function in Chapter 19.

Other parent functions

Three more parent functions may also come in handy on the ACCUPLACER. These are the parent functions for the absolute value function, the rational function, and the radical (square root) function (see Figure 17-4).

Absolute value parent function

The absolute value parent function is

$$f(x) = |x|$$

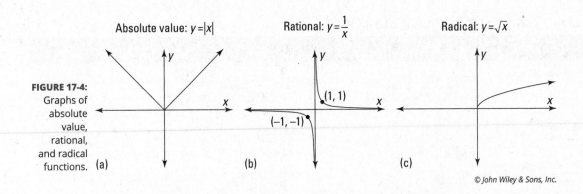

FIGURE 17-4: Graphs of absolute value, rational, and radical functions.

Absolute value: $y = |x|$

(a)

Rational: $y = \dfrac{1}{x}$

(1, 1)

(−1, −1)

(b)

Radical: $y = \sqrt{x}$

(c)

© John Wiley & Sons, Inc.

The domain of the absolute value function is $(-\infty, \infty)$, so all real numbers are permitted as x-values. The range, however, is $[0, \infty)$, so the function produces only non-negative y-values.

Rational parent function

The rational parent function is

$$f(x) = \frac{1}{x}$$

This curve, called a *hyperbola*, is shown in Figure 17-4b. Note that the domain of this function excludes the value 0 — you cannot input an x-value of 0. Similarly, the range of this function also excludes 0 — no x-value produces a y-value of 0. You can discover more about rational functions in Chapter 18.

Radical parent function

The radical parent function is

$$f(x) = \sqrt{x}$$

This function produces the curve shown in Figure 17-4c. Note that the domain of this function is $[0, \infty)$, so you cannot input a negative number into the function. Similarly, the range of this function is $[0, \infty)$, so output values are also limited to the non-negative numbers.

You can discover more about radical functions in Chapter 18.

Transformations of Functions

In a sense, function transformations are the big-kid version of the transformations that I discuss in Chapter 14. A transformation is a change made to a basic function $f(x)$. In this section, you work with the three most important types of function transformations: translations, dilations, and reflections.

Table 17-3 provides an overview of these transformation types.

TABLE 17-3   Transformations of Functions

Transformation	Description	Form	Example	Moves resulting function
Vertical translations	Up	$f(x)+n$	$f(x)+2$	Up 2
	Down		$f(x)-3$	Down 3
Horizontal translations	Left	$f(x-n)$	$f(x+1)$	Left 1
	Right		$f(x-2)$	Right 2
Vertical dilations	Stretch	$nf(x)$	$4f(x)$	Stretch by a factor of 4
	Compress	$\frac{1}{n}f(x)$	$\frac{1}{3}f(x)$	Compress by a factor of 3
Reflections	Across x-axis	$-f(x)$	$-f(x)$	Across x-axis
	Across y-axis	$f(-x)$	$f(-x)$	Across y-axis

Understanding transformations

Transformations are a set of variations to functions that show up predictably on the xy-graph. In this section, I discuss three types of transformations that you may see on the ACCUPLACER: translations, dilations, and reflections.

Translations

A *translation* moves a function from one place on the graph to another without changing its basic shape. A *vertical translation* moves a function either up or down on the graph, and a *horizontal translation* moves a function either left or right. In this section, you work with both of these types of translations.

VERTICAL TRANSLATIONS

The transformation $f(x)+n$ translates a function vertically in the xy-plane. When n is positive, the function moves up; when n is negative, the function moves down.

For example, $f(x)+2$ moves the function *up 2*, and $f(x)-3$ moves it *down 3*. Figure 17-5 shows these two examples of vertical translation using the parent function $f(x) = x^2$.

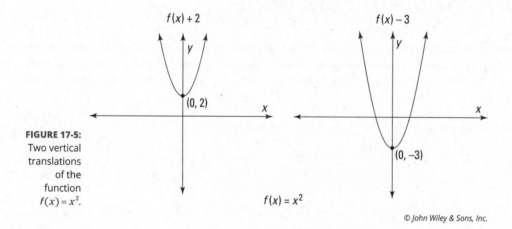

FIGURE 17-5: Two vertical translations of the function $f(x) = x^2$.

© John Wiley & Sons, Inc.

HORIZONTAL TRANSLATIONS

The transformation $f(x-n)$ translates a function horizontally in the xy-plane. When n is positive, the function moves to the left — that is, in the negative direction on the graph. When n is negative, the function moves to the right — that is, in the positive direction.

For example, $f(x+1)$ moves a function *left 1*, and $f(x-2)$ moves it *right 2*. Figure 17-6 shows these examples of horizontal translation using the parent function $f(x) = x^3$.

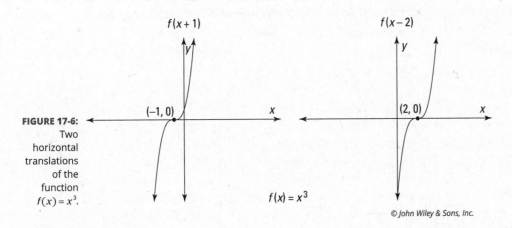

FIGURE 17-6: Two horizontal translations of the function $f(x) = x^3$.

© John Wiley & Sons, Inc.

Dilations

A *dilation* changes the shape of a function without substantially moving it on the graph, either stretching or compressing the function on the graph. The transformation $nf(x)$ stretches a function, and $\frac{1}{n}f(x)$ compresses a function.

For example, $4f(x)$ stretches a function by a factor of 4, and $\frac{1}{3}f(x)$ compresses it by a factor of 3. Figure 17-7 shows examples of dilations using the parent function $f(x) = |x|$.

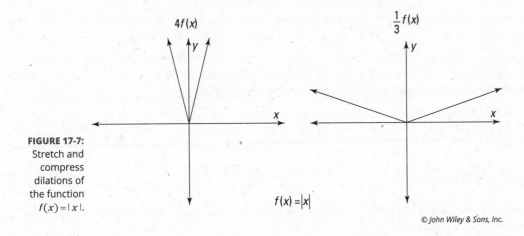

FIGURE 17-7: Stretch and compress dilations of the function $f(x) = |x|$.

© John Wiley & Sons, Inc.

Reflections

A *reflection* flips a function across either the x-axis or the y-axis.

The *vertical reflection* is $-f(x)$, which reflects a function vertically across the x-axis. For example, when $f(x) = x^2$, $-f(x) = -x^2$, as shown in Figure 17-8.

The *horizontal reflection* is $f(-x)$, which reflects a function horizontally across the y-axis. For example, when $f(x) = 2^x$, $f(-x) = 2^{-x}$. This particular reflection is very common with exponential functions, for changing exponential growth to exponential decay (see Chapter 18).

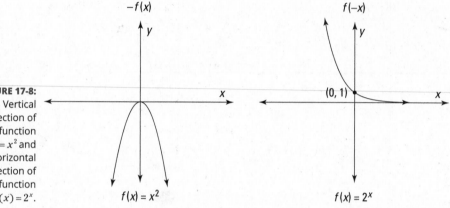

FIGURE 17-8: Vertical reflection of the function $f(x) = x^2$ and horizontal reflection of the function $f(x) = 2^x$.

© John Wiley & Sons, Inc.

Working with transformations

In practice, an ACCUPLACER question may ask you to apply one or more transformations to a particular function. For example:

EXAMPLE

Which transformation of the function $f(x) = x^2$ moves its vertex to the point (2, 5)?

The vertex of the function $f(x) = x^2$ is at (0, 0), so a transformation that moves it to (2, 5) moves it *up 5* and *right 2*:

$$f(x-2) + 5 = (x-2)^2 + 5$$

You can simplify this result by FOILing and combining like terms:

$$= (x-2)(x-2) + 5$$
$$= x^2 - 4x + 4 + 5$$
$$= x^2 - 4x + 9$$

Here's another example:

EXAMPLE

Which transformation of $f(x) = x^3$ reflects the function across the y-axis and then compresses it vertically by a factor of 2?

Using the transformations for compression and reflection results in the following:

$$\frac{1}{2}(-x)^3 = -\frac{1}{2}x^3$$

Here's a final example:

What is the result when you transform the function $f(x) = x^2 - 5x$ down 1 and left 3, then stretch it vertically by a factor of 2?

Apply the following transformations, using the rules described in the section, "Function Notation":

$$2[f(x+3)-1] = 2[(x+3)^2 - 5(x+3) - 1]$$

Now, simplify:

$$= 2(x^2 + 6x + 9 - 5x - 15 - 1)$$
$$= 2x^2 + 12x + 18 - 10x - 30 - 2$$
$$= 2x^2 + 2x - 14$$

Chapter **18**

Polynomials and Other Functions

In this chapter, you dive head first into some of the most important math on the ACCUPLACER Advanced Algebra and Functions Test. Each section of this chapter gives you an in-depth look at one of the many functions introduced in Chapter 17.

You start by looking at some general information about polynomial functions. Then you apply this knowledge to the quadratic (2nd-degree polynomial) function. After that, you work with exponential and logarithmic equations, using the exponential growth and decay functions to solve word problems.

From there, you discover how to calculate with radicals (square roots of non-square numbers). I then show you how to find the domain and range of both radical and rational functions.

Polynomial Functions

REMEMBER

In Chapter 17, I introduce *polynomials* as a class of functions of the following form:

$$f(x) = a_n x^n + a_{n-1} x^{n-1} + \ldots + a_2 x^2 + a_1 x + a_0$$

A good way to clarify this bewildering mess of letters and numbers is to focus on the parent functions and general forms for the first four polynomials — the linear, quadratic, cubic, and quartic functions — as shown in Table 18-1.

TABLE 18-1 **The First Four Polynomial Functions**

Degree	Name	Parent function	General function
1	Linear	$f(x) = x$	$f(x) = ax + b$
2	Quadratic	$f(x) = x^2$	$f(x) = ax^2 + bx + c$
3	Cubic	$f(x) = x^3$	$f(x) = ax^3 + bx^3 + cx + d$
4	Quartic	$f(x) = x^4$	$f(x) = ax^4 + bx^3 + cx^2 + dx + e$

Like all functions, a polynomial includes an input variable (usually x). Every term in a polynomial may include x raised to the power of a non-negative number, and then multiplied by a real number. Here are three examples of polynomials:

$$\frac{3}{4}x - 7 \qquad 3x^2 + 5x - \frac{1}{2} \qquad -x^{68} + 3x^{13} + 2.5x^7 - 11x^5 - 104x$$

As you can see, the terms of a polynomial are normally written in decreasing order based on the exponent of that term. If the polynomial includes a constant (a number without a variable), this term appears last.

When placed in this order, the *leading term* of a polynomial is its first term — that is, the term with the greatest exponent. And the *degree* of a polynomial is simply the exponent of the leading-term polynomial. So, the three polynomials shown here are, in order, a first-degree, a second-degree, and a 68th-degree polynomial.

Typically, polynomials are classified by their degree. Linear functions are the simplest, followed by quadratics, cubics, and so forth. Refer to Table 18-1 for a look at the four lowest-degree polynomials.

Distinguishing odd and even polynomials

A polynomial is odd or even depending upon its degree. Odd and even polynomials display different *end behavior* — that is, the direction that the function ultimately goes toward as x becomes infinitely large or infinitely small:

An *odd polynomial* has a degree that is an odd number. For example, linear and cubic functions are odd polynomials. The end behavior of an odd polynomial always goes in opposite directions.

An *even polynomial* has a degree that is an even number. For example, quadratic and quartic functions are even polynomials. The end behavior of an even polynomial always goes in the same direction.

Understanding positive and negative polynomials

The end behavior of a polynomial is also determined by whether it's *positive* or *negative* — that is, the sign of its leading term.

A *positive polynomial* has a positive leading term, and its positive end behavior (its direction as x gets very large) is also positive. Figure 18-1 shows four examples of positive polynomials.

A *negative polynomial* has a negative leading term, and its positive end behavior is negative. Figure 18-2 shows four examples of negative polynomials.

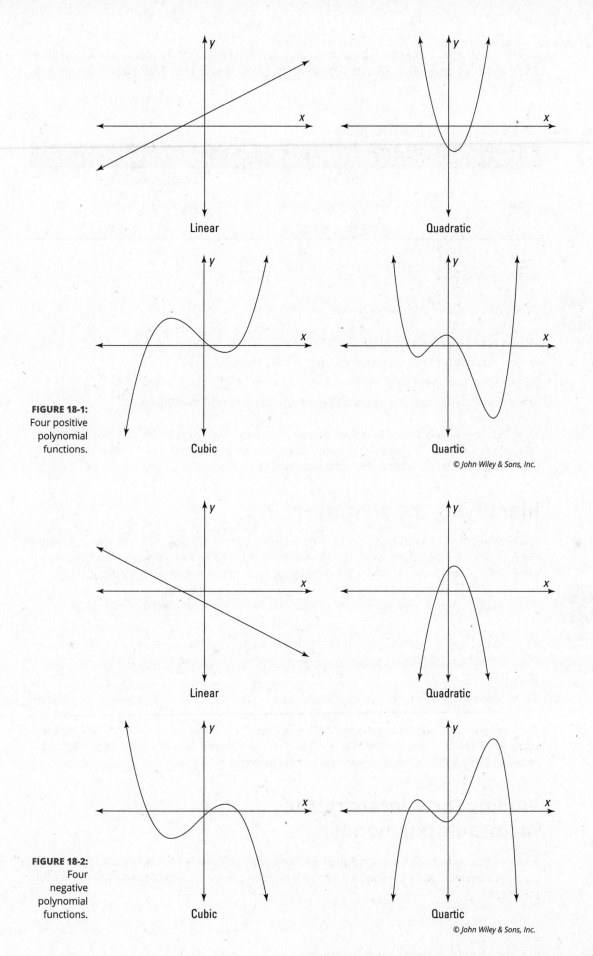

FIGURE 18-1:
Four positive polynomial functions.

Linear

Quadratic

Cubic

Quartic

© John Wiley & Sons, Inc.

FIGURE 18-2:
Four negative polynomial functions.

Linear

Quadratic

Cubic

Quartic

© John Wiley & Sons, Inc.

As you can see from Table 18-2, when you know whether a polynomial is odd or even and whether it's positive or negative, you can determine the end behavior of that polynomial in both directions.

TABLE 18-2 ## End Behavior of Polynomials

Polynomial type	Positive polynomials	Negative polynomials
Even	All end behavior: Positive	All end behavior: Negative
Odd	Very large x: Positive	Very large x: Negative
	Very small x: Negative	Very small x: Positive

Here's an example:

Which of the following is true of the end behavior of polynomial $3x^5 - 5x^3 - 7x$?

EXAMPLE

(A) It's positive for both very large and very small values of x.

(B) It's negative for both very large and very small values of x.

(C) It's positive for very large values of x and negative for very small values of x.

(D) It's negative for very large values of x and positive for very small values of x.

This fifth-degree polynomial is odd, so its end behavior is to go in opposite directions; therefore, you can rule out Answers A and B. And it's a positive polynomial, so its positive end behavior is positive; therefore, the correct answer is C.

Identifying the y-intercept

The constant term of a polynomial is the *y-intercept* of that polynomial — that is, the point where the function intersects the y-axis. You already know this fact from your work with linear functions of the form $y = mx + b$, where the constant term, b, is the y-intercept. For example:

Which of the following functions includes the point $(0, -9)$ when graphed on the xy-plane?

EXAMPLE

(A) $y = x^3 - 5x^2 - x + 9$

(B) $y = -9x^3 + 4x^2 - 6x - 2$

(C) $y = 6x^3 - 11x^2 + 8x - 9$

(D) $y = 3x^3 + 4x^2 - 9x$

The long way to do this problem is to plug in the values $x = -9$ and $y = 0$, and see which equation is correct. A faster way is to check the constant term of each polynomial to find its y-intercept. In order, these are 9, -2, -9, and 0, so the correct answer is C.

Finding the x-intercepts of factorable polynomials

So far in this section, all the polynomials you've worked with have been in *standard form* — that is, simplified down to their terms, without parentheses. Every polynomial can be written in standard form.

Some polynomials are *factorable*, which means that they can be written in an alternative form called *factored form*. Table 18-3 shows four examples of polynomials in standard form that can be factored — each by a different method discussed in Chapter 17 — changing each to its factored form.

TABLE 18-3 ## Polynomials in Standard and Factored Forms

Type of factoring	Standard form	Factored form
GCF factoring	$3x^2 - 6x$	$3x(x-2)$
Quadratic factoring	$x^2 + 8x + 7$	$(x+1)(x+7)$
Difference of 2 squares	$x^2 - 25$	$(x+5)(x-5)$
Factorable cubic	$x^3 - 2x^2 + 3x - 6$	$(x^2+3)(x-2)$

When a polynomial is factorable, you have an opportunity to find its x-intercepts — that is, the point or points (if any) where the function intersects the x-axis.

REMEMBER

Every polynomial has exactly one y-intercept, but

>> An even polynomial may have any even number of x-intercepts up to its degree — for example, a sixth-degree polynomial may have any number of x-intercepts from 0 to 6.

>> An odd polynomial always has at least one x-intercept and may have any number of them up to its degree — for example, a seventh-degree polynomial may have any number of x-intercepts from 1 to 7.

>> When a polynomial is factorable, you can often find some or all of its x-intercepts by setting each factor to 0 and solving for x.

REMEMBER

Mathematicians call x-intercepts by at least three other names: *roots*, *solutions*, and *zeros*. Don't be thrown off your game, because they all mean exactly the same thing: x-*intercepts*. Throughout the examples that follow, I use these four words interchangeably to give you practice recognizing them.

EXAMPLE

Which of the following is a root of the function $y = (x+3)(x-1)(2x-5)$?

(A) $\dfrac{2}{5}$

(B) $\dfrac{5}{2}$

(C) $-\dfrac{2}{5}$

(D) $-\dfrac{5}{2}$

To find the roots (x-intercepts) of this function, set each of its factors to 0 and solve for x:

$$x + 3 = 0 \qquad x - 1 = 0 \qquad 2x - 5 = 0$$
$$x = -3 \qquad\quad x = 1 \qquad\quad 2x = 5$$
$$x = \frac{5}{2}$$

Therefore, the correct answer is B.

Quadratic Equations and Functions

The standard form of a two-variable *quadratic function* is $y = ax^2 + bx + c$. On the xy-graph, this function takes the shape of a parabola.

You can change this function into the single-variable *quadratic equation* $ax^2 + bx + c = 0$ by setting y to 0. In this section, I discuss quadratic equations and functions, focusing on the types of questions you're likely to encounter on the ACCUPLACER.

Solving quadratic equations

The standard form of a quadratic equation is $ax^2 + bx + c = 0$. The presence of the x^2 term makes this a quadratic equation, and can sometimes be troublesome when you're trying to solve for x. In this section, I show you how to solve the most common ACCUPLACER problems that involve quadratic equations.

Solving 2-term quadratic equations

Although every quadratic equation is of the form $ax^2 + bx + c = 0$, when either b or c equals 0, the result is a quadratic equation with missing terms. These types of equations are simpler to solve than 3-term equations, but knowing how to solve them is key.

SOLVING QUADRATICS WHEN B=0

REMEMBER

A 2-term quadratic equation with a missing b term takes the form $ax^2 + c = 0$. Its general solution is $x = \pm\sqrt{-\dfrac{c}{a}}$.

For example, to solve $-3x^2 + 5 = 0$, subtract 5, divide by -3, and then take the square root of both sides:

$$-3x^2 + 5 = 0$$
$$-3x^2 = -5$$
$$x^2 = \frac{-5}{-3}$$
$$x = \pm\sqrt{\frac{5}{3}}$$

SOLVING QUADRATICS WHEN C=0

REMEMBER

A 2-term quadratic equation with a missing c term takes the form $ax^2 + bx = 0$. Its general solution is $x = 0$ and $x = -\dfrac{b}{a}$.

For example, to solve $7x^2 + 2x = 0$, first notice that $x = 0$ is a solution. To find the remaining solution, divide both sides by x, subtract 2, and divide by 7:

$$\frac{7x^2 + 2x}{x} = \frac{0}{x}$$
$$7x + 2 = 0$$
$$7x = -2$$
$$x = -\frac{2}{7}$$

Solving factorable quadratic equations

When a quadratic equation is factorable, you can solve it relatively quickly by factoring and then splitting it into two linear equations, each of which is easy to solve. You can't solve all quadratic equations in this way, but on the ACCUPLACER you'll find most problems yield to this method.

Earlier in this chapter (see "Solving quadratic equations"), I show you how to factor quadratic polynomials. Here, you put this information to work.

EXAMPLE

For example, consider the quadratic equation $x^2 - 5x - 14 = 0$. You can factor the left side of this equation as follows (as I show you in Chapter 17):

$$(x + 2)(x - 7) = 0$$

In this form, the equation states that when you multiply the two values $x + 2$ and $x - 7$, the result is 0. This result can only occur when one of these two values *already* equals 0. So you can split this equation into two separate equations, and solve each one easily:

$$x + 2 = 0 \qquad x - 7 = 0$$
$$x = -2 \qquad x = 7$$

So the solution to this equation is $x = -2$ or $x = 7$.

EXAMPLE

As another example, consider the equation $6x^2 + 13x = 5$. This time, begin by subtracting the 5 over to the left side of the equation, and then factor the resulting polynomial as follows (as I discuss in Chapter 17):

$$6x^2 + 13x - 5 = 0$$
$$(3x - 1)(2x + 5) = 0$$

Again, you can solve this equation by setting each factor to 0 and solving:

$$3x - 1 = 0 \qquad 2x + 5 = 0$$
$$3x = 1 \qquad 2x = -5$$
$$x = \frac{1}{3} \qquad x = -\frac{5}{2}$$

So the solution this time is $x = \frac{1}{3}$ or $x = -\frac{5}{2}$.

Solving quadratic equations with the quadratic formula

REMEMBER

The *quadratic formula* allows you to solve all quadratic equations (factorable and non-factorable) of the form $ax^2 + bx + c = 0$, as follows:

$$x = \frac{-b \pm \sqrt{b^2 - 4ac}}{2a}$$

On the ACCUPLACER, use the quadratic formula only as a last resort, when you're sure that you *really* need to solve a non-factorable 3-term quadratic equation.

EXAMPLE

For example, at first glance, the equation $x^2 - 6x + 7 = 0$ may look factorable, but play with it for a few moments and you'll find that you're stuck. To solve it using the quadratic formula, substitute in the values $a = 1$, $b = -6$, and $c = 7$, then simplify:

$$x = \frac{6 \pm \sqrt{(-6)^2 - 4(1)(7)}}{2(1)}$$
$$= \frac{6 \pm \sqrt{36 - 28}}{2}$$
$$= \frac{6 \pm \sqrt{8}}{2}$$

You can simplify this further by changing $\sqrt{8}$ to $2\sqrt{2}$, and then factoring a 2 in both the numerator and denominator:

$$= \frac{6 \pm 2\sqrt{2}}{2}$$
$$= 3 \pm \sqrt{2}$$

Although this looks like only one solution, it's actually two separate solutions: $3 + \sqrt{2} \approx 4.414$ and $3 - \sqrt{2} \approx 1.586$.

Solving systems that include a quadratic equation

Generally speaking, solving a system of equations in two variables requires you to reduce it to one equation with one variable (refer to Chapter 14). When one or both equations are quadratic, you may need to pull out your quadratic solving skills (refer to Chapter 17). For example,

$$y = 5x^2 + 10x - 4$$
$$y = 4x^2 + 12x - 1$$

EXAMPLE

Because the right side of each equation equals y, you can convert this to a quadratic equation in one variable by setting these two sides equal to each other, simplifying, and factoring:

$$5x^2 + 10x - 4 = 4x^2 + 12x - 1$$
$$x^2 - 2x - 3 = 0$$
$$(x + 1)(x - 3) = 0$$

You can probably see that the two solutions here are $x = -1$ and $x = 3$. To find the two corresponding y-values, plug these x-values into either equation and evaluate:

$$y = 5x^2 + 10x - 4 = 5(-1)^2 + 10(-1) - 4 = 5 - 10 - 4 = -9$$
$$y = 5x^2 + 10x - 4 = 5(3)^2 + 10(3) - 4 = 45 + 30 - 4 = 71$$

Therefore, the two solutions for this system of equations are $(-1, -9)$ and $(3, 71)$.

Quadratic functions

Every quadratic function can be written in three forms:

>> Standard form $y = ax^2 + bx + c$

>> Vertex form $y = a(x - h)^2 + k$

>> Factored form $y = (px + q)(rx + s)$

Technically speaking, the factored form of a quadratic equation may include complex numbers. On the ACCUPLACER, however, you probably won't have to worry about this type of equation.

TECHNICAL
STUFF

Quadratic functions in standard form: $y = ax^2 + bx + c$

The quadratic function in standard form is

$$y = ax^2 + bx + c.$$

REMEMBER

You can squeeze a lot of information out of this function with just a little bit of work. In this section, I show you how it's done.

IDENTIFYING A, B, AND C

The first step to understanding a specific quadratic function is to identify the values a, b, and c. These are the coefficients of the x^2, x, and constant terms. For example:

EXAMPLE

What are the a, b, and c values in the quadratic function $y = x^2 - 2x - 8$?

In this function, $a = 1$, $b = -2$, and $c = -8$.

USING A TO FIND THE DIRECTION OF CONCAVITY

When graphed on the xy-plane, every quadratic function takes the shape of a parabola that's either concave up or concave down:

When the a value is positive, the graph of the parabola is *concave up* — that is, the end behavior for both very large and very small x values is positive.

When the a value is negative, the graph of the parabola is *concave down* — that is, the end behavior for both very large and very small x values is negative.

Figure 18-3 shows two parabolas with different concavities, the first concave up and the second concave down.

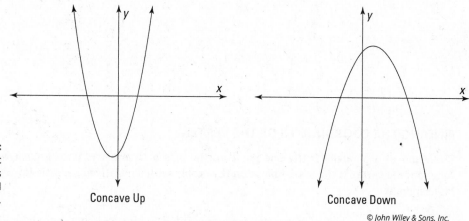

FIGURE 18-3: Quadratic functions and concavity.

Concave Up   Concave Down

© John Wiley & Sons, Inc.

USING C TO FIND THE Y-INTERCEPT

Earlier in this chapter, you discover that the constant term of a polynomial is its y-intercept. For a quadratic function in standard form, this is the c value.

EXAMPLE

What is the y-intercept of the function $y = x^2 - 6x - 7$?

The c value here is -7, so the point $(0, -7)$ is the y-intercept.

FINDING THE AXIS OF SYMMETRY WITH $x = \dfrac{-b}{2a}$

The *axis of symmetry* for a quadratic function is the vertical line on the xy-graph that passes symmetrically through it. You can think of the axis of symmetry as a mirror that reflects the same image to both its left and right.

The equation for the axis of symmetry of any quadratic function is $x = \dfrac{-b}{2a}$. For example:

What is the axis of symmetry for $y = -3x^2 + 12x + 1$?

EXAMPLE

In this function, $a = -3$ and $b = 12$. Plug these numbers into the formula for the axis of symmetry:

$$x = \frac{-b}{2a} = \frac{-12}{2(-3)} = \frac{-12}{-6} = 2$$

You can see how the axis of symmetry looks on the resulting graph, shown here.

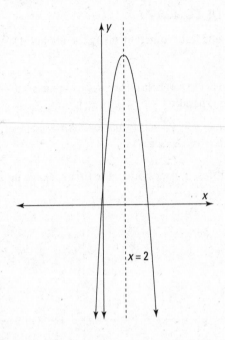

FINDING THE COORDINATES OF THE VERTEX

The *vertex* of a parabola is the one point on the axis of symmetry. With a positive quadratic function, the vertex is the lowest point on the graph; with a negative quadratic function, it's the highest point.

When you know the axis of symmetry for a function, you already know the x-coordinate of the vertex. You can plug this x-value into the function to find the y-coordinate of the vertex.

What are the coordinates of the vertex of the quadratic function $y = x^2 - 10x + 24$?

EXAMPLE

Begin by calculating the axis of symmetry:

$$x = \frac{-b}{2a} = \frac{10}{2(1)} = 5$$

Now, plug this value into the function:

$$y = (5)^2 - 10(5) + 24 = 25 - 50 + 24 = -1$$

So the coordinates of the vertex are $(5, -1)$.

Quadratic functions in vertex form: $y = a(x - h)^2 + k$

A second way to express a quadratic function is in vertex form:

$$y = a(x - h)^2 + k$$

In this formula, the constant a is the same as in standard form. The constants h and k indicate that the vertex of the parabola is the point (h, k).

If you graph $y = (x + 4)^2 + 6$ on the xy-plane, what's the vertex of the resulting parabola?

EXAMPLE The function is in vertex form, so the answer is $(-4, 6)$. Note that to find the vertex, I flip the sign of the value 4 to –4 and keep the value 6 unchanged.

CHANGING VERTEX FORM TO STANDARD FORM

To change a quadratic function from vertex form to standard form, use algebra to simplify the right side of the function. For example,

Rewrite $y = 2(x - 3)^2 + 1$ in standard form.

EXAMPLE Use algebra to simplify $2(x - 3)^2 + 1$:

$$y = 2(x - 3)^2 + 1$$
$$y = 2(x - 3)(x - 3) + 1$$
$$y = 2(x^2 - 6x + 9) + 1$$
$$y = 2x^2 - 12x + 18 + 1$$
$$y = 2x^2 - 12x + 19$$

So the answer is $y = 2x^2 - 12x + 19$. Note that the a value of 2 in vertex form remains the same in standard form.

CHANGING STANDARD FORM TO VERTEX FORM

To change a standard form quadratic to vertex form, most teachers use a method called *completing the square*. And most students find this method confusing for a very simple reason: It's confusing.

But in the earlier section, "Quadratic functions in standard form," I give you all the tools you need to make this conversion without completing the square. To make this conversion, remember that vertex form is $y = a(x - h)^2 + k$, so you need to find the three values a, h, and k. Follow these steps:

REMEMBER

1. **Use the a-value from the standard form function.**

2. **To get the h-value, calculate the axis of symmetry: $x = \dfrac{-b}{2a}$.**

3. **To get the k-value, plug in the value you just calculated into the function as x and solve for y.**

An example should make this clear:

Convert $y = 3x^2 + 12x + 5$ to vertex form.

EXAMPLE

To begin, note that $a = 3$ in both forms of the function. Next, to calculate the h-value, use the axis of symmetry formula:

$$x = \frac{-b}{2a} = \frac{-12}{2(3)} = \frac{-12}{6} = -2$$

Thus, $h = -2$. Now, plug this value as x into the function and solve for y:

$$y = 3(-2)^2 + 12(-2) + 5 = 12 - 24 + 5 = -7$$

So $k = -7$. To complete the problem, plug these three values into $y = a(x - h)^2 + k$:

$$y = 3(x + 2)^2 - 7$$

This result is the vertex-form version of $y = 3x^2 + 12x + 5$. You can verify this by following the steps outlined in "Changing vertex form to standard form" to reverse the process:

$$y = 3(x + 2)^2 - 7$$
$$y = 3(x + 2)(x + 2) - 7$$
$$y = 3(x^2 + 4x + 4) - 7$$
$$y = 3x^2 + 12x + 12 - 7$$
$$y = 3x^2 + 12x + 5$$

Quadratic functions in factored form

The third and final way to express a quadratic function is in factored form:

$$y = (px + q)(rx + s)$$

If this form seems overly complex, remember that this is just the result of factoring a quadratic expression, which I explain earlier in this section. And here's some more good news: In most cases, the quadratic functions that you'll need to factor on the ACCUPLACER will be the simpler version, with a leading coefficient of 1. Here are some examples:

Standard form	Factored form
$y = x^2 + 5x + 6$	$y = (x + 2)(x + 3)$
$y = x^2 - 8x + 12$	$y = (x - 2)(x - 6)$
$y = x^2 + 2x - 15$	$y = (x + 5)(x - 3)$
$y = x^2 - 4x - 21$	$y = (x + 3)(x - 7)$

REMEMBER

Expressing a quadratic equation in factored form allows you to find its x-intercepts. And before I proceed to the main event, may I take a moment to call out mathematicians everywhere for giving the world so many words that also mean x-intercepts! I count four so far, and it seems that every couple of years I discover a new one to add to this list:

    x-intercepts   roots   zeros   solutions

Let me make myself perfectly clear: All four of these words mean exactly the same thing. I prefer to use x-intercepts whenever possible, because most students know that that this means "the point(s) where the function crosses the x-axis."

Keep this definition in mind as you continue to read. In this section, I use all four terms interchangeably to get you used to the harsh realities of math out on the mean streets.

USING FACTORED FORM TO FIND INTEGER ROOTS

REMEMBER

EXAMPLE

When a quadratic equation in standard form has a leading coefficient of 1 and can be factored, its roots (x-intercepts) are integers. For example:

What are the roots of the function $y = x^2 + 14x - 32$?

Begin by factoring the right side:

$$y = (x + 16)(x - 2)$$

Now to find the roots, set y equal to 0 and solve for x:

$$0 = (x + 16)(x - 2)$$
$$x = -16, 2$$

Therefore, this function intersects the x-axis at the points $(-16, 0)$ and $(2, 0)$.

TIP

When you know the roots of the function, you can find the axis of symmetry easily: Just find the midpoint of the two roots by averaging the roots; the axis of symmetry intersects this point. In this case, the roots are at −16 and 2, so their midpoint is

$$\frac{-16 + 2}{2} = \frac{-14}{2} = -7$$

Therefore, the equation of the axis of symmetry is $x = -7$.

FRACTIONAL ROOTS

Fractional roots can arise from the factored form of a quadratic that has a leading coefficient other than 1. For example:

EXAMPLE

What are the zeros of the equation $y = (3x + 1)(2x - 5)$?

Remember that the zeros are the roots, or x-intercepts. To find them, set y equal to 0:

$$0 = (3x + 1)(2x - 5)$$

Solve this equation by splitting it into two equations and solving each one separately:

$$0 = 3x + 1 \qquad 0 = 2x - 5$$
$$-1 = 3x \qquad 5 = 2x$$
$$-\frac{1}{3} = x \qquad \frac{2}{5} = x$$

Therefore, this function intersects the x-axis at the points $(-\frac{1}{3}, 0)$ and $(\frac{2}{5}, 0)$.

IRRATIONAL ROOTS

REMEMBER

Irrational roots arise when a quadratic function isn't factorable. As I discuss earlier in this chapter (in the section, "Solving quadratic equations with the quadratic formula"), when a quadratic equation cannot be solved by factoring, your trusty nuclear option is the quadratic formula:

$$x = \frac{-b \pm \sqrt{b^2 - 4ac}}{2a}$$

EXAMPLE

What are the roots of the quadratic function $y = 3x^2 + 2x - 6$?

To solve this non-factorable quadratic equation, use the quadratic formula, plugging in 3 for a, 2 for b, and –6 for c:

$$x = \frac{-2 \pm \sqrt{2^2 - 4(3)(-6)}}{2(3)} = \frac{-2 \pm \sqrt{76}}{6}$$

You can simplify the radical and then reduce the resulting fraction by a factor of 2:

$$= \frac{-2 \pm \sqrt{4}\sqrt{19}}{6} = \frac{-2 \pm 2\sqrt{19}}{6} = \frac{-1 \pm \sqrt{19}}{3}$$

To be clear, this function has a pair of roots at $\frac{-1+\sqrt{19}}{3}$ and $\frac{-1-\sqrt{19}}{3}$, but the compact form $\frac{-1 \pm \sqrt{19}}{3}$ is commonly used.

Functions with fewer than 2 roots

So far in this section, all the functions you've worked with have had two solutions — that is, all have intersected the x-axis at two distinct points.

Although a quadratic function can never have more than two x-intercepts, it can have fewer than two — that is, one root or no roots. For example, here are two such functions:

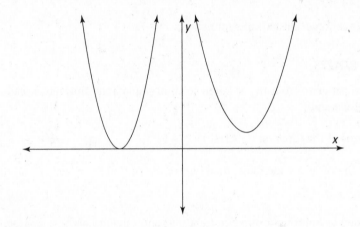

In this section, you explore how these two cases arise.

UNDERSTANDING FUNCTIONS WITH 1 ROOT

When a quadratic function can be expressed as the square of a binomial, it has only one root.

EXAMPLE

For example, consider the function $y = x^2 - 6x + 9$. Watch what happens when you factor this function:

$$y = (x - 3)(x - 3)$$
$$y = (x - 3)^2$$

This function's only x-intercept is at $(3, 0)$.

A quadratic function can also have its only root at a point that is a rational number but not an integer. For example, the function $y = 4x^2 + 4x + 1$ is factorable as follows:

$$y = (2x + 1)(2x + 1)$$
$$y = (2x + 1)^2$$

This function has only a single root, which you can find by solving the following equation:

$$0 = 2x + 1$$
$$-1 = 2x$$
$$-\frac{1}{2} = x$$

So this function's only zero is located at $(-\frac{1}{2}, 0)$.

Finally, a quadratic function's only x-intercept may be at an irrational. For example, the function $y = (x - \sqrt{5})^2$ has just one root at $(\sqrt{5}, 0)$. The standard form of this function isn't pretty, but you can find it by expanding this function and then FOILing:

$$y = (x - \sqrt{5})(x - \sqrt{5})$$
$$y = x^2 - 2\sqrt{5}x + 5$$

Not a very pretty function, but on the graph it looks relatively simple:

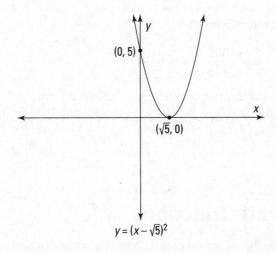

$$y = (x - \sqrt{5})^2$$

UNDERSTANDING FUNCTIONS WITH NO REAL ROOTS

Functions that have no real roots simply float over the x-axis without touching it. The simplest example I know is $y = x^2 + 1$, shown in the following figure.

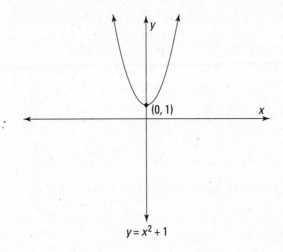

$$y = x^2 + 1$$

When you set y to 0 and solve an equation like this, something weird happens. Here, I make use of the solving technique that I show you earlier in this section, in "Solving quadratics when $b = 0$":

$$0 = x^2 + 1$$
$$-1 = x^2$$
$$\pm\sqrt{-1} = x$$
$$\pm i = x$$

This seemingly simple quadratic function forces you to reckon with those strange little numbers known as imaginary numbers. The imaginary number i is equivalent to $\sqrt{-1}$.

Thus, when a quadratic function appears to have no roots, it's common to say that it has no *real* roots. On the ACCUPLACER, you won't need to delve too deeply into imaginary numbers. But just in case you encounter this type of function, I'll present one more example, this time solvable with the quadratic equation:

What are the zeros of the function $y = 5x^2 - 7x + 3$?

EXAMPLE

To solve this problem, use the quadratic equation:

$$x = \frac{-(-7) \pm \sqrt{(-7)^2 - 4(5)(3)}}{2(5)} = \frac{7 \pm \sqrt{49 - 60}}{10} = \frac{7 \pm \sqrt{-11}}{10}$$

The negative radical $\sqrt{-11}$ earmarks this pair of solutions as *complex numbers* — that is, numbers that include a real value plus an imaginary value. Values of this kind are commonly expressed by factoring out $\sqrt{-1}$ and then simplifying this value to i:

$$= \frac{7 \pm \sqrt{11}\sqrt{-1}}{10} = \frac{7 \pm \sqrt{11}i}{10}$$

Systems of quadratic functions

In some cases, a problem may include a system of equations that includes one or more quadratic functions. In most cases, the best way to solve the problem is to reduce the system to a single quadratic equation in one variable.

If $x > 0$, what value of x in the following system of equations?

EXAMPLE

$$y = -x^2 + 4$$
$$4x - y = 8$$

To solve, substitute $-x^2 + 4$ for y into the second equation and simplify:

$$4x - (-x^2 + 4) = 8$$
$$4x + x^2 - 4 = 8$$

Put this equation into standard form and factor:

$$x^2 + 4x - 12 = 0$$
$$(x + 6)(x - 2) = 0$$

The two solutions are $x = -6$ and $x = 2$. However, the problem tells you that $x > 0$, so the correct answer is 2.

Exponential Equations and Functions

In Chapter 13, I fill you in on enough basic algebra involving exponents to pull you through the Algebra Test on the ACCUPLACER. Here are some key exponent facts to keep in mind:

$$x^a \cdot x^b = x^{a+b} \qquad \frac{x^a}{x^b} = x^{a-b} \qquad (x^a)^b = x^{ab} \qquad x^{-a} = \frac{1}{x^a}$$

In this section, I build upon that information to help you solve the types of problems you may face in the Advanced Algebra Test on the ACCUPLACER.

Rewriting fractional exponents

In Chapter 13, you discover that an exponent of $\frac{1}{2}$ is equivalent to a square root. For example,

$$25^{\frac{1}{2}} = \sqrt{25} = 5 \qquad 49^{\frac{1}{2}} = \sqrt{49} = 7 \qquad 121^{\frac{1}{2}} = \sqrt{121} = 11$$

You can extend this idea to other fractions. For example, an exponent of $\frac{1}{3}$ is equivalent to a cube root, so,

$$8^{\frac{1}{3}} = \sqrt[3]{8} = 2 \quad 27^{\frac{1}{3}} = \sqrt[3]{27} = 3 \quad 125^{\frac{1}{3}} = \sqrt[3]{125} = 5$$

Similarly, an exponent of $\frac{2}{3}$ equals the square of a cube root. For example,

$$125^{\frac{2}{3}} = (\sqrt[3]{125})^2 = 5^2 = 25$$

Generally speaking, you can express any fractional exponent as a radical in either of the following two ways (use whichever is most helpful):

$$x^{\frac{a}{b}} = \sqrt[b]{x^a} \quad x^{\frac{a}{b}} = (\sqrt[b]{x})^a$$

For example:

What does $8^{\frac{4}{3}}$ equal?

EXAMPLE

Rewrite the fractional exponent as a radical and simplify:

$$8^{\frac{4}{3}} = (\sqrt[3]{8})^4 = 2^4 = 16$$

Here's a trickier problem:

Which of the following is equivalent to $x^{\frac{3}{5}} x^{\frac{1}{4}}$?

EXAMPLE

(A) $\sqrt[3]{x^{20}}$

(B) $\sqrt[20]{x^3}$

(C) $\sqrt[17]{x^{20}}$

(D) $\sqrt[20]{x^{17}}$

Multiply the two values by adding their exponents, then rewrite the fractional exponent as a radical:

$$x^{\frac{3}{5}} x^{\frac{1}{4}} = x^{\frac{3}{5}+\frac{1}{4}} = x^{\frac{17}{20}} = \sqrt[20]{x^{17}}$$

Therefore, the correct answer is D.

Finding a common base to solve exponential problems

When a problem includes a pair of exponent expressions with different bases, you're faced with the challenge of expressing every exponent using a common base. The best way to understand this type of problem is with an example:

EXAMPLE

Which of the following is equivalent to the expression $\frac{9^n}{3^{n-1}}$?

(A) 3^n

(B) 9^n

(C) 3^{n+1}

(D) 9^{n-1}

Your first task is to think of a way to rewrite the bases 3 and 9 using a common base. The lower whole number here is 3, so use this number as your common base. You can rewrite 9 as 3^2, so plug in this value for 9:

$$\frac{9^n}{3^{n-1}} = \frac{(3^2)^n}{3^{n-1}}$$

Now, you need to remove the parentheses from the expression in the numerator. Use the rule $(x^a)^b = x^{ab}$ for this purpose:

$$= \frac{3^{2n}}{3^{n-1}}$$

Now, you can simplify this expression using $\frac{x^a}{x^b} = x^{a-b}$:

$$= 3^{2n-(n-1)}$$

To complete the problem, simplify the exponent by distributing and combining like terms:

$$= 3^{2n-n+1} = 3^{n+1}$$

Therefore, the answer is C.

When you know the basic moves, you're ready to take on an exponential equation.

What is the value of k if $(\frac{1}{5})^{3k} = \sqrt{5}^{k-1}$?

This problem is a bit trickier. You can probably see, however, that a good common base for both sides of the equation is 5. Now, you need to draw upon your understanding of exponents to express both $\frac{1}{5}$ and $\sqrt{5}$ as exponents with a base of 5. Recall that $\frac{1}{5} = 5^{-1}$ and that $\sqrt{5} = 5^{\frac{1}{2}}$, so,

$$(5^{-1})^{3k} = (5^{\frac{1}{2}})^{k-1}$$

As you make this change, don't forget those important parentheses. They should remind you that on the next step, you need to remove the parentheses by multiplying exponents:

$$5^{-1(3k)} = 5^{\frac{1}{2}(k-1)}$$

At this point, you have an equation in which the bases are the same but the exponents are different. To solve it, drop the bases and set the exponents equal:

$$-1(3k) = \frac{1}{2}(k-1)$$
$$-3k = \frac{1}{2}(k-1)$$
$$-6k = k-1$$
$$-7k = -1$$
$$k = \frac{1}{7}$$

Exponential growth and decay functions

Technically speaking, exponential growth and decay functions take different forms:

>> Exponential growth: $y = ab^x$

>> Exponential decay: $y = ab^{-x}$

When solving word problems on the ACCUPLACER, however, you'll most likely want to take a different approach to distinguishing exponential growth and decay:

>> Exponential decay: $y = ab^x$ with $b > 1$

>> Exponential growth: $y = ab^x$ with $0 < b < 1$

In this section, I show you how to think about percent increase and percent decrease in a way that will set you up for success when working with exponential functions. Then, you use the exponential growth and exponential decay functions to model the most common forms of ACCUPLACER word problems.

Reviewing percent increase and percent decrease

In Chapter 11, I discuss percent increase and percent decrease. There, I show you how to do these kinds of problems in one step rather than two.

For example, suppose you want to calculate a 20% tip added onto a restaurant bill of $42. Instead of finding 20% of $42 and then adding this amount to the bill, simply calculate 120% of $42:

$$1.2 \times 42 = 50.4$$

So the bill with a 20% tip comes to $50.40.

As another example, suppose you want to calculate 15% off the price of a computer that normally sells for $1350? This time, instead of finding 15% of $1350 and then subtracting this amount from the price, simply calculate 85% of $1350:

$$0.85 \times 1350 = 1147.5$$

So the price with a 15% discount comes to $1147.50.

This strategy is useful when working with percent increase and percent decrease problems, but it's *essential* when working with exponential growth and decay.

Exponential growth

The exponential growth function is

$$y = ab^x \text{ with } b > 1$$

In a typical ACCUPLACER problem, a is the initial value, b is the constant indicating the rate of growth, and x is the input variable indicating time. (Sometimes, the variable t replaces x, but this change doesn't affect the basic structure of the function.)

Here is a typical ACCUPLACER exponential growth problem:

EXAMPLE

A biologist observes that an initial population of 1000 bacteria doubles every 3 hours. Which of the following expressions best predicts the size of the population in 2 days?

(A) $1000(2)^3$

(B) $1000(3)^2$

(C) $1000(2)^{16}$

(D) $1000(3)^{48}$

Plug 1000 for a (the initial number of bacteria) and 2 for b (the constant indicating the rate of growth) into the function for exponential growth:

$$y = 1000(2)^x$$

Now, consider that 2 days equals 48 hours, and population doubles every 3 hours. Thus, the population doubles 16 times in 48 hours, so plug in this value for x:

$$y = 1000(2)^{16}$$

Therefore, the answer is C.

EXAMPLE

The value of a company is predicted to grow at 11% per year for each of the next 3 years. If the current value of the company is $2,400,000, which of the following expressions represents the predicted value of the company in 3 years, in millions of dollars?

(A) $2.4(0.1)^3$

(B) $2.4(0.11)^3$

(C) $2.4(1.11)^3$

(D) $2.4(1.011)^3$

All four answers already have 2.4 plugged in for a and 3 for x. The key to solving this problem is knowing how to represent 11% growth as the value b. In a percent increase problem, an 11% increase is represented as 111%, which is equivalent to the decimal 1.11. You use this same procedure when finding the base in an exponential growth problem, so the answer is C.

Exponential decay

The exponential decay function is

$$y = ab^x \text{ with } 0 < b < 1$$

As with exponential growth (see the previous section), once again *a* is the initial value, *b* is the constant indicating the rate of growth, and *x* is the input variable indicating time. (Sometimes, the variable *t* replaces *x*, but this change doesn't affect the basic structure of the function.)

A $200,000 investment lost 8% of its value every year for 4 years. How much was the investment worth at the end of this time?

EXAMPLE

Using the function $y = ab^x$, plug in 200,000 for *a* and 4 for *x*. For *b*, remember that a percent decrease of 8% is calculated as 92% of the value, so plug in 0.92 for *b*:

$$y = 200,000(0.92)^4 \approx 200,000(0.72) \approx 143,278.59$$

So the value of the investment after 4 years was approximately $143,278.59.

Logarithmic Equations (and Functions)

Logarithms (logs for short) are an alternative way to think about information more commonly expressed using exponents. In fact, log functions are the inverse of exponential functions (see "Exponential growth and decay functions).

Logarithms can be confusing, even for me (and I've been teaching this stuff since Abraham Lincoln was a boy). So, whenever I start working with logs, I tend to jot down the following two equations:

$$10^2 = 100 \qquad \log_{10}100 = 2$$

The first equation, which is an exponential equation, means that when you start with a base of 10 and multiply it by itself 2 times, the result is 100.

The second equation, which is a logarithmic equation, means essentially the same thing, but in a mixed-up way. This equation also has a base of 10: the base of the logarithm. The use of this word in both equations isn't a coincidence, and you can use this fact as a starting point to help you turn any log equation into an equivalent exponential equation:

1. Write down the base.

2. Move across the equals sign and use the value you find there as the exponent.

3. Write an equals sign.

4. Move back across the equals sign in the original equation and write down the remaining value.

For example, consider this log problem:

What is the value of *x* if $\log_4 x = 3$?

EXAMPLE

This is bound to be confusing until you rewrite it as an exponential equation. Starting with the base of 4, cross the equals sign and pick up 3 as the exponent, write an equals sign, and finish by picking up the *x* value from the original equation:

$$4^3 = x$$

In this form, the equation is a lot easier to solve:

$$64 = x$$

Here's another example:

EXAMPLE

If $\log_x 8 = 3$, what does x equal?

Again, this equation is confusing until you untangle it:

$$x^3 = 8$$

You can probably tell by inspection that $x = 2$, because $2^3 = 8$. One more example:

EXAMPLE

What is the value of x given that $\log_3 81 = x$?

Rewrite this equation as follows:

$$3^x = 81$$

Now, you need to find the exponent that makes this equation true. This time, the answer is $x = 4$, because $3^4 = 81$.

Here's a slightly tougher problem that relies on a reversal of this strategy:

EXAMPLE

If $2^{9x} = 5$, which of the following is the value of x?

(A) $\dfrac{\log_2 5}{9}$

(B) $\dfrac{\log_5 2}{9}$

(C) $\dfrac{\log_5 9}{2}$

(D) $\dfrac{\log_2 9}{5}$

To solve this problem, you need to turn the exponential equation into a log equation. Remember, always start with the base, cross the equals sign to pick up the next value, and then cross back to pick up the final value:

$$\log_2 5 = 9x$$

This result may look more complicated, but now you can divide both sides by 9 to isolate x:

$$\frac{\log_2 5}{9} = x$$

Therefore, the correct answer is A.

Radical Functions

The radical parent function is $f(x) = \sqrt{x}$. The graph of this function is shown here:

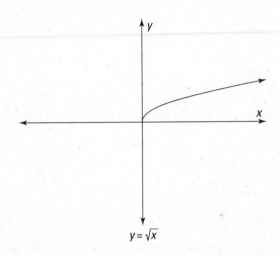

$$y = \sqrt{x}$$

Radicals may arise in other types of problems, so you'll need to know how to do some basic computation with radicals. Additionally, ACCUPLACER questions about a radical function may require you to find the domain or the range of that function. In this section, you build on your knowledge of radicals.

Calculating with radicals

Radicals arise in a variety of algebra problems. The ACCUPLACER also includes questions testing your specific knowledge of how to work with radicals. In this section, I give you some information about calculating with radicals that is most likely to help you on the test.

Estimating the value of a radical

When you take the square root of a square number, the result is a whole number. I place these values in their proper positions on the following number line:

Now, consider a few radicals and their irrational equivalents:

$$\sqrt{2} \approx 1.414 \qquad \sqrt{3} \approx 1.732 \qquad \sqrt{5} \approx 2.236$$

When you place these three values on the number line, they arrange themselves in order with the radicals that equal whole numbers:

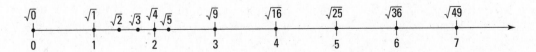

Every radical of a non-square number also falls on the number line in this way. This understanding allows you to find the approximate value of a radical. For example:

When expressed as a decimal, $\sqrt{42}$ falls between which two whole numbers?

EXAMPLE

(A) 3 and 4

(B) 4 and 5

(C) 5 and 6

(D) 6 and 7

First, notice that $\sqrt{36} < \sqrt{42} < \sqrt{49}$. You can rewrite this inequality by changing $\sqrt{36}$ and $\sqrt{49}$ to whole numbers:

$$6 < \sqrt{42} < 7$$

Thus, $\sqrt{42}$ is between 6 and 7.

Multiplying and dividing radicals

Multiplying and dividing radicals is easy and intuitive: Just multiply or divide the numbers inside the radicals. Here are some examples of radical multiplication:

$$\sqrt{2}\sqrt{3} = \sqrt{6} \qquad \sqrt{5}\sqrt{7} = \sqrt{35} \qquad \sqrt{10}\sqrt{13} = \sqrt{130}$$

Radical division is typically expressed as a rational expression, but the process is just as easy. For example:

$$\frac{\sqrt{15}}{\sqrt{3}} = \sqrt{5} \qquad \frac{\sqrt{22}}{\sqrt{11}} = \sqrt{2} \qquad \frac{\sqrt{70}}{\sqrt{7}} = \sqrt{10}$$

Simplifying radicals

Some radicals can be simplified by factoring out the square root of a square number, such as 4, 9, 16, 25, and so on. For example, here are three radicals that can be simplified by factoring out the square root of 4 and then changing this value to 2:

$$\sqrt{8} = \sqrt{4}\sqrt{2} = 2\sqrt{2} \qquad \sqrt{12} = \sqrt{4}\sqrt{3} = 2\sqrt{3} \qquad \sqrt{20} = \sqrt{4}\sqrt{5} = 2\sqrt{5}$$

And here are three examples of radicals that can be simplified by factoring out the square root of 9:

$$\sqrt{18} = \sqrt{9}\sqrt{2} = 3\sqrt{2} \qquad \sqrt{27} = \sqrt{9}\sqrt{3} = 3\sqrt{3} \qquad \sqrt{45} = \sqrt{9}\sqrt{5} = 3\sqrt{5}$$

Here are a few examples of radicals that can be simplified by factoring out the square roots of larger numbers:

$$\sqrt{32} = \sqrt{16}\sqrt{2} = 4\sqrt{2} \qquad \sqrt{50} = \sqrt{25}\sqrt{2} = 5\sqrt{2} \qquad \sqrt{98} = \sqrt{49}\sqrt{2} = 7\sqrt{2}$$

Adding and subtracting with radicals

You can add a pair of radical expressions that have the same radical part as follows:

$$\sqrt{3} + \sqrt{3} = 2\sqrt{3} \qquad 6\sqrt{5} + \sqrt{5} = 7\sqrt{5} \qquad 5\sqrt{19} + 10\sqrt{19} = 15\sqrt{19}$$

You can subtract a pair of radicals in a similar manner:

$$8\sqrt{2} - 7\sqrt{2} = \sqrt{2} \qquad 10\sqrt{6} - 2\sqrt{6} = 8\sqrt{6} \qquad 100\sqrt{14} - 90\sqrt{14} = 10\sqrt{14}$$

Generally speaking, you cannot add or subtract radical expressions that have different radical parts. For example, the following expressions cannot be simplified:

$$\sqrt{2} + \sqrt{3} \qquad 3\sqrt{5} + 6\sqrt{11} \qquad 8\sqrt{5} - 7\sqrt{2} \qquad 10\sqrt{15} - 10\sqrt{17}$$

In some cases, however, you can simplify one or more radicals and then add or subtract them. For example:

EXAMPLE

Which of the following is the value of $2\sqrt{18} - \sqrt{8}$?

(A) $4\sqrt{2}$

(B) $2\sqrt{6}$

(C) $4\sqrt{8}$

(D) $2\sqrt{10}$

These two radical expressions have different radical parts, so you can't subtract them as they are now. However, both radicals can be simplified by factoring out the square root of a square number:

$$2\sqrt{18} = 2\sqrt{9}\sqrt{2} = 6\sqrt{2}$$
$$\sqrt{8} = \sqrt{4}\sqrt{2} = 2\sqrt{2}$$

These two values have the same radical part ($\sqrt{2}$), so you can subtract them:

$$2\sqrt{18} - \sqrt{8} = 6\sqrt{2} - 2\sqrt{2} = 4\sqrt{2}$$

Therefore, the answer is A.

Rationalizing the denominator

For a variety of esoteric reasons, mathematicians don't like radicals in the denominator of a rational expression, and prefer to remove them. This process, called *rationalizing the denominator*, simply requires you to multiply the numerator and the denominator of the expression you're working with by the radical that's already in the denominator.

For example, the expression $\dfrac{1}{\sqrt{2}}$ has a radical in the denominator. To rationalize this denominator, multiply both the numerator and the denominator by $\sqrt{2}$:

$$\frac{1}{\sqrt{2}} = \frac{1\sqrt{2}}{\sqrt{2}\sqrt{2}} = \frac{1\sqrt{2}}{\sqrt{4}} = \frac{\sqrt{2}}{2}$$

This new form is simply an equivalent version of the same value, but without the radical in the denominator. Similarly, you can rationalize the expression $\dfrac{6\sqrt{7}}{5\sqrt{3}}$ by multiplying both the numerator and the denominator by $\sqrt{3}$:

$$\frac{6\sqrt{7}}{5\sqrt{3}} = \frac{6\sqrt{7}\sqrt{3}}{5\sqrt{3}\sqrt{3}} = \frac{6\sqrt{21}}{5\sqrt{9}} = \frac{6\sqrt{21}}{15}$$

In this case, you can take one final step to cancel out a factor of 3 from both the numerator and the denominator:

$$= \frac{2\sqrt{21}}{5}$$

Domain and range of a radical function

REMEMBER

The value inside a radical expression (called the *radicand*) cannot be negative, because when working with real numbers, you can't take the square root of a negative number. So, to find the domain of a radical function (the set of x-values that can be inputted into it), create an inequality using $\geq$ with the value inside the radical on the left side and 0 on the right side. For example:

EXAMPLE

What's the domain of the function $f(x) = 3\sqrt{2x-5} + 1$?

The value $2x - 5$ is inside the radical, so create and solve the following inequality:

$$2x - 5 \geq 0$$
$$2x \geq 5$$
$$x \geq \frac{5}{2}$$

So the domain of the function is $x \geq \frac{5}{2}$.

The range of a function is the set of y-values that can be outputted by that function. The range of a radical function can be expressed as an inequality. To find it, keep two things in mind:

REMEMBER

When the function is positive, this inequality includes $\geq$; when the function is negative, it includes $\leq$.

The value added to or subtracted from the radical part of the function goes on the right side of the inequality. (This works because the radical part of the expression is set to 0, so it drops out of the expression.)

A few examples should clarify how to find the range of a radical function:

Radical function	Range
$y = 3\sqrt{2x-5} + 1$	$y \geq 1$
$y = 3\sqrt{2x-5} - 1$	$y \geq -1$
$y = -3\sqrt{2x-5} + 1$	$y \leq 1$
$y = -3\sqrt{2x-5} - 1$	$y \leq -1$

Finding the Domain and Range of a Rational Function

Recall from Chapter 17 that the parent rational function is $y = \frac{1}{x}$. The graph of this function is shown here:

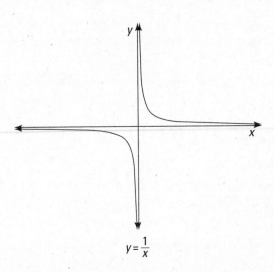

$$y = \frac{1}{x}$$

Also, remember that the domain of a function is the set of x-values that can be inputted into the function. A rational function has *at least one* (and sometimes more than one) vertical asymptote, which is a value excluded from the domain. In the parent function, this value is 0, but changes to this function can alter the domain.

A rational function also has *exactly one* horizontal asymptote, which is a value excluded from the range. In the parent function, this value is 0, but changes to this function can alter the range.

REMEMBER

EXAMPLE

The value in the denominator of a rational expression cannot equal zero. So, to find the domain of a radical function, set the denominator equal to 0 and solve. The solutions to this equation are *excluded* from the domain. For example:

What's the domain of the function $f(x) = \dfrac{3}{x^2 - 6x - 27} - 4$?

The value in the denominator is $x^2 - 6x - 27$, so set this expression equal to 0 and solve for x:

$$x^2 - 6x - 27 = 0$$
$$(x+3)(x-9) = 0$$
$$x = -3, 9$$

Therefore, the domain of the function includes all real values *except* –3 and 9. You can state this domain formally as $(-\infty, -3) \cup (-3, 9) \cup (9, \infty)$.

REMEMBER

EXAMPLE

The range of a function is the set of y-values that can be outputted by that function. Finding the range of a rational function is easier: Simply note the value that is separately either added to or subtracted from the fractional part of the function, and exclude this value. (If no such value exists, then this value is 0.)

What is the range of the function $f(x) = \dfrac{3}{x^2 - 6x - 27} - 4$?

The value –4 is separated from the fractional part of the function, so the range is all real values *except* –4. You can write this formally as $(-\infty, -4) \cup (-4, \infty)$.

IN THIS CHAPTER

» **Understanding triangle congruency and similarity**

» **Graphing circles on the *xy*-plane**

» **Applying the six trigonometric ratios to 45-45-90 and 30-60-90 right triangles**

» **Calculating arc length with radian measure**

» **Solving trig equations**

Chapter **19**

Geometry and Trigonometry

T his chapter picks up where Chapter 14 leaves off in its discussion of geometry, and extends this topic to include trigonometry.

You start off working with a few formulas for the volume of solid objects: spheres, pyramids, and cones. Then you work with a variety of theorems for measuring angles, as well as triangle congruency and similarity. The section concludes with a look at how to write the equations for circles on the *xy*-plane.

You then move on to trigonometry, where I introduce you to the six trigonometric ratios, and show you a variety of ways to work with them. You apply this understanding to two important special right triangles — the 45-45-90 degree triangle and the 30-60-90 degree triangle. You also work with radian measure and arc length. The chapter concludes with a look at some more advanced concepts in trigonometry: proving identities, solving trig equations, and the laws of sines and cosines.

Geometry

In Chapter 14, I present the geometry that you'll need to know for the second ACCUPLACER test, the QAS Math Test. For the AAF Math Test, you'll need all this *plus* some additional information. In this section, I bring you up to speed on the geometry you need to know, so that you'll be ready for trigonometry later in this chapter.

Determining volume of non-prism objects

The following sections cover the formulas for determining the volume of different types of non-prism objects.

Volume of a sphere

Here's the formula for the volume of a sphere:

$$V = \frac{4\pi r^3}{3}$$

This formula requires only the radius of the sphere.

For example, here's how to find the volume of a ball that has a radius of 6 inches:

EXAMPLE

$$V = \frac{4\pi 6^3}{3} = \frac{4\pi(216)}{3} = 4\pi(72) = 288\pi$$

You can approximate the value of π to be 3.14 to find the measurement as a decimal:

$$\approx 904.32$$

So the volume is approximately 904.32 cubic inches.

Volume of a pyramid

REMEMBER

The formula for the volume of a pyramid is

$$V = \frac{A_b h}{3} = \frac{s^2 h}{3}$$

I give this formula in two ways so that, depending on the problem, you'll be able to find the answer. Here, A_b stands for *the area of the base*. And because the base of a pyramid is a square, if you know how long the side of the base is, you can find the area of the base as s^2 and then proceed to the answer. For example:

The Great Pyramid of Giza has a height of 481 feet, and the side of its base measures 756 feet. What is the volume of the pyramid in cubic feet?

EXAMPLE

Plug these numbers into the formula and simplify:

$$V = \frac{756^2(481)}{3} = \frac{571,536(481)}{3} = 190,512(481) = 91,636,272$$

So the volume of the pyramid is approximately 91,636,272 cubic feet.

Volume of a cone

REMEMBER

Calculate the volume of a cone as follows:

$$V = \frac{A_b h}{3} = \frac{\pi r^2 h}{3}$$

As with the formula for the volume of a pyramid, I give you two versions of this formula. The first includes A_b, which stands for *the area of the base*, and the second replaces this value with the area formula for a circle, which is πr^2. Here's an example:

EXAMPLE

A large ice cream cone is 18 centimeters in height, and its base has a radius of 3 centimeters. What is its volume?

Find the answer by plugging this information into the formula:

$$V = \frac{\pi 3^2 (18)}{3} = 9\pi(6) = 54\pi$$

Approximate the value of π to be 3.14 to find the measurement as a decimal:

$$\approx 169.56$$

So the volume is about 169.56 cubic cm.

Using the intersecting line theorems

Geometry includes a relatively short list of theorems — or rules — for calculating the measurements of angles. In this section, I discuss these useful theorems and show you how to apply them to answer a variety of typical ACCUPLACER questions.

Vertical angles are equal

REMEMBER

When two lines intersect, two pairs of *vertical angles* — angles that are opposite from each other — are formed. A pair of vertical angles always have the same angle measurement.

For example, in the following figure, the two pairs of vertical angles are *w* and *z*, and *x* and *y*. Here, *w* measures 130 degrees, so by the vertical angles rule, *z* also measures 130 degrees.

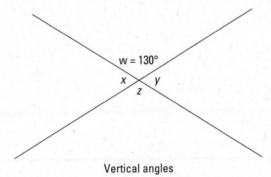

Vertical angles

Two supplementary angles form a line equal to 180°

REMEMBER

When two lines intersect, any pair of adjacent angles are *supplementary angles* — that is, their sum equals 180°. Combining this information with the fact that vertical angles are equal allows you to find all the angles formed when two lines intersect, if you know any one of those angles.

For example, in this figure, you already know that w and z both measure 130 degrees. Additionally, y and z both measure 50 degrees, because $130 + 50 = 180$.

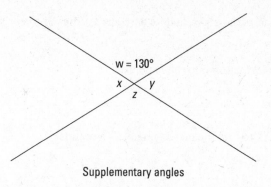

Supplementary angles

Corresponding angles are equal

REMEMBER

When a line intersects two parallel lines, a pair of angles that appear in the same relative position, called *corresponding angles*, always have the same angle measurement.

Combining this information with the rules for vertical angles and supplementary angles gives you a lot of power for determining the measurements of angles in figures.

For example, in the following figure, e measures 145 degrees, because a and e are corresponding angles. Additionally, d and h also measure 145 degrees, because they are vertical angles to a and e, respectively. And b, c, f, and g all measure 35 degrees, because $145° + 35° = 180°$.

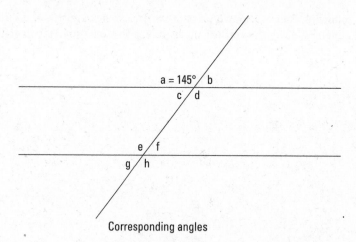

Corresponding angles

Three interior angles of a triangle equal 180°

REMEMBER

The sum of the three interior angles of a triangle is always 180 degrees. This rule implies that when you know the measures of any two angles in a triangle, you can determine the measure of the third angle. For example:

EXAMPLE

What is the measure of angle *a* in this figure?

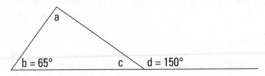

Interior angles of a triangle

The measure of angle *d* is 150 degrees, and *c* is supplementary to *d*, so the measure of *c* is 30 degrees. You also know that the measure of *b* is 65 degrees, so you can conclude that the measure of *a* is 85 degrees, because $85° + 65° + 30° = 180°$.

A pair of complementary angles equals 90°

REMEMBER

By definition, a pair of *complementary angles* equals 90 degrees. On the ACCUPLACER, complementary angles tend to come in two varieties:

A pair of angles that form a right angle

The two smaller angles of a right triangle

EXAMPLE

In this figure, the measure of angle *r* is 20 degrees greater than the measure of angle *q*. How many degrees is angle *p*?

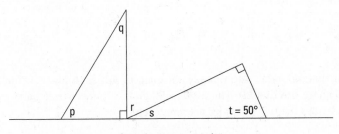

Complementary angles

Angle *t* is 50 degrees, and angle *s* is complementary to angle *t*, so the measure of angle *s* is 40 degrees. Angle *r* is complementary to angle *s*, so angle *r* is 50 degrees. Thus, because angle *r* is 20 degrees greater than angle *q*, angle *q* is 30 degrees. Angle *p* is complementary to angle *q*, so angle *p* is 60 degrees.

Using triangle congruency and similarity theorems

In this section, I discuss two ideas from geometry that loom large on the ACCUPLACER (and other standardized tests, such as the SAT and ACT): triangle congruency and similarity.

Some ACCUPLACER questions may ask you to prove that a pair of triangles are *congruent* — that is, identical in size and shape. These questions usually require you to identify which of the four congruency theorems is needed to prove that the triangles are congruent.

Other ACCUPLACER questions may ask you to recognize that a pair of triangles are *similar* — that is, identical in shape, and proportional in size. This type of question often requires you to perform a calculation by creating and solving a proportion (see Chapter 13 for more details).

Proving congruent triangles

When a pair of triangles have three identical sides and three identical angles, they're called *congruent triangles*. Geometry features a short list of ways to prove that a pair of triangles are congruent, based on whether some of their sides and/or angles are congruent.

In this section, I show you the four possible ways to prove that two angles are congruent: SSS, SAS, ASA, and SAA. I also show you two mistakes to be avoided: *SSA* and *AAA*.

Three sides (SSS)

REMEMBER

When two triangles have three pairs of sides that are equal in length, the two triangles are congruent. This rule is called *SSS* (*side-side-side*). For example:

EXAMPLE

Triangle *ABC* is isosceles where *AB* = *BC*, and *BD* bisects *AC*. Show that triangle *ABD* is congruent to triangle *CBD*.

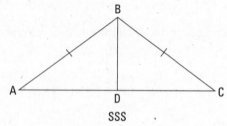

SSS

The problem tells you that *AB* = *BC*. Additionally, *BD* bisects *AC*, so *AD* = *DC*. Finally, the two triangles *ABC* and *BCD* share the line *BD*. Therefore, *ABC* and *BCD* have three pairs of sides that have equal lengths, so by *SSS* the two triangles are congruent.

Two sides plus the angle between them (SAS)

REMEMBER

When two triangles have two pairs of sides that are equal in length, and the angles *between these two sides* are equal in measure, the two triangles are congruent. This rule is called *side-angle-side* (abbreviated *SAS*). Notice the very important provision that the angle in common must be the angle between the two sides whose lengths you know.

EXAMPLE

Triangle *ABC* is isosceles with *AB* = *BC*, and *BD* bisects the angle *ABC*. Show that triangle *ABD* is congruent to triangle *CBD*.

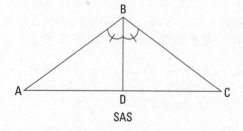

SAS

The problem tells you that $AB = BC$. Additionally, BD bisects the angle ABC, so angle ABD has the same measure as angle DBC. Finally, the two triangles ABD and CBD share the line BD. Therefore, ABD and CBD share two pairs of sides that have equal lengths, and the angles between these two sides are equal to each other. Therefore, by SAS the two triangles are congruent.

ONE SIDE PLUS ANY TWO ANGLES (ASA AND SAA)

REMEMBER

When two triangles have one pair of sides that are equal in length, and any two pairs of angles are equal in measure, the two triangles are congruent. This rule is called either *angle-side-angle* or *side-angle-angle* (*ASA* or *SAA*), depending upon whether the side whose length you know lies between the two angles that you know.

Both of these rules are essentially the same. To see why, recall that when you know the measures of two angles in a triangle, you can find the measure of the third angle (see "Three interior angles of a triangle equal 180°"). So, maybe this rule should just be called *ASAA*.

EXAMPLE

In triangle ABC, BD bisects the angle ABC and BD is perpendicular to AC. Show that triangle ABD is congruent to triangle CBD.

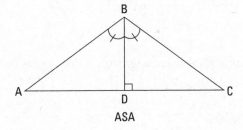

ASA

The line BD bisects the angle ABC, so angle ABD has the same measure as angle DBC. Additionally, BD is perpendicular to AC, so angles ADB and CDB are both right angles. Thus, two angles in triangle ABC have the same measures as two angles in CBD, so the third angles in each triangle are also equivalent. Finally, the two triangles ABD and CBD share the line BD. Therefore, ABD and CBD share one pair of sides that have equal lengths, and three pairs of angles that are equal in measure. As a result, by either ASA or SAA, the two triangles are congruent.

TWO WAYS TO MAKE MISTAKES WITH NON-CONGRUENT TRIANGLES (SSA AND AAA)

WARNING

Two close cases exist that look like they prove congruency but actually don't:

Side-side-angle (*SSA*)

Angle-angle-angle (*AAA*)

To see why *SSA* fails to prove that a pair of triangles are congruent, look at the following figure. Clearly, triangles ABC and BCD are not congruent: In fact, ABC includes all of BCD and more. Now, consider that these two triangles both share angle C and side BC. Furthermore, $AB = BD$, so both of these triangles have additional sides of equivalent length. Therefore, triangles ABC and BCD are non-congruent, and yet *SSA* holds for them.

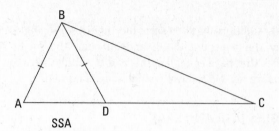

SSA

To understand why *AAA* fails to prove that two triangles are congruent, consider the following figure. Triangles *PQR* and *STU* are obviously different sizes, so they're not congruent. Yet the two triangles have three pairs of angles that are equal in measure.

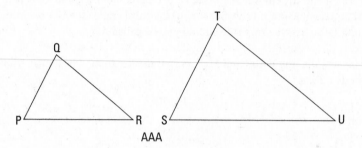

AAA

Although *AAA* doesn't prove that a pair of triangles are congruent, it does prove that they are similar. I discuss this concept in the next section.

Calculating with similar triangles

When a pair of triangles have equal corresponding angles, they are called *similar triangles*. Similar triangles have corresponding sides that are all proportionate. A typical ACCUPLACER question will ask you to identify a pair of similar triangles and then use their proportionality to find the length of one side. For example:

EXAMPLE

In this figure, the lines containing *AB* and *DE* are parallel, with *AE* intersecting *BD* at point *C*. If the lengths of the line segments are as shown, what is the length of *DE*?

The trick here is to recognize that triangles *ABC* and *CDE* are similar. To see why this is so, note that angles *BCA* and *DCE* are vertical angles, so they're equal. Angles *A* and *E* are equal, as are angles *B* and *D*. Thus, the two triangles are similar by *AAA*, so their corresponding sides are proportional. *AC* = 4 and *CE* = 10, so corresponding sides of the two triangles are in a 2:5 ratio. So you can make a proportion as follows:

$$\frac{2}{5} = \frac{6}{x}$$
$$2x = 30$$
$$x = 15$$

Using circle equations in the coordinate plane

REMEMBER

The simplest equation for a circle in the coordinate plane is as follows, for a circle centered at the origin $(0,0)$ with a radius of r:

$$x^2 + y^2 = r^2$$

EXAMPLE

So, for example, the following three equations produce circles centered at the origin with radii of 3, 6, and 10, respectively:

$$x^2 + y^2 = 9 \qquad x^2 + y^2 = 36 \qquad x^2 + y^2 = 100$$

REMEMBER

This formula generalizes to a circle centered anywhere on the xy-plane. Here is the general formula, for a circle centered at the point (h,k) with a radius of r:

$$(x-h)^2 + (y-k)^2 = r^2$$

EXAMPLE

To see how this works, suppose you want to find the equation of a circle centered at $(3,-4)$ with a radius of 5. Plug these values into the equation, then simplify:

$$(x-3)^2 + [y-(-4)]^2 = 5^2$$
$$(x-3)^2 + (y+4)^2 = 25$$

The resulting circle is shown in the following figure:

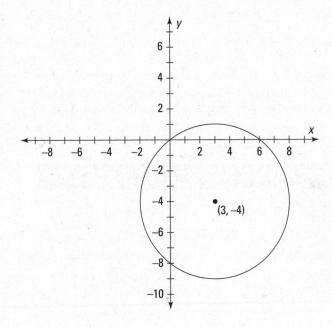

Trigonometry

Trigonometry is the study of triangles, especially right triangles. Trigonometry introduces a new set of ideas called the trigonometric ratios, or trig ratios for short. Trig ratios are a powerful tool for solving a wide variety of problems in math, and trigonometry is pretty much the most advanced math you'll face on the ACCUPLACER.

Knowing the trig ratios

The six *trigonometric ratios* are functions of a reference angle in a right triangle. For example, consider the 3–4–5 right triangle shown in this figure:

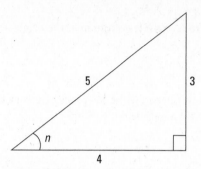

I've labeled the smaller of the two acute angles in this triangle as the reference angle n. The position of this angle allows me to label each of the three sides of the triangle:

The *opposite* side is the leg of the triangle that doesn't touch angle n.

The *adjacent* side is the leg of the triangle that touches n.

The *hypotenuse* is the longest side of the triangle.

With the sides of the triangle labeled in this way, there are six possible ratios that can be made:

Each of these trig ratios is a function that takes an angle x as its input and produces a real number as its output, as shown in the third column of Table 19-1. Using the angle n from the 3–4–5 right triangle as an input produces the six values shown in the fourth column of the table.

TABLE 19-1 ## The Six Trig Ratios

Name	Ratio	General function	Function with input n
Sine x	$\dfrac{\text{Opposite}}{\text{Hypotenuse}}$	$\sin x = \dfrac{O}{H}$	$\sin n = \dfrac{3}{5}$
Cosine x	$\dfrac{\text{Adjacent}}{\text{Hypotenuse}}$	$\cos x = \dfrac{A}{H}$	$\cos n = \dfrac{4}{5}$
Tangent x	$\dfrac{\text{Opposite}}{\text{Adjacent}}$	$\tan x = \dfrac{O}{A}$	$\tan n = \dfrac{3}{4}$
Cotangent x	$\dfrac{\text{Adjacent}}{\text{Opposite}}$	$\cot x = \dfrac{A}{O}$	$\cot n = \dfrac{4}{3}$
Secant x	$\dfrac{\text{Hypotenuse}}{\text{Adjacent}}$	$\sec x = \dfrac{H}{A}$	$\sec n = \dfrac{5}{4}$
Cosecant x	$\dfrac{\text{Hypotenuse}}{\text{Opposite}}$	$\csc x = \dfrac{H}{O}$	$\csc n = \dfrac{5}{3}$

An important insight is that when you know the value of one trig ratio for a given angle, you can find the remaining five values. For example:

If $\sin p° = \dfrac{2}{3}$, what does the cotangent of $p°$ equal?

EXAMPLE

Because $\sin x = \dfrac{O}{H}$, you can use the value 2 for the opposite side and 3 for the hypotenuse. Using this information, draw a picture of the triangle you're working with:

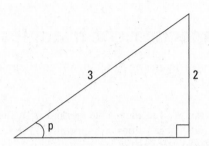

Now, use the Pythagorean theorem to find the length of the triangle's adjacent side:

$$a^2 + b^2 = c^2$$
$$2^2 + b^2 = 3^2$$
$$4 + b^2 = 9$$
$$b^2 = 5$$
$$b = \sqrt{5}$$

The length of the adjacent side is $\sqrt{5}$, so plug this value into the equation for the cotangent:

$$\cot p° = \dfrac{A}{O} = \dfrac{\sqrt{5}}{2}$$

Remembering not to forget with SOH-CAH-TOA

Most students who have done a little trigonometry remember the mnemonic device SOH–CAH–TOA as a trick for remembering the first three trig ratios:

REMEMBER

Abbreviation	Stands for
SOH	$\sin x = \dfrac{O}{H}$
CAH	$\cos x = \dfrac{A}{H}$
TOA	$\tan x = \dfrac{O}{A}$

TIP

If you can remember SOH-CAH-TOA, you can build these three trig ratios. And from there, you can use the three reciprocal identities, which I discuss later in this section, to recall the remaining three trig ratios. For now, simply remember that the following pairs are reciprocals of each other:

Sine and cosecant

Cosine and secant

Tangent and cotangent

Getting friendly with the special right triangles

The two special right triangles are the 45-45-90 triangle and the 30-60-90 triangle. Both of these triangles yield results for the six trig ratios.

If you've already had some exposure to trigonometry, you may have worked with these values before, and perhaps have some of them memorized. These values — especially those for the sine, cosine, and tangent functions — are useful on the ACCUPLACER.

If you don't have them memorized, the next best thing is to understand how to build them from scratch by sketching and labeling the triangle you need and then calculating the trig function. In this section, I show you how to do this.

The 45-45-90 right triangle

The 45-45-90 right triangle has sides whose lengths are in the proportion $1:1:\sqrt{2}$.

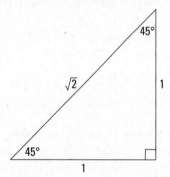

TIP

Before moving on, take a moment to notice the angles and sides of this triangle. Better yet, close the book and see if you can redraw it from memory.

Using this triangle (with an assist from trusty SOH-CAH-TOA), you can list all the trig ratios for an input of 45°, as shown in Table 19-2.

These values are very useful when solving trig problems, so do your best to memorize them — or, at least, the values for sine, cosine, and tangent.

TABLE 19-2 **Trig Functions for Special Right Triangles**

30°	45°	60°
$\sin 30° = \dfrac{1}{2}$	$\sin 45° = \dfrac{\sqrt{2}}{2}$	$\sin 60° = \dfrac{\sqrt{3}}{2}$
$\cos 30° = \dfrac{\sqrt{3}}{2}$	$\cos 45° = \dfrac{\sqrt{2}}{2}$	$\cos 60° = \dfrac{1}{2}$
$\tan 30° = \dfrac{\sqrt{3}}{3}$	$\tan 45° = 1$	$\tan 60° = \sqrt{3}$
$\cot 30° = \sqrt{3}$	$\cot 45° = 1$	$\cot 60° = \dfrac{\sqrt{3}}{3}$
$\sec 30° = \dfrac{2\sqrt{3}}{3}$	$\sec 45° = \sqrt{2}$	$\sec 60° = 2$
$\csc 30° = 2$	$\csc 45° = \sqrt{2}$	$\csc 60° = \dfrac{2\sqrt{3}}{3}$

The 30-60-90 right triangle

The 30-60-90 right triangle has sides whose lengths are in the proportion $1 : \sqrt{3} : 2$.

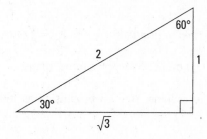

TIP

As with the 45-45-90 triangle, study this triangle for a minute, and then close the book and see if you can redraw it from memory.

Using this triangle (again, with SOH-CAH-TOA to help), you can list all the trig ratios for inputs of 30° and 60°, as shown in Table 19-2. As you think through this idea, remember that in trigonometry, only the Hypotenuse of a triangle is fixed — it's always the longest side of a triangle. In contrast, the Opposite and Adjacent sides of a triangle aren't fixed, but instead are *with respect to the angle*.

So, when you're working with the 30° angle, the Opposite side equals 1 and the Adjacent side equals $\sqrt{3}$. On the other hand, when you're working with the 60° angle, the Opposite side equals $\sqrt{3}$ and the Adjacent side equals 1.

This information is summarized in Table 19-2.

These values are worth knowing — especially the values for sine, cosine, and tangent — so committing them to memory will be time well spent.

Understanding radian measure

REMEMBER

Radian measure is an alternative way to measure angles, distinct from degrees. Radian measure derives in a natural way from facts about circles that you know from geometry.

To begin understanding radian measure, recall the circumference formula for a circle:

$$C = 2\pi r$$

Next, consider a unit circle — that is, a circle with a radius of 1. Plugging 1 for r into this formula results in the following simplified equation:

$$C = 2\pi$$

Next, remember that the circumference of a circle has 360°, so substitute this value in place of C:

$$360° = 2\pi$$

This formula provides the basis for radian measure.

$$360° = 2\pi \text{ radians}$$

REMEMBER

You can simplify it by dividing both sides by 2:

$$180° = \pi \text{ radians}$$

REMEMBER

Now, divide this value by 2, 3, 4, and 6 to find equivalences for the important angles that arise in trigonometry:

$$90° = \frac{\pi}{2}\text{radians} \qquad 60° = \frac{\pi}{3}\text{radians} \qquad 45° = \frac{\pi}{4}\text{radians} \qquad 30° = \frac{\pi}{6}\text{radians}$$

REMEMBER

You can use the equivalence $180° = \pi$ to convert any number of degrees into radians.

For example, suppose you want to convert 25° to radians. Write a proportion as follows, and solve for x:

$$\frac{180}{\pi} = \frac{25}{x}$$
$$180x = 25\pi$$
$$x = \frac{25\pi}{180} = \frac{5\pi}{36}$$

Therefore, $25° = \frac{5\pi}{36}$ radians.

Measuring arc length

Arc length is the distance around a portion of a circle, from one point on that circle to another. This formula for arc length, in my opinion, is the simplest to remember and use:

REMEMBER

$$\text{Arc length} = \text{Radius} \times \text{Radians}$$

EXAMPLE

In the following figure, Points X and Y are on the circle centered at O, which has a radius of 6, and $\angle XOY$ measures 100°. What is the arc length from X to Y?

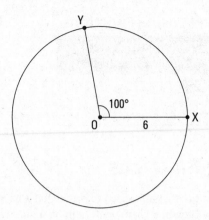

To begin, convert the measurement of $\angle XOY$ from degrees to radians:

$$100° \times \frac{\pi}{180°} = \frac{5\pi}{9}$$

Now, plug the radius and radians into the formula and simplify:

$$\text{Arc length} = 6 \times \frac{5\pi}{9} = \frac{30\pi}{9} = \frac{10\pi}{3}$$

So the arc length from X to Y is $\frac{10\pi}{3}$ units.

Graphing trig relationships

In Chapter 17, I introduce you to the two most important trig parent functions:

$$f(x) = \sin x \qquad f(x) = \cos x$$

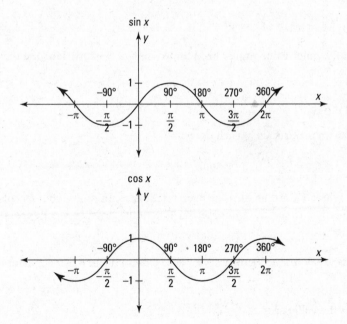

Table 19-3 provides a handy reference for the most important values for the sine and cosine functions from 0° to 360°.

TABLE 19-3 **Values of Key Trig Ratios for Sine and Cosine Functions**

$0° = 0$	$90° = \dfrac{\pi}{2}$	$180° = \pi$	$270° = \dfrac{3\pi}{2}$	$360° = 2\pi$
$\sin 0° = 0$	$\sin 90° = 1$	$\sin 180° = 0$	$\sin 270° = -1$	$\sin 360° = 0$
$\cos 0° = 1$	$\cos 90° = 0$	$\cos 180° = -1$	$\cos 270° = 0$	$\cos 360° = 1$

Evaluating equivalent trig functions

Trigonometry includes a wide variety of *trig identities* that allow you to prove that a pair of apparently different trig functions are equivalent. Knowing them can help you do well on the ACCUPLACER. At a minimum, here are the trig identities you should know cold.

Complementary angles identity

REMEMBER

If I had to spend the rest of my life on a desert island and could only take one trig identity with me, it would be this one:

$$\sin x = \cos(90° - x)$$

In my experience, this is by far the most common trig identity to appear on standardized tests, and it allows you to easily solve problems that look really tough. For example:

If $\cos 65° = \sin n°$, what is the value of n?

Because $90° - 65° = 25°$, the identity tells you that $\sin 25° = \cos 65°$, so $n = 25°$.

Analogous identities also exist for each of the other pairs of co-functions (tangent and cotangent, secant and cosecant).

$$\tan x = \cot(90° - x) \quad \sec x = \csc(90° - x)$$

The proviso here is that the input value x must be in the domain of the function, and the function itself must not equal 0.

Reciprocal identities

REMEMBER

Three pairs of trig ratios are reciprocals of each other:

$$\sin x = \frac{1}{\csc x} \qquad \cos x = \frac{1}{\sec x} \qquad \tan x = \frac{1}{\cot x}$$

The usefulness of these three identities can't be overstated. For one thing, your calculator may not even include the cotangent, secant, and cosecant functions, because the information they provide is redundant.

You can combine this identity with the identity $\sin x = \cos(90° - x)$ to answer more complex questions. For example:

EXAMPLE

If $\sin r° = \dfrac{1}{\sec 60°}$ what is the value of r?

Begin by using the reciprocal identity $\cos x = \dfrac{1}{\sec x}$ to simplify the right side of the equation:

$$\sin r° = \cos 60°$$

Now, the identity $\sin x = \cos(90° - x)$ (which I introduce earlier in this section) allows you to rewrite the equation as follows:

$$\sin 30° = \cos 60°$$

Therefore, $r = 30$.

Ratio identities

The two ratio identities are as follows:

$$\tan x = \frac{\sin x}{\cos x} \qquad \cot x = \frac{\cos x}{\sin x}$$

REMEMBER

These two identities are different only in that *tan x* and *cot x* are reciprocals (as I discuss in the previous section). They can also be surprisingly handy, especially when used in conjunction with the reciprocal identities (which I introduce earlier in this section). For example:

Which of the following is equivalent to $\tan x \csc x$?

EXAMPLE

(A) $\sin x$

(B) $\cos x$

(C) $\cot x$

(D) $\sec x$

Begin by applying both $\tan x = \frac{\sin x}{\cos x}$ and $\sin x = \frac{1}{\csc x}$, and then simplify by cross-canceling factors:

$$\tan x \csc x = \frac{\sin x}{\cos x} \cdot \frac{1}{\sin x} = \frac{1}{\cos x}$$

Now, apply the reciprocal identity $\cos x = \frac{1}{\sec x}$:

$$= \sec x$$

So the correct answer is D.

Pythagorean identities

The Pythagorean identities are as follows:

$$\sin^2 x + \cos^2 x = 1$$
$$\tan^2 x + 1 = \sec^2 x$$
$$\cot^2 x + 1 = \csc^2 x$$

REMEMBER

The first of these three identities is by far the most commonly used, so be sure to memorize it. The notation $\sin^2 x$ means $(\sin x)^2$ — that is, $\sin x \sin x$. Similarly, $\cos^2 x$ means $(\cos x)^2$ — that is, $\cos x \cos x$.

Which of the following is equivalent to $\frac{\sin x}{\csc x}$?

EXAMPLE

(A) $\sin^2 x - 1$

(B) $1 - \sin^2 x$

(C) $\cos^2 x - 1$

(D) $1 - \cos^2 x$

To begin, use the reciprocal identity $\sin x = \dfrac{1}{\csc x}$ as follows:

$$\frac{\sin x}{\csc x} = \sin x \sin x = \sin^2 x$$

Now, subtract $\cos^2 x$ from both sides of the Pythagorean identity:

$$\sin^2 x + \cos^2 x = 1$$
$$\sin^2 x = 1 - \cos^2 x$$

These calculations show that both $\dfrac{\sin x}{\csc x}$ and $1 - \cos^2 x$ are equal to $\sin^2 x$, so the correct answer is D.

Solving trig equations

Although solving trig equations can get pretty complicated, the equations that you see on the ACCUPLACER are likely to yield to a few basic tricks. In some cases, you can solve a trig equation by simply being familiar with the trig ratios for the special right triangles that I discuss earlier in this chapter. For example:

EXAMPLE

If $\cos x° = \dfrac{\sqrt{3}}{2}$, and $0 < x < 90°$, what is the value of x?

The familiar value $\dfrac{\sqrt{3}}{2}$ should remind you that $\cos 30° = \dfrac{\sqrt{3}}{2}$, so the answer is 30°.

TECHNICAL STUFF

Although the information $0 < x < 90$ seems unnecessary to solving the problem, it's needed here to rule out possible values of x outside this range. For example, adding multiples of 360 would also yield $\dfrac{\sqrt{3}}{2}$, so other possible values of x outside this range could be 390, 750, 1110, and so forth.

In a slightly more difficult problem, you may need to apply your knowledge of the basic trig identities. For example:

EXAMPLE

Given that $-90° \le x \le 90°$, what is the value of x if $\dfrac{\tan x°}{\sec x°} = 1$?

Begin by simplifying $\dfrac{\tan x°}{\sec x°}$ using identities:

$$\frac{\tan x°}{\sec x°} = \tan x° \cos x° = \frac{\sin x°}{\cos x°} \cos x° = \sin x°$$

Now, substitute $\sin x°$ back into the equation:

$$\sin x° = 1$$

At this point, you may remember that $\sin 90° = 1$, so $x = 90°$.

Another type of problem may require you to isolate the trig values on one side of the equation and the numerical values on the other. For example:

EXAMPLE

Given that $0 \le x \le 180°$, what is the value of x if $2\cot x° = \csc x°$?

A good first step here is to use algebra to separate the trig values from the value 2:

$$2\cot x° = \csc x°$$
$$\frac{\cot x°}{\csc x°} = \frac{1}{2}$$

Now, use identities to simplify the left side of the equation:

$$\frac{\cot x^\circ}{\csc x^\circ} = \sin x^\circ \cot x^\circ = \sin x^\circ \frac{\cos x^\circ}{\sin x^\circ} = \cos x^\circ$$

Substitute $\cos x^\circ$ back into the equation:

$$\cos x^\circ = \frac{1}{2}$$

Again, the last step is to recognize that $\cos 60^\circ = \frac{1}{2}$, so the answer is 60°.

Another complication to solving a trig equation may be the introduction of radian measure into the mix. For example:

EXAMPLE

Which of the following is a possible value for x in the equation $\frac{1}{\csc x} = \frac{\sqrt{2}}{2}$?

(A) $\frac{\pi}{2}$

(B) $\frac{\pi}{3}$

(C) $\frac{\pi}{4}$

(D) $\frac{\pi}{6}$

As usual, simplifying the trig function is always a good idea. Here, apply the ratio identity $\sin x = \frac{1}{\csc x}$:

$$\sin x = \frac{\sqrt{2}}{2}$$

Your knowledge of the trig ratios for the special triangle reminds you that $\sin 45^\circ = \frac{\sqrt{2}}{2}$. In radians, $45^\circ = \frac{\pi}{4}$, so the correct answer is C.

Applying inverse trig functions

In the previous section, you solve trig equations using your knowledge of special right triangles. *Inverse trig functions* provide a formal way of "undoing" trig functions, which can be useful when the angle you're working with is unfamiliar.

To show how this works, I'll start with a familiar angle:

$$\sin 30^\circ = \frac{1}{2}$$

In this trig function, an input value of 30° produces the output value of $\frac{1}{2}$. So, you can think of the inverse sine as a function that accepts an input value of $\frac{1}{2}$ and produces the output value of 30°:

$$\sin^{-1} \frac{1}{2} = 30^\circ$$

The notation $\sin^{-1}$ means "inverse sine." Be careful not to think of this notation as the reciprocal of sine, which you know from the reciprocal identities to be the secant function.

WARNING

This function provides a formal way to solve problems that involve unfamiliar angles. For example, here's a new question about a triangle you've seen before:

EXAMPLE

In this triangle, what is the value of $n°$ to the nearest whole number of degrees?

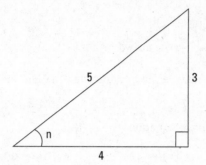

To begin, use your knowledge of trig to write an equation that uses $n°$ as its input. Here, I choose the sine function:

$$\sin n° = \frac{3}{5}$$

To solve this problem for $n°$, undo the sine function by applying the inverse sine function to both sides of the equation:

$$\sin^{-1}(\sin n°) = \sin^{-1}\frac{3}{5}$$

On the left side of the equation, the inverse sine function cancels out the sine function:

$$n° = \sin^{-1}\frac{3}{5}$$

To complete the problem, you need to evaluate the trig expression $\sin^{-1}\frac{3}{5}$ as a number of degrees. In practice on the ACCUPLACER, if you need to calculate an inverse trig function, you'll be given a calculator to do it:

$$\sin^{-1}\frac{3}{5} \approx 37°$$

Thus, the smaller angle in a 3–4–5 right triangle measures approximately 37°.

Using the law of sines and cosines

Earlier in this chapter, you discover four ways to show that a pair of triangles are congruent: *SSS*, *SAS*, *AAS*, and *ASA*. Another way to think of these four cases is that each *determines* all the sides and angles of a single triangle. For example, *SSS* tells you that if you know the lengths of all three sides of a triangle, the measurements of the three angles are determined.

The *law of sines* and the *law of cosines* are two formulas that allow you to plug in the information you know about a triangle whose dimensions are determined to find the information you don't know.

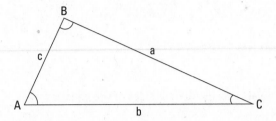

Notice that in this figure, each side (*a*, *b*, and *c*) is opposite its corresponding angle (*A*, *B*, and *C*). The following laws hold true for *all* such triangles:

Law of sines:

$$\frac{\sin A}{a} = \frac{\sin B}{b} = \frac{\sin C}{c}$$

Law of cosines:

$$a^2 + b^2 - 2ab\cos C = c^2$$

TIP

Here's a quick way to know which law to use first: When you have two or more sides (*SSS* and *SAS*), use the law of cosines; otherwise, use the law of sines. This information is shown in this table:

Triangle case	First law to use
SSS	Law of cosines
SAS	Law of cosines
SAA	Law of sines
ASA	Law of sines

To see how these formulas work, consider the following example:

EXAMPLE

In this triangle, what is the value of *x*° to the nearest whole number of degrees?

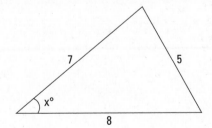

You know the lengths of three sides (*SSS*), so use the law of cosines. Using $x°$ for your *C* value, $c = 5$, and you can allow *a* and *b* to stand in for 7 and 8 in any order you like:

$$8^2 + 7^2 - 2(8)(7)\cos x° = 5^2$$
$$113 - 112\cos x° = 25$$
$$-112\cos x° = -88$$
$$\cos x° = \frac{-88}{-112}$$
$$\cos x° \approx 0.786$$

To complete the problem, apply an inverse cosine to both sides of the equation:

$$\cos^{-1}(\cos x°) = \cos^{-1} 0.786$$
$$x° \approx 38°$$

Chapter **20**

Advanced Algebra and Functions Practice Questions

Ready to practice the toughest math on the ACCUPLACER? The practice questions in this chapter focus on all the skills you gain from Chapters 17, 18, and 19. Detailed answers follow at the end of the chapter.

Advanced Algebra and Functions Practice Questions

These practice questions include both multiple choice and open-ended questions. If you get stuck on a question or just want to check you answer, flip to the end of the chapter for both the answer and a detailed explanation of how to find it.

Factoring polynomials

Fully factor each of the following polynomials.

1. $6x^5y^4 + 18x^3y^3 - 30xy^5$

2. $x^2 + 7x - 18$

3. $-2x^2 - 18x + 44$

4. $4x^2 - 15x + 9$

5. $x^4 - 25y^2$

6. $64x^3 + 27y^3$

7. $x^3 - 5x^2 + 3x - 15$

8. $x^3 - 11x^2 - 4x + 44$

9. $24x^3 - 54xy^2$

10. $16x^2 - 40x - 24$

11. $96x^4y^2z^3 - 12xy^5z^3$

12. $2x^3 - 4x^2 - 18x + 36$

Function notation

Evaluate each of these functions as either a value or an expression. Use the following function definitions for these questions:

$$f(x) = 5x - 2$$
$$g(x) = x^2 + 3$$

1. $f(3) =$

2. $g(-5) =$

3. $f(-4) + 3 =$

4. $-2g(7) =$

5. $\dfrac{2f(1) - 3}{-g(6)} =$

6. $f(k - 1) =$

7. $4g(k + 2) - 3 =$

8. $(f + g)(2t + 1) =$

9. $(fg)(-1) =$

10. $f(g(-3)) =$

11. $g(f(p - 1)) - 10 =$

12. $3(f \circ g)(5) + 1 =$

Parent functions and transformations

1. Write out the parent functions for the four polynomial functions from degree 1 to degree 4:

   (A) Linear function

   (B) Quadratic function

   (C) Cubic function

   (D) Quartic function

2. Write out the parent functions for the following functions:

   (A) Exponential function (with a base of 2)

   (B) Logarithmic function (with a base of 2)

3. Write out the parent functions for the two trig functions:

   (A) Sine function

   (B) Cosine function

4. Write out the parent functions for the following three functions:

   (A) Absolute value function

   (B) Rational function

   (C) Radical (square root) function

5. Transform the function $f(x) = x^2$, stretching it vertically by a factor of 2, and then translating it down 4 and right 1.

6. Transform the function $f(x) = x^2$, compressing it vertically by a factor of 3, translating it up 7, and then reflecting it vertically across the x-axis.

7. Transform the function $f(x) = x^2$, translating it left 2, stretching it vertically by a factor of 5, and reflecting it vertically across the x-axis.

8. Consider the function $g(x) = -3(x+2)^2 - 9$ as a transformation of the function $f(x) = x^2$:

   (A) What is the vertical translation for this function?

   (B) What is the horizontal translation?

   (C) What is the dilation (stretch or compression), and by what factor?

   (D) What is the reflection?

9. Consider the function $g(x) = -\frac{1}{7}(x-5)^3 + 10$ as a transformation of the function $f(x) = x^3$:

   (A) What is the vertical translation for this function?

   (B) What is the horizontal translation?

   (C) What is the dilation (stretch or compression), and by what factor?

   (D) What is the reflection?

Polynomial functions

1. Which of the following polynomials displays end behavior that is negative for both very large and very small values of x?

(A) $y = x^5 - x^2 + 3$

(B) $y = x^6 + 3x^4 - 9x$

(C) $y = -5x^8 + x^7 - 4x + 10$

(D) $y = -x^9 - 2x^7 + x^6 - 4x + 11$

2. If the polynomial $y = mx^n + x - 1$ displays positive end behavior for both very large and very small values of x, which of the following could be true.

(A) m is positive and n is even.

(B) m is positive and n is odd.

(C) m is negative and n is even.

(D) m is negative and n is odd.

3. Which of the following functions includes the point $(0, 5)$ when graphed on the xy-plane?

(A) $y = 3x^2 + x - 5$

(B) $y = 5x^3 - 4x - 1$

(C) $y = -3x^3 - 2x^2 + 5$

(D) $y = -7x^3 + 6x^2 + 5x$

4. Which of the following is NOT a root of the function $y = (x+3)(x+7)(x-4)(-3x+1)$?

(A) 3

(B) 4

(C) −7

(D) $\frac{1}{3}$

Quadratic equations and functions

1. What two values of x satisfy the equation $2x^2 - 7 = 0$?

2. Solve for x: $3x^2 + 10x = 0$.

3. What are the two solutions for $x^2 - 9x + 20 = 0$?

4. Find the two values of x that satisfy the equation $x^2 + 3x - 28 = 0$.

5. Use the quadratic formula to find the two solutions of $3x^2 - 4x - 6 = 0$.

6. Solve the following system of quadratic equations: $y = 3x^2 - 4x + 17$ and $y = 4x^2 - 13x - 5$.

7. Which of the following could be an equation of the function matching this figure?

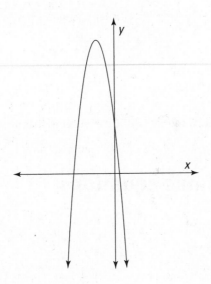

(A) $y = x^2 + 5x + 6$

(B) $y = 2x^2 - 4x - 6$

(C) $y = -x^2 - 8x - 6$

(D) $y = -2x^2 - 9x + 6$

8. What is the axis of symmetry for the function $y = x^2 - 10x + 21$?

9. What are the coordinates of the vertex for function $y = -5x^2 - 20x + 1$?

10. Convert the function $y = 5(x - 2)^2 - 11$ to standard form.

11. Rewrite the function $y = x^2 + 6x - 3$ in vertex form.

12. What are the roots of the function $y = x^2 - 5x - 36$?

13. What is the axis of symmetry of the function $y = (x + 3)(x - 9)$?

14. What are the solutions of the function $y = (4x - 5)(6x + 1)$?

15. What are the roots of the quadratic function $y = 4x^2 - x - 2$?

16. Which of the following is the complete list of solutions for the function $y = x^2 + 14x + 49$?

(A) 7 and −7

(B) 7 only

(C) −7 only

(D) No real solutions

17. Which of the following is a zero of the function $y = (3x + 5)^2$?

(A) $\frac{3}{5}$

(B) $\frac{5}{3}$

(C) $-\frac{3}{5}$

(D) $-\frac{5}{3}$

18. The function $y = x^2 - bx + 10$ has just one real root at $(\sqrt{10}, 0)$. Assuming that $b > 0$, what is the value of b?

Exponential and logarithmic equations

1. What does $9^{\frac{5}{2}}$ equal?

2. Which of the following is equivalent to $x^{\frac{1}{3}} x^{\frac{2}{7}}$?

(A) $\sqrt[21]{x^2}$

(B) $\sqrt[21]{x^{13}}$

(C) $\sqrt[13]{x^{21}}$

(D) $\sqrt[3]{x^{21}}$

3. Which of the following is equivalent to the expression $\dfrac{2^{5n+2}}{8^n}$?

(A) 2^{2n+1}

(B) 2^{2n+2}

(C) 4^{2n-1}

(D) 8^{2n-2}

4. What is the value of n if $(\frac{1}{9})^{2n} = \sqrt{3}^{\,n+4}$?

5. A video on IToob had just 150 views at the beginning of last year. In each of the next 12 months, the video tripled its viewership. Which of the following numbers is the closest to the number of views that the video had as the year ended?

(A) 80,000

(B) 800,000

(C) 8,000,000

(D) 80,000,000

6. The value of a company at the time of its IPO in 2013 grew by 17% per year for 5 years in a row. If the value of the company was $1.9 million in 2013, what was its approximate value in 2018?

(A) $2.5 million

(B) $3 million

(C) $3.5 million

(D) $4 million

7. An investment property valued at $455,000 in 2010 lost 14% of its value in each of the next three years. Approximately how much was the property worth in 2013?

   (A) $270,000

   (B) $280,000

   (C) $290,000

   (D) $300,000

8. What is the value of x given that $\log_9 x = 2$?

9. If $\log_k 8 = -1$, what does k equal?

10. What is the value of n when $\log_8 2 = n$?

11. If $7^{6x} = 10$, which of the following is the value of x?

   (A) $\dfrac{\log_6 7}{10}$

   (B) $\dfrac{\log_{10} 6}{7}$

   (C) $\dfrac{\log_{10} 7}{6}$

   (D) $\dfrac{\log_7 10}{6}$

Radical and rational equations

1. When expressed as a decimal, which of the following values falls between 7 and 8?

   (A) $\sqrt{60}$

   (B) $\sqrt{70}$

   (C) $\sqrt{80}$

   (D) $\sqrt{90}$

2. What is the simplified value of $\dfrac{\sqrt{6}\sqrt{15}}{\sqrt{10}}$?

3. Simplify $\sqrt{63}$.

4. When you simplify $\dfrac{\sqrt{35}}{\sqrt{20}}$, what is the result?

5. What is $\sqrt{24} + \sqrt{54}$?

6. When you rationalize the denominator of $\dfrac{15}{\sqrt{30}}$ and simplify, what is the result?

7. What is the domain of the function $f(x) = -\sqrt{4x-3} + 7$?

8. What is the range of the function $f(x) = -\sqrt{4x-3} + 7$?

9. What is the domain of $f(x) = \dfrac{1}{x^2 + 13x - 30} + 5$?

10. What is the range of $f(x) = \dfrac{1}{x^2 + 13x - 30} + 5$?

Geometry

1. A golf ball has a radius of about 1.7 inches. What is its volume in cubic inches, to the nearest tenth of a cubic inch? (Use 3.14 to approximate π.)

2. The Pyramid at Cholula in Mexico has a base whose side measures 450 meters and a height of 66 meters. To the nearest cubic meter, what is its volume?

3. A tent that has the shape of a cone has a base with a radius of 6 feet and a height of 9 feet. What is its volume in cubic feet, to the nearest whole cubic foot? (Use 3.14 to approximate π.)

4. In the following figure, what is the value of x?

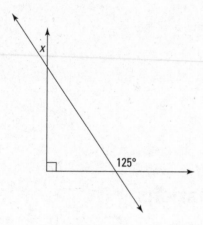

5. In the following figure, lines k and m are parallel, and $x + y = 215$. What is the value of z?

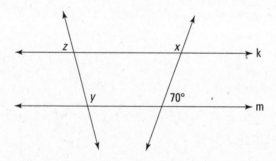

6. In the following figure, all the following relationships are sufficient to prove that triangle *ABD* and triangle *CBD* are congruent EXCEPT

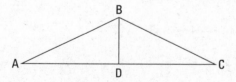

(A) $AB = BC$ and $AD = CD$

(B) $AB = BC$ and $\angle ABD \cong \angle CBD$

(C) $AD = CD$ and $\angle ABD \cong \angle CBD$

(D) $\angle ABD \cong \angle CBD$ and $\angle A \cong \angle C$

7. In the following figure, $\angle G \cong \angle T$ and $GF = RT$. To show that the two triangles are congruent, each of the following relationships would be sufficient EXCEPT

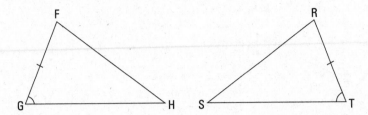

(A) $GH = ST$

(B) $HF = RS$

(C) $\angle H \cong \angle S$

(D) $\angle F \cong \angle R$

8. In the following figure, $\angle WXY = 90°$ and WY is perpendicular XZ. Which of the following pairs of triangles are similar?

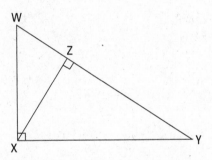

(A) Triangle WXY and triangle XZY

(B) Triangle WXY and triangle WZX

(C) Triangle WZX and triangle XZY

(D) All three pairs of triangles are similar.

9. What is the equation of a circle with a radius of 13 centered at the origin of the xy-plane?

10. What is the equation of a circle with a radius of 4 centered at the point $(-2,7)$?

Trigonometry

1. What are the following values in terms of O, A, and H:

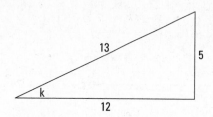

    (A) $\sin k =$

    (B) $\cos k =$

    (C) $\tan k =$

    (D) $\cot k =$

    (E) $\sec k =$

    (F) $\csc k =$

2. If $\cos q° = \frac{6}{7}$, what does the tangent of $q°$ equal?

3. What are the following values?

    (A) $\sin 45°$

    (B) $\cos 45°$

    (C) $\tan 45°$

4. What are the following values?

    (A) $\sin 30°$

    (B) $\cos 30°$

    (C) $\tan 30°$

    (D) $\sin 60°$

    (E) $\cos 60°$

    (F) $\tan 60°$

5. What is the value of $40°$ in radian measure?

6. What is the value of $\frac{7\pi}{9}$ radians in degrees?

7. In the following figure, Points A and B are on the circle centered at O, which has a radius of 18, and $\angle AOB$ measures 160°. What is the minor arc length from A to B, counterclockwise around the circle?

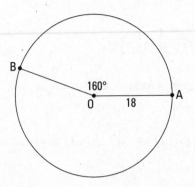

8. If $\cos k° = \sin 15°$, and $0 < k < 90$, what is the value of k?

9. The trig expression $\sin x \cos x \tan x$ equals which of the following?

(A) $1 + \sin^2 x$

(B) $1 - \sin^2 x$

(C) $1 + \cos^2 x$

(D) $1 - \cos^2 x$

10. In the trig expression $\sin x \cos x \cot x \sec x = \dfrac{\sqrt{3}}{2}$, x equals which of the following?

(A) $\dfrac{\pi}{2}$

(B) $\dfrac{\pi}{3}$

(C) $\dfrac{\pi}{4}$

(D) $\dfrac{\pi}{6}$

11. If $3 \sin n° \sec n° = \sqrt{3}$, and $0 < n < 90$, what is the value of n?

12. In the following triangle, what is the value of x, rounded to 1 decimal place? (You may use a calculator to solve this problem.)

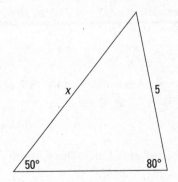

Answers

Here are the answers to all the questions in this chapter. Each answer includes a detailed explanation of how to find that solution.

Answers to factoring polynomials

1. **$6xy^3(x^4y + 3x^2 - 5y^2)$.** Factor out the greatest common factor of $6xy^3$:

$$6x^5y^4 + 18x^3y^3 - 30xy^5 = 6xy^3(x^4y + 3x^2 - 5y^2)$$

2. **$(x+9)(x-2)$.** Find a pair of numbers that multiply to –18 and add to 7 (that is, 9 and –2) to use for quadratic factoring:

$$x^2 + 7x - 18 = (x+9)(x-2)$$

3. **$-2(x+11)(x-2)$.** Begin by factoring out the GCF of –2, then find a pair of numbers that multiply to –22 and add to 9 (that is, 11 and –2) to do quadratic factoring:

$$-2x^2 - 18x + 44 = -2(x^2 + 9x - 22) = -2(x+11)(x-2)$$

4. **$(4x-3)(x-3)$.** The expression $4x^2 - 15x + 9$ is quadratic. To factor it, use the analogous case of $x^2 - 15x + 36$ (multiplying 4 by 9, and using this value as the constant, as I explain in Chapter 17):

$$x^2 - 15x + 36 = (x-3)(x-12)$$

Use this answer as a starting point to find the actual answer. Next, because the original expression begins with 4x, replace both xs in the result above with 4x, then factor out the GCF:

$$(4x-3)(4x-12) = 4(4x-3)(x-3)$$

Dropping the factor of 4 here leads to the correct answer of $(4x-3)(x-3)$.

5. **$(x^2+5y)(x^2-5y)$.** This is a difference of two squares, so follow the pattern:

$$x^4 - 25y^2 = (x^2+5y)(x^2-5y)$$

6. **$(4x+3y)(16x^2-12xy+9y^2)$.** Follow the pattern for the sum of two cubes:

$$64x^3 + 27y^3 = (4x+3y)(16x^2-12xy+9y^2)$$

7. **$(x^2+3)(x-5)$.** This is a factorable cubic expression. Find the GCF of the first two terms and the last two terms, then use the distributive property:

$$x^3 - 5x^2 + 3x - 15 = x^2(x-5) + 3(x-5) = (x^2+3)(x-5)$$

8. **$(x+2)(x-2)(x-11)$.** This is a factorable cubic expression, so find the GCF of the first two terms and the last two terms, then use the distributive property:

$$x^3 - 11x^2 - 4x + 44 = x^2(x-11) - 4(x-11) = (x^2-4)(x-11)$$

Now, $x^2 - 4$ is a difference of two squares, so factor according to this pattern:

$$= (x+2)(x-2)(x-11)$$

9. $6x(2x+3y)(2x-3y)$. Begin by factoring out the GCF of $6x$:

$$24x^3 - 54xy^2 = 6x(4x^2 - 9y^2)$$

The result inside the parentheses is a difference of two squares, so use the pattern for factoring:

$$= 6x(2x+3y)(2x-3y)$$

10. $8(2x+1)(x-3)$. Begin by factoring out the GCF of 8:

$$16x^2 - 40x - 24 = 8(2x^2 - 5x - 3)$$

The result inside the parentheses is quadratic. To factor it, use the analogous case of $8(x^2 - 5x - 6)$ (multiplying 2 by –3 and using this value as the constant, as I show you in Chapter 17):

$$8(x^2 - 5x - 6) = 8(x+1)(x-6)$$

Use this answer as a starting point to find the actual answer:

$$8(2x+1)(2x-6) = 8[2(2x+1)(x-3)]$$

Dropping the factor of 2 here leads to the correct answer:

$$8(2x+1)(x-3)$$

11. $12xy^2z^3(2x-y)(4x^2+2xy+y^2)$. Begin by factoring out the GCF of $12xy^2z^3$:

$$96x^4y^2z^3 - 12xy^5z^3 = 12xy^2z^3(8x^3 - y^3)$$

The result is a difference of two cubes, so use the pattern for factoring:

$$= 12xy^2z^3(2x-y)(4x^2+2xy+y^2)$$

12. $2(x+3)(x-3)(x-2)$. Begin by factoring out the GCF of 2:

$$2x^3 - 4x^2 - 18x + 36 = 2(x^3 - 2x^2 - 9x + 18)$$

Next, factor the cubic expression:

$$= 2[x^2(x-2) - 9(x-2)] = 2(x^2 - 9)(x-2)$$

The factor $x^2 - 9$ is the difference of two squares, so factor according to the pattern:

$$= 2(x+3)(x-3)(x-2)$$

Answers to function notation

1. 13. Substitute 3 for x into $5x - 2$, then evaluate:

$$f(3) = 5(3) - 2 = 15 - 2 = 13$$

2. 28. Substitute –5 for x into $x^2 + 3$, then evaluate:

$$g(-5) = (-5)^2 + 3 = 25 + 3 = 28$$

3. **−19.** Substitute −4 for x into $5x - 2$ and add 3:

$$f(-4) + 3 = 5(-4) - 2 + 3 = -20 - 2 + 3 = -19$$

4. **−104.** Substitute 7 for x into $x^2 + 3$, and multiply the resulting expression by −2:

$$-2g(7) = -2(7^2 + 3) = -2(49 + 3) = -2(52) = -104$$

5. $-\dfrac{1}{13}$. In the numerator, substitute 1 for x into $5x - 2$, multiply the resulting expression by 2, and then subtract 3. In the denominator, substitute 6 for x into $x^2 + 3$, and then multiply the resulting expression by −1:

$$\frac{2f(1) - 3}{-g(6)} = \frac{2[5(1) - 2] - 3}{-(6^2 + 3)}$$

Evaluate using PEMDAS:

$$= \frac{2[5 - 2] - 3}{-(36 + 3)} = \frac{2[3] - 3}{-39} = \frac{6 - 3}{-39} = \frac{3}{-39} = -\frac{1}{13}$$

6. $5k - 7$. Substitute $k - 1$ for x into $5x - 2$, then simplify:

$$f(k - 1) = 5(k - 1) - 2 = 5k - 5 - 2 = 5k - 7$$

7. $4k^2 + 16k + 25$. Substitute $k + 2$ for x into $x^2 + 3$, then simplify and subtract 3:

$$4((k + 2)^2 + 3) - 3$$

Simplify:

$$4((k + 2)(k + 2) + 3) = 4((k^2 + 4k + 4) + 3) - 3$$
$$= 4(k^2 + 4k + 7) - 3$$
$$= 4(k^2 + 16k + 28) - 3$$
$$= 4(k^2 + 16k + 25)$$

8. $4t^2 + 14t + 7$.

To begin, rewrite the problem in expanded form:

$$(f + g)(2t + 1) = f(2t + 1) + g(2t + 1)$$

Now, substitute $2t + 1$ for x into $5x - 2$ and $2t + 1$ into x for $x^2 + 3$:

$$= 5(2t + 1) - 2 + (2t + 1)^2 + 3$$

Simplify:

$$= 10t + 5 - 2 + 4t^2 + 4t + 1 + 3 = 4t^2 + 14t + 7$$

9. **−28.** To begin, rewrite as multiplication:

$$(fg)(-1) = f(-1)g(-1)$$

Next, substitute -1 for x into both functions and evaluate:

$$= [5(-1) - 2][(-1)^2 + 3] = (-5 - 2)(1 + 3) = (-7)(4) = -28$$

10. 58. To begin, evaluate $g(-3)$:

$$g(-3) = (-3)^2 + 3 = 9 + 3 = 12$$

Now, substitute 12 for $g(-3)$ into $f(g(-3))$:

$$f(g(-3)) = f(12)$$

Now, substitute 12 for x into $5x - 2$:

$$= 5(12) - 2 = 60 - 2 = 58$$

So $f(g(-3)) = 58$.

11. $25p^2 - 70p + 42$. To begin, evaluate $f(p-1)$:

$$f(p-1) = 5(p-1) - 2 = 5p - 5 - 2 = 5p - 7$$

Now, substitute $5p - 7$ for $f(p-1)$ into $g(f(p-1)) - 10$:

$$g(f(p-1)) - 7 = g(5p - 7) - 10$$

Now, substitute $5p - 7$ for x into $x^2 + 3$ and simplify:

$$= (5p - 7)^2 + 3 - 10 = 25p^2 - 70p + 49 - 7 = 25p^2 - 70p + 42$$

12. 415. To begin, rewrite as a function of a function:

$$3(f \circ g)(5) + 1 = 3f(g(5)) + 1$$

Now, evaluate $g(5)$:

$$g(5) = 5^2 + 3 = 25 + 3 = 28$$

Next, substitute 28 for $g(5)$ in $3f(g(5)) + 1$:

$$3f(g(5)) + 1 = 3f(28) + 1$$

Now, substitute 28 for x in $5x - 2$, multiply this expression by 3, add 1, and evaluate:

$$3[5(28) - 2] + 1 = 3(140 - 2) + 1 = 3(138) + 1 = 414 + 1 = 415$$

Answers to parent functions and transformations

1. (A) $f(x) = x$
   (B) $f(x) = x^2$
   (C) $f(x) = x^3$
   (D) $f(x) = x^4$

2. (A) $f(x) = 2^x$
   (B) $f(x) = \log_2 x$

3. **(A)** $f(x) = \sin x$

 (B) $f(x) = \cos x$

4. **(A)** $f(x) = |x|$

 (B) $f(x) = \dfrac{1}{x}$

 (C) $f(x) = \sqrt{x}$

5. $2x^2 - 4x - 2$. The transformation is as follows:

$$2f(x-1) - 4 = 2(x-1)^2 - 4$$

Now, simplify:

$$= 2(x^2 - 2x + 1) - 4$$
$$= 2x^2 - 4x + 2 - 4$$
$$= 2x^2 - 4x - 2$$

6. $-\dfrac{1}{3}x^2 - 7$. Transform the function $f(x) = x^2$, translating it up 7, compressing it by a factor of 3, and reflecting it vertically across the x-axis. The transformation is as follows:

$$-\dfrac{1}{3}f(x) - 7 = -\dfrac{1}{3}x^2 - 7$$

7. $-5x^2 - 20x - 20$. Transform the function $f(x) = x^2$, translating it left 2, stretching it vertically by a factor of 5, and reflecting it vertically across the x-axis.

$$-5f(x+2) = -5(x+2)^2$$

Simplify as follows:

$$= -5(x^2 + 4x + 4)$$
$$= -5x^2 - 20x - 20$$

8. **(A)** 9 units down.

 (B) 2 units left.

 (C) Stretch factor of 3 vertically.

 (D) Across the x-axis.

9. **(A)** 10 units up.

 (B) 5 units right.

 (C) Compression vertically factor of 7.

 (D) Across the x-axis.

Answers to polynomial functions

1. **C.** $y = -5x^8 + x^7 - 4x + 10$. A polynomial that displays end behavior that is negative for both very large and very small values of x must be even and negative — that is, it must have a leading term with an even exponent and a negative coefficient. Therefore, the correct answer is C.

2. A. *m* is positive and *n* is even. A polynomial that displays positive end behavior for both very large and very small values of x must be an even polynomial that's positive. Therefore, the correct answer is A.

3. C. $y = -3x^3 - 2x^2 + 5$. The point (0, 5) is a y-intercept of 5, so the constant c must be 5; therefore, the correct answer is C.

4. A. 3. To find the roots of the function, set each factor equal to 0 and solve for x.

$$\begin{array}{llll} x+3=0 & x+7=0 & x-4=0 & -3x+1=0 \\ x=-3 & x=-7 & x=4 & -3x=-1 \\ & & & x=\dfrac{-1}{-3}=\dfrac{1}{3} \end{array}$$

Therefore, the correct answer is A.

Answers to quadratic equations and functions

1. $\pm\sqrt{\dfrac{7}{2}}$. Add 7, divide by 2, and then take the square root of both sides:

$$\begin{aligned} 2x^2 - 7 &= 0 \\ 2x^2 &= 7 \\ x^2 &= \frac{7}{2} \\ x &= \pm\sqrt{\frac{7}{2}} \end{aligned}$$

2. 0 and $-\dfrac{10}{3}$. First, notice that $x = 0$ is a solution. To find the remaining solution, divide both sides by x, subtract 10, and divide by 3:

$$\begin{aligned} \frac{3x^2 + 10x}{x} &= \frac{0}{x} \\ 3x + 10 &= 0 \\ 3x &= -10 \\ x &= -\frac{10}{3} \end{aligned}$$

3. 4 and 5. Begin by factoring the left side of the equation:

$$\begin{aligned} x^2 - 9x + 20 &= 0 \\ (x-4)(x-5) &= 0 \end{aligned}$$

Now set each factor to 0:

$$\begin{array}{ll} x-4=0 & x-5=0 \\ x=4 & x=5 \end{array}$$

4. 4 and –7. Factor the left side of the equation:

$$\begin{aligned} x^2 + 3x - 28 &= 0 \\ (x+7)(x-4) &= 0 \end{aligned}$$

Set each factor to 0:

$$x+7=0 \qquad x-4=0$$
$$x=-7 \qquad x=4$$

5. $\dfrac{2\pm\sqrt{22}}{3}$. Substitute the values $a=3$, $b=-4$, and $c=-6$ into the quadratic formula, then simplify:

$$x = \frac{4\pm\sqrt{4^2-4(3)(-6)}}{2(3)}$$
$$= \frac{4\pm\sqrt{16+72}}{6}$$
$$= \frac{4\pm\sqrt{88}}{6}$$
$$= \frac{4\pm2\sqrt{22}}{6}$$
$$= \frac{2\pm\sqrt{22}}{3}$$

6. **(−2, 37) and (11, 336).** Begin by setting the right sides of both equations equal to each other and moving all terms to one side of the equation; then factor:

$$3x^2-4x+17 = 4x^2-13x-5$$
$$x^2-9x-22 = 0$$
$$(x+2)(x-11) = 0$$

Thus, $x=-2$ and $x=11$. Plug both of these values back into either of the original equations to find the corresponding y-values:

$$y = 3x^2-4x+17 = 3(-2)^2-4(-2)+17 = 12+8+17 = 37$$
$$y = 3x^2-4x+17 = 3(11)^2-4(11)+17 = 363-44+17 = 336$$

So the two answers are $(-2, 37)$ and $(11, 336)$.

7. **D. $y=-2x^2-9x+6$.** The sketch shows a parabola that is concave down, so the leading coefficient is negative, which rules out answers A and B. And the sketch also shows a parabola with a positive y-intercept, so the constant value c is positive, which rules out answer C; therefore, answer D is correct.

8. **$x=5$.** Calculate the axis of symmetry for $y=x^2-10x+21$, plugging in 1 for a and -10 for b, as follows:

$$x = -\frac{b}{2a} = -\frac{-10}{2(1)} = \frac{10}{2} = 5$$

9. **(−2, 21).** To begin, calculate the axis of symmetry for $y=-5x^2-20x+1$, plugging in -5 for a and -20 for b, as follows:

$$x = -\frac{b}{2a} = -\frac{-20}{2(-5)} = -\frac{20}{10} = -2$$

So -2 is the x-value of the vertex. Plug -2 for x into the equation to find the y-value:

$$y = -5(-2)^2-20(-2)+1 = -20+40+1 = 21$$

So the answer is $(-2, 21)$.

10. $y = 5x^2 - 20x + 9$. Expand the exponent, FOIL the result, distribute, and simplify:

$$y = 5(x-2)^2 - 11$$
$$y = 5(x-2)(x-2) - 11$$
$$y = 5(x^2 - 4x + 4) - 11$$
$$y = 5x^2 - 20x + 20 - 11$$
$$y = 5x^2 - 20x + 9$$

11. $y = (x+3)^2 - 12$. To begin, calculate the axis of symmetry for $y = x^2 + 6x - 3$, plugging in 1 for a and 6 for b, as follows:

$$x = -\frac{b}{2a} = -\frac{6}{2(1)} = -\frac{6}{2} = -3$$

So -3 is the x-value of the vertex. Plug -3 for x into the equation to find the y-value:

$$y = (-3)^2 + 6(-3) - 3 = 9 - 18 - 3 = -12$$

So the vertex is at $(-3, -12)$. Now, plug in 1 for a, -3 for h, and -12 for k in the vertex form of the quadratic function:

$$y = a(x-h)^2 + k$$
$$y = (x+3)^2 - 12$$

12. −4 **and** 9. Begin by factoring the right side:

$$y = (x+4)(x-9)$$

Now, to find the roots, set y equal to 0 and solve for x:

$$0 = (x+4)(x-9)$$
$$x = -4, 9$$

13. $x = 3$. Begin by finding the roots of the function:

$$0 = (x+3)(x-9)$$
$$x = -3, 9$$

Locate the midpoint of these two points by finding their average:

$$\frac{-3+9}{2} = \frac{6}{2} = 3$$

So the equation of the axis of symmetry is $x = 3$.

14. $\frac{5}{4}$ **and** $-\frac{1}{6}$. Begin by setting the y to 0:

$$0 = (4x - 5)(6x + 1)$$

Now, split this equation into two separate equations and solve both:

$$0 = 4x - 5$$
$$5 = 4x$$
$$\frac{5}{4} = x$$

$$0 = 6x + 1$$
$$-1 = 6x$$
$$-\frac{1}{6} = x$$

So the solutions are $\frac{5}{4}$ and $-\frac{1}{6}$.

15. $\frac{1 \pm \sqrt{33}}{8}$. To solve this non-factorable quadratic equation, use the quadratic formula, plugging in 4 for a, -1 for b, and -2 for c:

$$x = \frac{1 \pm \sqrt{(-1)^2 - 4(4)(-2)}}{2(-4)} = \frac{1 \pm \sqrt{1 + 32}}{8} = \frac{1 \pm \sqrt{33}}{8}$$

16. **C.** -7. Begin by setting y to 0 and factoring:

$$0 = x^2 + 14x + 49$$
$$0 = (x + 7)(x + 7)$$
$$0 = (x + 7)^2$$

This equation can be expressed as the square of a binomial, so it has only one solution, which rules out answers A and D. Find this value by setting this binomial equal to 0:

$$0 = x + 7$$
$$-7 = x$$

The solution is -7, so answer C is correct.

17. **D.** $-\frac{5}{3}$. Set the binomial $3x + 5$ equal to 0 and solve for x:

$$0 = 3x + 5$$
$$-5 = 3x$$
$$-\frac{5}{3} = x$$

So answer D is correct.

18. $2\sqrt{10}$. In order to have one real root, the function $y = x^2 - bx + 10$ must be expressible as the square of a binomial. A binomial takes the form $px + q$, so the factored form of the function is:

$$y = (px + q)^2$$
$$y = (px + q)(px + q)$$
$$y = p^2x^2 + 2pqx + q^2$$

Now, set the right side of this equation equal to the right side of the original function:

$$p^2x^2 + 2pqx + q^2 = x^2 - bx + 10$$

For this equation to be true, each of the coefficients on the left side must be equal to its corresponding coefficient on the right side. Thus:

$$p^2 = 1 \quad \text{and} \quad q^2 = 10$$
$$p = \pm 1 \qquad\qquad q = \pm\sqrt{10}$$

Therefore:

$$-b = \pm 2pq = \pm 2(1)(\sqrt{10}) = \pm 2\sqrt{10}$$

Answers to exponential and logarithmic equations

1. 243. Use $x^{\frac{a}{b}} = \sqrt[b]{x^a}$:

$$9^{\frac{5}{2}} = \sqrt{9}^5 = 3^5 = 243$$

2. B. $\sqrt[21]{x^{13}}$. Begin by adding the exponents:

$$x^{\frac{1}{3}} x^{\frac{2}{7}} = x^{\frac{1}{3} + \frac{2}{7}} = x^{\frac{7+6}{21}} = x^{\frac{13}{21}}$$

Now, use $x^{\frac{a}{b}} = \sqrt[b]{x^a}$:

$$= \sqrt[21]{x^{13}}$$

Therefore, the answer is B.

3. B. 2^{2n+2}. Rewrite the expression in the denominator using a base of 2, then remove parentheses:

$$\frac{2^{5n+2}}{8^n} = \frac{2^{5n+2}}{(2^3)^n} = \frac{2^{5n+2}}{2^{3n}}$$

Divide by subtracting exponents, and then simplify:

$$= 2^{5n+2-3n} = 2^{2n+2}$$

Therefore, the answer is B.

4. $-\dfrac{4}{9}$. Rewrite both sides of the equation using a base of 3, then remove parentheses:

$$(\tfrac{1}{9})^{2n} = \sqrt{3}^{n+4}$$

$$(3^{-2})^{2n} = (3^{\frac{1}{2}})^{n+4}$$

$$3^{-4n} = 3^{\frac{1}{2}(n+4)}$$

Set the exponents equal and solve for n:

$$-4n = \frac{1}{2}(n+4)$$

$$-8n = n + 4$$

$$-9n = 4$$

$$n = -\frac{4}{9}$$

5. D. 80,000,000. Using the exponential function $y = ab^x$, plug in 150 for a, 3 for b, and 12 for x, then solve for y:

$$y = ab^x = 150(3)^{12} = 150(531,441) = 79,716,150$$

Therefore, answer D is correct.

6. D. $4 million. Starting with $y = ab^x$, plug in 1,900,000 for a and 5 for x. Next, calculate the percent increase of 17% as 117%, which equals 1.17, and set b equal to this value:

$$y = ab^x = 1,900,000(1.17)^5 \approx 1,900,000(2.19) \approx 4,161,000$$

So the company was worth Therefore, D. . .

7. C. $290,000. An investment property valued at $455,000 in 2010 lost 14% of its value in each of the next three years. How much was the property worth in 2013?

Model the problem using the function $y = ab^x$. Plug in 455,000 for a and 3 for x. Now, calculate the percent decrease of 14% as 86% of the price, which equals 0.86, and set b equal to this value:

$$y = ab^x = 455,000(0.86)^3 \approx 455,000(0.64) \approx 290,000$$

Therefore, the answer is $290,000.

8. 81. Rewrite as an exponential equation and solve:

$$9^2 = x$$
$$81 = x$$

9. $\frac{1}{8}$. Rewrite as an exponential equation:

$$k^{-1} = 8$$
$$\frac{1}{k} = 8$$
$$k = \frac{1}{8}$$

10. $\frac{1}{3}$. Rewrite as an exponential equation:

$$8^n = 2$$

Now, rewrite both sides of the equation using a base of 2 and remove parentheses:

$$(2^3)^n = 2^1$$
$$2^{3n} = 2^1$$

Set the exponents equal and solve for n:

$$3n = 1$$
$$n = \frac{1}{3}$$

11. D. $\frac{\log_7 10}{6}$. Rewrite as a logarithmic equation, then isolate x:

$$\log_7 10 = 6x$$
$$\frac{\log_7 10}{6} = x$$

Therefore, D. . .

Answers to radical and rational equations

1. **A. $\sqrt{60}$**. Begin by noticing that $7 = \sqrt{49}$ and $8 = \sqrt{64}$. Of the four choices, the only one that falls between these two values is $\sqrt{60}$, so answer A is correct.

2. **3**. Begin by multiplying the two values in the numerator, then divide by the denominator, and simplify:

$$\frac{\sqrt{6}\sqrt{15}}{\sqrt{10}} = \frac{\sqrt{90}}{\sqrt{10}} = \sqrt{9} = 3$$

3. **$3\sqrt{7}$**. Factor out $\sqrt{9}$ and then simplify this value:

$$\sqrt{63} = \sqrt{9}\sqrt{7} = 3\sqrt{7}$$

4. **$\frac{\sqrt{7}}{2}$**. Begin by factoring out a 5 from both the numerator and the denominator, then cancel these factors and simplify:

$$\frac{\sqrt{35}}{\sqrt{20}} = \frac{\sqrt{5}\sqrt{7}}{\sqrt{5}\sqrt{4}} = \frac{\sqrt{7}}{\sqrt{4}} = \frac{\sqrt{7}}{2}$$

5. **$5\sqrt{6}$**. Begin by simplifying both radicals, then add the whole number parts and keep the radical part:

$$\sqrt{24} + \sqrt{54} = \sqrt{4}\sqrt{6} + \sqrt{9}\sqrt{6} = 2\sqrt{6} + 3\sqrt{6} = 5\sqrt{6}$$

6. **$\frac{\sqrt{30}}{2}$**. Multiply both the numerator and denominator by $\sqrt{30}$, then simplify the denominator and cancel out a common factor of 15:

$$\frac{15}{\sqrt{30}} = \frac{15\sqrt{30}}{\sqrt{30}\sqrt{30}} = \frac{15\sqrt{30}}{30} = \frac{\sqrt{30}}{2}$$

7. **$x \geq \frac{3}{4}$**. To find the domain of a radical function, create an inequality in which the radicand (number inside the radical) is greater than or equal to 0, then simplify:

$$4x - 3 \geq 0$$
$$4x \geq 3$$
$$x \geq \frac{3}{4}$$

8. **$y \leq 7$**. This function is negative, and the number 7 is added outside the radical portion, so the range is $y \leq 7$.

9. **$(-\infty, -15) \cup (-15, 2) \cup (2, \infty)$**. To find the domain of a rational function, set the denominator equal to 0:

$$x^2 + 13x - 30 = 0$$
$$(x + 15)(x - 2) = 0$$
$$x = -15, 2$$

Therefore, the domain is all real values *except* −15 and 2: $(-\infty, -15) \cup (-15, 2) \cup (2, \infty)$.

10. **$(-\infty, 5) \cup (5, \infty)$**. To find the range of a rational function, exclude the value that is added to or subtracted from the fractional portion of the function. In this case, this number is 5, so the range is all real values *except* 5: $(-\infty, 5) \cup (5, \infty)$.

Answers to geometry

1. **20.6 cubic inches.** Plug 1.7 as the radius and 3.14 as π into the formula for the volume of a sphere:

$$V = \frac{4\pi r^3}{3} = \frac{4(3.14)(1.7)^3}{3} = \frac{4(3.14)(4.913)}{3} \approx 20.6$$

2. **4,455,000 cubic meters.** Plug 450 as the side and 66 as the height into the formula for the volume of a pyramid:

$$V = \frac{s^2 h}{3} = \frac{450^2 (66)}{3} = 202{,}500 \times 22 = 4{,}455{,}000$$

3. **339 cubic feet.** Plug 6 as the radius, 9 as the height, and 3.14 as π into the formula for the volume of a cone:

$$V = \frac{\pi r^2 h}{3} = \frac{(3.14)(6^2)(9)}{3} 3.14(36)(3) \approx 339$$

4. **35°.** The angle that is supplementary to 125° is 55°. The 55° angle and the other small angle in the triangle are complementary, so the latter angle is 35°. And this angle and angle x are vertical angles, so they are equal.

5. **75°.** The angle that is supplementary to 70° is 110°. This angle and angle x are corresponding angles, so they are equal. Because $x + y = 215$, you know that $y = 105°$. The angle supplementary to y is 75°. This angle is a corresponding angle to z, so $z = 75°$.

6. **C. $AD = CD$ and $\angle ABD \cong \angle CBD$.** The two triangles share side BD. Answer C only gives you $AD = CD$ and $\angle ABD \cong \angle CBD$, which is sufficient to show SSA (side-side-angle). However, this is not sufficient to prove congruency.

7. **B. $HF = RS$.** With the information given, answer B is only sufficient to show SSA (side-side-angle), which is not sufficient to prove congruency.

8. **D.** All three triangles are right triangles, so $\angle W$ and $\angle Y$ are complementary. Also, $\angle W$ and $\angle WXZ$ are complementary, so $\angle Y = \angle WXZ$. By the same reasoning, $\angle ZXY$ and $\angle Y$ are complementary, so $\angle W = \angle ZXY$ and. Thus, each of the three triangles includes one right angle and a pair of equal angles, which is sufficient to prove AAA for all three triangles. Therefore, all three pairs of triangles are similar, so D is the correct answer.

9. **$x^2 + y^2 = 169$.** Use the formula for a circle centered at the origin, plugging in 13 for the radius:

$$x^2 + y^2 = 13^2$$
$$x^2 + y^2 = 169$$

10. **$(x + 2)^2 + (y - 7)^2 = 16$.** Use the formula for a circle centered at the point (h, k) with a radius of r:

$$(x - h)^2 + (y - k)^2 = r^2$$
$$(x - (-2))^2 + (y - 7)^2 = 4^2$$
$$(x + 2)^2 + (y - 7)^2 = 16$$

Answers to trigonometry

1. **(A)** $\sin k = \dfrac{5}{13}$

   (B) $\cos k = \dfrac{12}{13}$

   (C) $\tan k = \dfrac{5}{12}$

   (D) $\cot k = \dfrac{12}{5}$

   (E) $\sec k = \dfrac{13}{12}$

   (F) $\csc k = \dfrac{13}{5}$

2. $\dfrac{\sqrt{13}}{6}$. Because $\cos x = \dfrac{A}{H}$, you can use the value 6 for the adjacent side and 7 for the hypotenuse. Using this information, draw a picture of the triangle you're working with:

   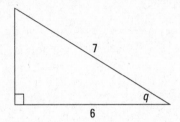

   Now, use the Pythagorean theorem to find the length of the triangle's opposite side:

   $$a^2 + b^2 = c^2$$
   $$6^2 + b^2 = 7^2$$
   $$36 + b^2 = 49$$
   $$b^2 = 13$$
   $$b = \sqrt{13}$$

   The length of the opposite side is $\sqrt{13}$, so plug this value into the equation for the tangent:

   $$\tan p = \dfrac{O}{A} = \dfrac{\sqrt{13}}{6}$$

3. **(A)** $\sin 45° = \dfrac{\sqrt{2}}{2}$

   (B) $\cos 45° = \dfrac{\sqrt{2}}{2}$

   (C) $\tan 45° = 1$

4. **(A)** $\sin 30° = \dfrac{1}{2}$

   (B) $\cos 30° = \dfrac{\sqrt{3}}{2}$

   (C) $\tan 30° = \dfrac{\sqrt{3}}{3}$

   (D) $\sin 60° = \dfrac{\sqrt{3}}{2}$

   (E) $\cos 60° = \dfrac{1}{2}$

   (F) $\tan 60° = \sqrt{3}$

5. $\frac{2\pi}{9}$. Using the equivalence $180° = \pi$, write a proportion as follows, and solve for x:

$$\frac{180}{\pi} = \frac{40}{x}$$
$$180x = 40\pi$$
$$x = \frac{40\pi}{180} = \frac{2\pi}{9}$$

6. **140°**. Using the equivalence $180° = \pi$, write a proportion as follows, and solve for x:

$$\frac{180}{\pi} = \frac{x}{\frac{7\pi}{9}}$$
$$180 \frac{7\pi}{9} = \pi x$$
$$180 \frac{7}{9} = x$$
$$140 = x$$

7. **16π**. Begin by converting 160° to radians:

$$\frac{180}{\pi} = \frac{160}{x}$$
$$180x = 160\pi$$
$$x = \frac{160\pi}{180} = \frac{8\pi}{9}$$

Now, use the formula for arc length:

$$\text{Arc length} = \text{Radius} \times \text{Radians} = 18 \times \frac{8\pi}{9} = 16\pi \text{ units}$$

8. **75**. Use the identity $\sin x = \cos(90° - x)$:

$$\sin 15° = \cos(90 - 15)° = \cos 75°$$

9. **D. $1 - \cos^2 x$**. Apply the identity $\tan x = \frac{\sin x}{\cos x}$ and simplify as follows:

$$\sin x \cos x \tan x = \sin x \cos x \frac{\sin x}{\cos x} = \sin^2 x$$

Now, apply algebra to the Pythagorean identity:

$$\sin^2 x + \cos^2 x = 1$$
$$\sin^2 x = 1 - \cos^2 x$$

Therefore, the answer is D.

10. **D. $\frac{\pi}{6}$**. Apply the ratio identity to cot x and the reciprocal identity to sec x, then simplify by canceling factors:

$$\sin x \cos x \cot x \sec x = \sin x \cos x \frac{\cos x}{\sin x} \cdot \frac{1}{\cos x} = \cos x$$

Thus, you can rewrite the equation as:

$$\cos x = \frac{\sqrt{3}}{2}$$

You know from the special right triangles that $\cos 30° = \frac{\sqrt{3}}{2}$, and $30° = \frac{\pi}{6}$, so the answer is D.

11. **30°.** Begin by isolating the trig expression on the left side of the equation:

$$3\sin n° \sec n° = \sqrt{3}$$

$$\sin n° \sec n° = \frac{\sqrt{3}}{3}$$

Now, use identities to simplify the left side:

$$\frac{\sin n°}{\cos n°} = \frac{\sqrt{3}}{3}$$

$$\tan n° = \frac{\sqrt{3}}{3}$$

Your knowledge of special right triangles tells you that $\tan 30° = \frac{\sqrt{3}}{3}$, so the answer is 30°.

12. **6.4.** The problem provides you with two angle measurements and the length of one side, so use the law of sines. Notice that the 50° angle is opposite the side of length 5, and that the 80° angle is opposite the side of length x:

$$\frac{\sin 50°}{5} = \frac{\sin 80°}{x}$$

Use your calculator to replace the two trig functions with approximate values, then solve for x:

$$\frac{0.766}{5} = \frac{0.985}{x}$$

$$0.766x = 4.952$$

$$x \approx 6.4$$

7

ACCUPLACER
Practice Tests

Test your knowledge with two complete ACCUPLACER practice tests.

Find the answer to every question, including a detailed explanation, so that you know what you need to work on before taking the ACCUPLACER for real.

Chapter 21
Practice Test 1

This practice exam features five tests, just like the real ACCUPLACER. As you may have guessed, the sample tests in this book are paper-based tests. When you take the actual ACCUPLACER, it will be a computer-based exam. Remember that with the computer-based test you can't skip a question and go back to it, and you can't change an answer after you enter it into the computer. You can find out all you need to know about the ACCUPLACER in Part 1.

TIP

To get the most out of this sample test, take it under the same conditions as the real ACCUPLACER:

>> Find a quiet place where you won't be interrupted.

>> Feel free to take the five separate sections of the test in any order you choose. There is no time limit, so take as long as you need. (But bear in mind that when you really take the test, you'll need to start and finish each section of the test in one sitting.)

After you complete the entire sample test, or any section, check your answers against the answers and explanations in Chapter 22.

Answer Sheet for Practice Exam 1

Reading Test

1. Ⓐ Ⓑ Ⓒ Ⓓ	5. Ⓐ Ⓑ Ⓒ Ⓓ	9. Ⓐ Ⓑ Ⓒ Ⓓ	13. Ⓐ Ⓑ Ⓒ Ⓓ	17. Ⓐ Ⓑ Ⓒ Ⓓ
2. Ⓐ Ⓑ Ⓒ Ⓓ	6. Ⓐ Ⓑ Ⓒ Ⓓ	10. Ⓐ Ⓑ Ⓒ Ⓓ	14. Ⓐ Ⓑ Ⓒ Ⓓ	18. Ⓐ Ⓑ Ⓒ Ⓓ
3. Ⓐ Ⓑ Ⓒ Ⓓ	7. Ⓐ Ⓑ Ⓒ Ⓓ	11. Ⓐ Ⓑ Ⓒ Ⓓ	15. Ⓐ Ⓑ Ⓒ Ⓓ	19. Ⓐ Ⓑ Ⓒ Ⓓ
4. Ⓐ Ⓑ Ⓒ Ⓓ	8. Ⓐ Ⓑ Ⓒ Ⓓ	12. Ⓐ Ⓑ Ⓒ Ⓓ	16. Ⓐ Ⓑ Ⓒ Ⓓ	20. Ⓐ Ⓑ Ⓒ Ⓓ

Writing Test

1. Ⓐ Ⓑ Ⓒ Ⓓ	6. Ⓐ Ⓑ Ⓒ Ⓓ	11. Ⓐ Ⓑ Ⓒ Ⓓ	16. Ⓐ Ⓑ Ⓒ Ⓓ	21. Ⓐ Ⓑ Ⓒ Ⓓ
2. Ⓐ Ⓑ Ⓒ Ⓓ	7. Ⓐ Ⓑ Ⓒ Ⓓ	12. Ⓐ Ⓑ Ⓒ Ⓓ	17. Ⓐ Ⓑ Ⓒ Ⓓ	22. Ⓐ Ⓑ Ⓒ Ⓓ
3. Ⓐ Ⓑ Ⓒ Ⓓ	8. Ⓐ Ⓑ Ⓒ Ⓓ	13. Ⓐ Ⓑ Ⓒ Ⓓ	18. Ⓐ Ⓑ Ⓒ Ⓓ	23. Ⓐ Ⓑ Ⓒ Ⓓ
4. Ⓐ Ⓑ Ⓒ Ⓓ	9. Ⓐ Ⓑ Ⓒ Ⓓ	14. Ⓐ Ⓑ Ⓒ Ⓓ	19. Ⓐ Ⓑ Ⓒ Ⓓ	24. Ⓐ Ⓑ Ⓒ Ⓓ
5. Ⓐ Ⓑ Ⓒ Ⓓ	10. Ⓐ Ⓑ Ⓒ Ⓓ	15. Ⓐ Ⓑ Ⓒ Ⓓ	20. Ⓐ Ⓑ Ⓒ Ⓓ	25. Ⓐ Ⓑ Ⓒ Ⓓ

Arithmetic Test

1. Ⓐ Ⓑ Ⓒ Ⓓ	5. Ⓐ Ⓑ Ⓒ Ⓓ	9. Ⓐ Ⓑ Ⓒ Ⓓ	13. Ⓐ Ⓑ Ⓒ Ⓓ	17. Ⓐ Ⓑ Ⓒ Ⓓ
2. Ⓐ Ⓑ Ⓒ Ⓓ	6. Ⓐ Ⓑ Ⓒ Ⓓ	10. Ⓐ Ⓑ Ⓒ Ⓓ	14. Ⓐ Ⓑ Ⓒ Ⓓ	18. Ⓐ Ⓑ Ⓒ Ⓓ
3. Ⓐ Ⓑ Ⓒ Ⓓ	7. Ⓐ Ⓑ Ⓒ Ⓓ	11. Ⓐ Ⓑ Ⓒ Ⓓ	15. Ⓐ Ⓑ Ⓒ Ⓓ	19. Ⓐ Ⓑ Ⓒ Ⓓ
4. Ⓐ Ⓑ Ⓒ Ⓓ	8. Ⓐ Ⓑ Ⓒ Ⓓ	12. Ⓐ Ⓑ Ⓒ Ⓓ	16. Ⓐ Ⓑ Ⓒ Ⓓ	20. Ⓐ Ⓑ Ⓒ Ⓓ

Quantitative Reasoning, Algebra, and Statistics Test

1. Ⓐ Ⓑ Ⓒ Ⓓ	5. Ⓐ Ⓑ Ⓒ Ⓓ	9. Ⓐ Ⓑ Ⓒ Ⓓ	13. Ⓐ Ⓑ Ⓒ Ⓓ	17. Ⓐ Ⓑ Ⓒ Ⓓ
2. Ⓐ Ⓑ Ⓒ Ⓓ	6. Ⓐ Ⓑ Ⓒ Ⓓ	10. Ⓐ Ⓑ Ⓒ Ⓓ	14. Ⓐ Ⓑ Ⓒ Ⓓ	18. Ⓐ Ⓑ Ⓒ Ⓓ
3. Ⓐ Ⓑ Ⓒ Ⓓ	7. Ⓐ Ⓑ Ⓒ Ⓓ	11. Ⓐ Ⓑ Ⓒ Ⓓ	15. Ⓐ Ⓑ Ⓒ Ⓓ	19. Ⓐ Ⓑ Ⓒ Ⓓ
4. Ⓐ Ⓑ Ⓒ Ⓓ	8. Ⓐ Ⓑ Ⓒ Ⓓ	12. Ⓐ Ⓑ Ⓒ Ⓓ	16. Ⓐ Ⓑ Ⓒ Ⓓ	20. Ⓐ Ⓑ Ⓒ Ⓓ

Advanced Algebra and Functions Test

1. Ⓐ Ⓑ Ⓒ Ⓓ	5. Ⓐ Ⓑ Ⓒ Ⓓ	9. Ⓐ Ⓑ Ⓒ Ⓓ	13. Ⓐ Ⓑ Ⓒ Ⓓ	17. Ⓐ Ⓑ Ⓒ Ⓓ
2. Ⓐ Ⓑ Ⓒ Ⓓ	6. Ⓐ Ⓑ Ⓒ Ⓓ	10. Ⓐ Ⓑ Ⓒ Ⓓ	14. Ⓐ Ⓑ Ⓒ Ⓓ	18. Ⓐ Ⓑ Ⓒ Ⓓ
3. Ⓐ Ⓑ Ⓒ Ⓓ	7. Ⓐ Ⓑ Ⓒ Ⓓ	11. Ⓐ Ⓑ Ⓒ Ⓓ	15. Ⓐ Ⓑ Ⓒ Ⓓ	19. Ⓐ Ⓑ Ⓒ Ⓓ
4. Ⓐ Ⓑ Ⓒ Ⓓ	8. Ⓐ Ⓑ Ⓒ Ⓓ	12. Ⓐ Ⓑ Ⓒ Ⓓ	16. Ⓐ Ⓑ Ⓒ Ⓓ	20. Ⓐ Ⓑ Ⓒ Ⓓ

Reading Test

DIRECTIONS: This test contains items that measure your ability to understand what you read. Read each of the following passages and then answer the question(s) based on what is implied or stated directly in that passage. Mark your choice on your answer sheet, using the correct letter with each question number.

Hazel Morse was a large, fair woman of the type that incites some men when they use the word "blonde" to click their tongues and wag their heads roguishly. She prided herself upon her small feet and suffered for her vanity, boxing them in snub-toed, high-heeled slippers of the shortest bearable size.

She was not a woman given to recollections. At her middle thirties, her old days were a blurred and flickering sequence, an imperfect film, dealing with the actions of strangers.

In her twenties, after the deferred death of a hazy widowed mother, she had been employed as a model in a wholesale dress establishment. Her job was not onerous, and she met numbers of men and spent numbers of evenings with them, laughing at their jokes and telling them she loved their neckties. Men liked her, and she took it for granted that the liking of many men was a desirable thing. Popularity seemed to her to be worth all the work that had to be put into its achievement.

No other form of diversion, simpler or more complicated, drew her attention. She never pondered if she might not be better occupied doing something else.

When she had been working in the dress establishment some years she met Herbie Morse. He was thin, quick, attractive, with shifting lines about his shiny, brown eyes and a habit of fiercely biting at the skin around his finger nails.

She liked him immediately upon their meeting. He was as promptly drawn to her. They were married six weeks after they had met.

She was delighted at the idea of being a bride; coquetted with it, played upon it. Other offers of marriage she had had, and not a few of them, but it happened that they were all from stout, serious men who had visited the dress establishment as buyers; men from Des Moines and Houston and Chicago and, in her phrase, even funnier places. There was always something immensely comic to her in the thought of living elsewhere than New York. She could not regard as serious proposals that she share a western residence.

She wanted to be married. She was nearing thirty now, and she did not take the years well. She spread and softened, and her darkening hair turned her to inexpert dabblings with peroxide. There were times when she had little flashes of fear about her job. And she had had a couple of thousand evenings of being a good sport among her male acquaintances. She had come to be more conscientious than spontaneous about it. (Excerpt from *Big Blonde*, by Dorothy Parker)

1. Hazel Morse might best be described as which of the following?

(A) Discontent

(B) Innocent

(C) Scheming

(D) Unreflective

2. The author most likely includes the detail in Paragraph 3 that Hazel spent many evenings "laughing at [men's] jokes and telling them she loved their neckties" in order to

(A) imply that she had ulterior motives in dating these men.

(B) present these relationships as essentially superficial.

(C) provide evidence that she had an uncanny eye for detail.

(D) demonstrate that she had ample reason to be afraid of the men she dated and that she needed to placate them constantly.

3. We can reasonably infer that all of the following impelled Hazel to marry Herbie EXCEPT

(A) They felt a mutual attraction for each other.

(B) She was concerned about a lack of opportunity in her chosen career as she aged.

(C) She had had no other offers of marriage and had begun to grow desperate.

(D) She had begun to grow weary of her dating life as a single woman.

4. An accurate paraphrasing of the words "more conscientious than spontaneous" might be

(A) more faithful than independent

(B) more forced than natural

(C) more instinctive than rehearsed

(D) more mature than impulsive

Passage 1

TV advertising has a wide *reach*, which means that it can be seen by a lot of people from a wide variety of demographics (a "demographic" describes the characteristics of an audience in terms of age, gender, or life stage). As a result, TV advertising tends to be expensive in terms of professional production costs, and the most sought-after media slots (the ones in the commercial breaks of the most popular TV shows) are also expensive to occupy.

TV advertisements (known as "adverts") need to appeal to a large majority of viewers. They are, therefore, more suited to retailers and brands that have a wide geographic and demographic spread, such as Marks and Spencer (M&S), Gap, Nike, and Levi's. TV advertising can utilize music and movement to good effect: It is three dimensional and can attract attention but normally it only lasts between 30 and 60 seconds. It cannot show a whole range of products but needs to focus on a narrow presentation and on image building. Levi's iconic advert, set in a launderette with the background music "Heard It Through the Grapevine", only showed the 501 style of jeans. It created such a demand that stores ran out of 501s.

TV adverts can be cost effective in an integrated campaign where "stills" are replicated on billboards, in magazines, and in stores. However, that can saturate the media environment and the consumer can become bored or tired with seeing the same pictures. (*Fashion Marketing Communications*, by Gaynor Lea-Greenwood — Pages 19–20)

Passage 2

Ten minutes of work at a real advertising agency should be enough to convince even the most cynical that an agency's involvement in a client's business is anything but superficial. Every cubicle on every floor at an agency is occupied by someone intensely involved in improving the client's day-to-day business, shepherding its assets more wisely, sharpening its business focus, widening its market, improving its product, and creating new products.

GO ON TO NEXT PAGE ▶

Advertising isn't just some mutant offspring of capitalism. It isn't a bunch of caffeine junkies dreaming up clever ways to talk about existing products. Advertising is one of the main gears in the machinery of a huge economy, responsible in great part for creating and selling products that contribute to one of the highest standards of living the world's ever seen. That three-mile run you just clocked on your Nike+ GPS watch was created in part by an ad agency: R/GA. The Diet Coke you had when you cooled down at home was co-created with an agency called SSCB. These are just two of tens of thousands of stories out there where marketer and agency worked together to bring a product — and with it, jobs and industry — to life. (*Hey, Whipple, Squeeze This: The Classic Guide to Creating Great Ads*, by Luke Sullivan, Sam Bennett, and Edward Boches — Pages 19–20)

5. Which of the following is the main purpose of Passage 1?

(A) Provide an overview of a variety of aspects of TV advertising, especially as it relates to fashion merchandising.

(B) Warn that when it comes to TV advertising, there so many down sides that the value for fashion products is entirely overrated.

(C) Persuade people in the fashion industry who might be reluctant to advertise on TV that is has many unimagined benefits.

(D) Lead the reader to an understanding of how an integrated campaign utilizes an array of media, including billboards, magazines, and in-store photographs.

6. Which of the following offers the best explanation of why the authors of Passage 2 use the words "mutant offspring" and "caffeine junkies."

(A) To express their disgust at the field of advertising, which they view as unredeemably shallow and without merit.

(B) To question the sincerity of some people in advertising without entirely dishonoring all of its practitioners or the field as a whole.

(C) To discredit the argument that advertising is inconsequential by casting it as crude and uninformed.

(D) To shock readers out of their complacency and encourage them to take a specific recommended action.

7. Which of the following do both Passage 1 and Passage 2 have in common? They both

(A) assert that advertising is not just a tool for selling products but a vehicle for creating them.

(B) mention the potential high cost of advertising as a consideration that must be reckoned with in an advertising campaign.

(C) discuss advertising specifically in the context of a variety of different media.

(D) cite specific brands and products by name as examples of successful advertising campaigns.

8. Which of the following statements is most accurate?

(A) Both passages are more informative than persuasive.

(B) Passage 1 is more informative than persuasive, and Passage 2 is more persuasive than informative.

(C) Passage 1 is more persuasive than informative, and Passage 2 is more informative than persuasive.

(D) Both passages are more persuasive than informative.

In most software and hardware markets, the latest and greatest product is the one that everyone wants. People like all the bells and whistles in the new product, and they gobble up the marketing literature that gives you 101 reasons why this product is the answer to all of your prayers. In the world of cryptography, almost the exact opposite is true — nothing new is trusted until it has been extensively tested by the outside world. (*Cryptography For Dummies*, by Chey Cobb — Page 34)

9. It can be reasonably inferred from this passage that

(A) Many people who read marketing literature fail to understand the importance of cryptography.

(B) Most users judge cryptography products by standards that are similar to those by which they judge other types of software and hardware products.

(C) Cryptography is unlike most other software and hardware products in at least one important respect.

(D) In the future, cryptography markets will continue to distinguish themselves even more sharply from other software and hardware markets.

Where do antibiotic-resistant bacteria come from? The answer lies in the idea of natural selection, or survival of the fittest. Bacteria reproduce very quickly, and little changes in the traits of individuals occur with each generation so that even all the bacteria of one species aren't the same as each other. When people use antibiotics, the susceptible bacteria die first, leaving behind the most resistant cells. These resistant cells multiply and take over the available space. As this scenario repeats over time, populations of bacteria eventually become super-resistant to antibiotics, explaining why sometimes doctors don't have the drugs to help people who are infected with an antibiotic resistant bacteria. (*Biology for Dummies* by Rene Fester Kratz — Page 373)

10. The author uses the question that begins this passage to

(A) Express the central idea of the passage and then answer it.

(B) Express a supporting idea within the passage and then answer it.

(C) Encourage the reader to figure out their own answer to the question.

(D) Imply that the question cannot be answered simply or conclusively.

Why was radio's early history ignored? The "radio boom" of the 1920s seemed so sudden, the diffusion of the device so dramatic, and the entry of voice and music into people's homes without any connecting wires so miraculous (the word most frequently used at the time), that it eclipsed what came before. By contrast, the exchange of the Morse code between wireless operators could seem like an irrelevant prehistory with not nearly the cultural, political, and economic impact of radio. (*A Companion to the History of American Broadcasting*, edited by Aniko Bodroghkozy — Page 27)

11. According to the passage, all of the following events would be considered part of the "radio boom" EXCEPT

(A) An evening of orchestral music is broadcast locally from a small radio station in Phoenix, Arizona.

(B) A politician broadcasts a speech about the shortcomings of his rival in an upcoming mayoral election in New York.

(C) A sinking ship in the North Atlantic issues a distress call in Morse code, prompting a nearby ship to rescue all on board.

(D) A radio advertisement for a new brand of women's cold cream causes the product to become a topic of conversation in beauty parlors throughout New England.

We do not know the identity of the person who translated *Everyman* into English from the original Dutch version of the play, *Elckerlijc*. But we do know that the author of *Le Morte Darthur* (completed 1469 or 1470), Sir Thomas Malory, was a layman, a knight of the gentry class. With its interest in the chivalric exploits of Arthurian knighthood, the *Morte*, we might think, is untouched by the dogma that the Church in 15th-century England expected people like Malory to swallow. And yet Malory's "Tale of the Sankgreal" might be said to complement that dogma, in that it shows how Sir Lancelot, who is initially a failure on the Grail Quest, proceeds to obtain a partial vision of the Grail after he confesses his sins to a priest and does penance for them. (*A Concise Companion to Middle English Literature*, Marilyn Corrie, editor — Pages 49–50)

12. The author implies that the response to Malory's "Tale of the Sankgreal" by Church elders of 15th-century England would likely lean toward

(A) Approval, because the tale accepts Church dogma.

(B) Approval, because the tale rejects Church dogma.

(C) Disapproval, because the tale accepts Church dogma.

(D) Disapproval, because the tale rejects Church dogma.

Reading well out loud is one of the most essential traits of a voice actor. Even some of the most enthusiastic bookworms have trouble articulating a well-phrased passage when asked to do so out loud. A skilled voice actor can read aloud with ease. If you stumble over your words and need to start over again, then you need more practice. Try reading books and newspaper and magazine articles aloud. Doing so is a great way to practice reading a variety of writing styles, which is something you'll encounter in the voice-over industry.

Read everything you can find and interpret in various ways. Finding material doesn't have to be difficult. You can read the back of a cereal box or leaf through your favorite book and focus on a particular passage. (*Voice Acting For Dummies*, by Stephanie Ciccarelli and David Ciccarelli — Page 21)

13. The main purpose of this passage is to

(A) Describe a key skill that voice actors need and explain how to hone it.

(B) Discourage people with poor reading skills from becoming voice actors.

(C) Distinguish between people who enjoy reading and those who are skilled at reading aloud.

(D) Explain to would-be voice actors how practicing with one form of text is superior to practicing with other forms.

The processes that formed the infant Earth set the stage for its subsequent evolution into the dynamic and habitable planet we know today. Probing these earliest events is problematic in the extreme, as over four billion years of subsequent evolution have obscured most of Earth's primary features. The challenge is to unravel the series of complicated and bewildering events that began with condensation in the solar nebula, proceeded through the cataclysm of planetary accretion from which emerged a hot and molten proto-Earth, and ended with the large-scale differentiation including core formation, mantle differentiation, and proto-crust formation. (*The Early Earth: Accretion and Differentiation*, by James Badro and Michael J. Walter — Page 1)

14. Why is understanding what the Earth was like at the time of its formation so difficult to achieve?

(A) Earth has changed considerably in the billions of years since it was formed, concealing the topography of that time.

(B) Planetary accretion had disastrous effects on key evidence that might tell scientists how the planet originally formed.

(C) Three important scientific theories that describe three distinct periods in the evolution of the Earth are all at odds with one another.

(D) The most promising early theory about how proto-Earth formed was disproved when new information about large-scale differentiation was discovered.

Herbivores (animals that can digest the fibrous tissue of plants) subsist primarily on plants and plant materials, converting grassland products to high-quality meat and milk foods that complement the nutritive value of plant products for humans. Ruminants and other herbivores contribute to human well-being by producing meat and milk products that are rich sources of proteins, fats, vitamins, and minerals. In addition, ruminants provide non-food products of value, such as hides, wool, and horns for clothing, implements, and adornments. (*Forages, Volume 1: An Introduction to Grassland Agriculture*, edited by Michael Collins et al. — Page 4)

15. According to the passage, all of the following statements about ruminants are true EXCEPT

(A) Ruminants are herbivores, but herbivores are not necessarily ruminants.

(B) Some ruminants produce both food and non-food products.

(C) People who do not eat animals do not eat ruminants.

(D) Ruminants cannot directly digest the fibrous tissue of plants.

(1) Like the rest of the modern world, the USA and Canada have had a history of bed bug infestations. (2) Although the eventual return of the bed bug might have been predicted, Americans were taken by surprise when bed bugs began to reappear in the late 1990s. (3) Major urban centers like New York, Chicago, and Cincinnati were among the first cities to become re-infested with this pest. (4) Bed bug infestations spread quickly and infestations soon began to be reported all over the country. (5) By 2006, a national survey documented bed bug infestations within all 50 states.

(6) Because bed bugs had not been a problem for almost 40 years, most pest management companies (PMCs) had no experience in controlling infestations, and did not know which products would kill bed bugs. (7) The EPA and CDC joined forces to officially declare bed bugs a public health pest in 2007. (8) This declaration was intended to allow national research funding agencies (like the National Institute of Health) to justify the funding of bed bug research. (Excerpt from *Advances in the Biology and Management of Modern Bed Bugs*, edited by Stephen L. Doggett, Dini M. Miller, and Chow-Yang Lee)

GO ON TO NEXT PAGE

16. Which sentence helps to explain why the resurgence of bed bugs spiraled out of control for about 10 years?

(A) Sentence 4

(B) Sentence 5

(C) Sentence 6

(D) Sentence 8

Paleontology is fundamental to geology. From the study of environmental tolerances of their living relatives, fossils provide the clearest insight into the nature and development of ancient Earth environments. Paleontology also has a pivotal role in biology, in providing proof of the evolution and diversification of life on Earth. Despite this, paleontology is not popular with students. Dinosaurs have universal appeal, but simple invertebrate fossils appear insignificant and dull. The most common accusation is the "plethora of long names" pervading the subject. In reality, paleontology is more than a catalogue of fusty-sounding names: it is a living subject of fundamental importance to both geology and evolutionary biology. (*Understanding Fossils: An Introduction to Invertebrate Palaeontology* by Peter Doyle — Page 2)

17. Which of the following is the most likely reason the writer chose to place quotation marks around the words "plethora of long names"?

(A) To dispute the accusation that paleontology has an overabundance of long names.

(B) To defend the overabundance of long names in paleontology as necessary to appreciating its importance.

(C) To highlight how often he has heard this particular complaint about paleontology.

(D) To celebrate the fact that paleontology is, in essence, a catalogue of fusty-sounding names.

We know that through global technologies children access a largely global fare consisting of cartoons, situation comedies, soap operas, action-adventure serials, as well as Disney-style and Hollywood movies mostly produced in Euro-American cultures in the West. In addition, they also watch programs that come from other parts of the world, such as Latin-American tele-novellas; Japanese and Korean animated series; or local co-productions of the American series *Sesame Street*. Thus, in shifting from a focus on global access to technologies to the contents distributed, we know that children around the world are entertained through technologies that bring them popular culture-style productions, originating primarily in the USA but also in other parts of the world. These productions are diffused through a process theorized as *The Megaphone Effect*. The USA, according to this thesis, collects and adopts cultural artifacts from around the world, adapts them to the American "palate," and then USA-based media conglomerates serve as a megaphone, spreading them to other markets and turning them into a global phenomenon. (*Children and Media: A Global Perspective* by Dafna Lemish — Page 6)

18. In reference to *The Megaphone Effect*, the author uses the words "theorized" and "according to this thesis" in order to

(A) Underscore that she personally agrees with this idea.

(B) Underscore that she personally doesn't agree with this idea.

(C) Distinguish this idea from an objective fact that must be taken as true.

(D) Denounce this idea as having no factual basis whatsoever.

Directions for Questions 19–20: Each of the following sentences has a blank indicating that something has been left out. Beneath each sentence are four words or phrases. Choose the response that best fits the meaning of the sentence.

19. Finding that his friends had already left the party by the time he arrived, Erik _____ his plans and joined them at the beach.

    (A) accustomed

    (B) amended

    (C) misrepresented

    (D) regulated

20. Despite her disappointment at losing her first two tennis tournaments, Jasmyn _____ and in her senior year became the tennis champion in her division.

    (A) capitulated

    (B) concurred

    (C) persevered

    (D) tolerated

STOP

Writing Test

DIRECTIONS: This test, which contains items that measure your ability to write and improve English sentences in a given context, includes five early drafts of essays. Read each essay and then choose the best answer to the questions that follow.

(1) Tokyo is certainly on the short list of world cities that many Americans would love to visit. (2) With its mix of ancient and modern culture — not to mention its virtually endless supply of opportunities for shopping, dining, and entertainment, Japans capital city is an ideal tourist destination. (3) In summary, despite its many attractive features, some potential American travelers are intimidated by such a journey. (4) Otherwise bold vacationers are daunted by a variety of perceived deterrents; cultural differences, a formidable language barrier, and a city whose sheer magnitude may seem impenetrable.

(5) This is unfortunate, because these obstacles are for the most part quite surmountable. (6) Since the 1950s, Japanese culture has actively embraced the West, and nowhere is this more true than in Tokyo. (7) English is commonly spoken among hotel workers, restaurant staff, and others who work in the city's ever-burgeoning tourism industry. (8) Additionally, the Internet offers a wide supply of resources to help potential visitors narrow down their options from among the many available and choose a location that suits their needs and preferences.

1. Which is the best version of the underlined portion of Sentence 2 (reproduced here)?

With its mix of ancient and modern culture — not to mention its virtually endless supply of opportunities for shopping, dining, and <u>entertainment, Japans</u> capital city is an ideal tourist destination.

   (A) (as it is now)
   (B) entertainment — Japans
   (C) entertainment, Japan's
   (D) entertainment — Japan's

2. Which is the best version of the underlined portion of Sentence 3 (reproduced here)?

<u>In summary</u>, despite its many attractive features, some potential American travelers are intimidated by such a journey.

   (A) (as it is now)
   (B) Consequently
   (C) However
   (D) Finally

3. Which is the best version of the underlined portion of Sentence 4 (reproduced here)?

Otherwise bold vacationers are daunted by a variety of perceived <u>deterrents</u>; cultural differences, a formidable language barrier, and a city whose sheer magnitude may seem impenetrable.

   (A) (as it is now)
   (B) deterrents —
   (C) deterrents,
   (D) deterrents:

4. Which is the best version of the underlined portion of Sentence 5 (reproduced here)?

This is unfortunate, because these obstacles are for the most part quite surmountable.

(A) (as it is now)

(B) This makes me sad

(C) This is a downer

(D) I disagree

5. The writer would like to end the essay with a sentence that best sums up their viewpoint with an appropriate tone. Which choice best serves this purpose?

(A) All in all, while Tokyo may not be the ideal vacation spot, these prospects can help to make a difficult stay more pleasant.

(B) Last but not least, Tokyo is equally attractive to business travelers, who find its stable economy and openness to foreign exchange a potentially lucrative investment of their time and effort.

(C) To counter this trend, an increasing number of airlines are now adding non-stop options for flying to Tokyo's Narita and Haneda Airports from a widening choice of U.S. cities.

(D) In short, while Tokyo is still an adventurous destination, most Americans who venture there relate that the many positive experiences far outweigh the few hardships.

(1) Pasteurization is a process by which foods (especially liquid foods, such as milk, fruit juices, and beer) are heated to destroy potentially dangerous microorganisms. (2) Invented by scientist Louis Pasteur, for whom it is named, pasteurization not only makes a wide variety of foods safer to eat, but also increases their shelf life by eliminating bacteria that propagate spoilage.

(3) Yet some opponents of pasteurization are concerned that while the procedure indisputably eliminates harmful pathogens, it can also eliminate important nutrients and natural probiotics ("good bacteria") that human's bodies need to survive and thrive. (4) Others claim that food simply tastes better when not subjected to heating and cooling that is resounding with the pasteurization process.

(5) Proponents of pasteurization fire back that more than a century of scientific research has failed to unearth any reasonable evidence that unpasteurized products are safer than those having been pasteurized. (6) They also challenge naysayers to provide proof that even the most discerning palate can taste the difference between pasteurized and unpasteurized foods.

6. Which is the best version of the underlined portion of Sentence 1 (reproduced here)?

Pasteurization is a process by which foods (especially liquid foods, such as milk, fruit juices, and beer) are heated to destroy potentially dangerous microorganisms.

(A) (as it is now)

(B) foods, especially

(C) foods — especially

(D) foods: especially

7. Which is the best version of the underlined portion of Sentence 2 (reproduced here)?

Invented by scientist Louis Pasteur, for whom it is named, pasteurization not only makes a wide variety of foods safer to eat, but also increases their shelf life by eliminating bacteria that propagate spoilage.

(A) (as it is now)

(B) food safety and shelf life are both increased by the elimination of bacteria that propagate spoilage in pasteurization.

(C) the elimination of bacteria that propagates spoilage causes pasteurization to make a wide variety of foods safer to eat while increasing shelf life.

(D) by pasteurization are not only a wide variety of foods made safer to eat, but also shelf life is increased by eliminating bacteria that propagates spoilage.

8. Which is the best version of the underlined portion of Sentence 3 (reproduced here)?

Yet some opponents of pasteurization are concerned that while the procedure indisputably eliminates harmful pathogens, it can also eliminate important nutrients and natural probiotics ("good bacteria") that human's bodies need to survive and thrive.

(A) (as it is now)

(B) a human's body need

(C) humans' bodies needs

(D) the human body needs

9. Which is the best version of the underlined portion of Sentence 4 (reproduced here)?

Others claim that food simply tastes better when not subjected to heating and cooling that is resounding with the pasteurization process.

(A) (as it is now)

(B) forthcoming from

(C) inherent in

(D) unavoidable on

10. Which is the best version of the underlined portion of Sentence 5 (reproduced here)?

Proponents of pasteurization fire back that more than a century of scientific research has failed to unearth any reasonable evidence that unpasteurized products are safer than those having been pasteurized.

(A) (as it is now)

(B) those that are pasteurized

(C) the ones that have now been pasteurized

(D) those that have been subjected to the process of pasteurization

(1) Some people say, "Less is more." (2) Well, in an Italian-American family, when it comes to Sunday dinner, more is more — and then some! (3) At least, that's how my grandmother saw it.

(4) The traditional Italian big meal of the day is served midday, with a two-hour break from work to accommodate travel time to and from work, not to mention the meal itself. (5) But with the longer commutes and one-hour lunchtime more common in the United States, many Italian families adopted a more typical American routine, postponing dinner until after work. (6) Although Italy fought against the U.S. during the Second World War, my family considered themselves citizens of their adopted country, and did not support Mussolini's aggression in Europe. (7) Because my grandfather took the train each day from the suburbs to New York's Garment District, where he worked long hours to support his family, a traditional midday dinner would have been out of the question.

(8) Sundays, however were reserved for the traditional Italian trifecta of family, food, and fun. (9) Each week, my grandmother would spend hours upon hours — often, late into Saturday night — cooking and baking in anticipation of the big event, most often assisted by one of her sisters or another family member. (10) Although the apartment was small, at least one or two relatives were almost always staying for the weekend in the spare room; my sister and I, however, were small enough to sleep in the living room. (11) We were cozy, to be sure, but the warmth and closeness we felt as a family more than made up for any inconvenience.

(12) Dinner started soon after noon, as course after course was served: appetizers, soup, salad, bread, pasta, vegetables, and usually at least two varieties of meat. (13) Coffee and dessert were often postponed to give us all a chance to work up an appetite again. (14) Meanwhile, the table was cleared for a long game of poker with a maximum bet of two pennies.

(15) All of the relatives would leave at the end of the evening with their hearts and bellies equally full. (16) More often than not, at least one proclamation could be heard to the effect that "The diet starts tomorrow!" (17) Yet it seemed that before the last dish was washed and the leftovers placed precariously into the overstuffed fridge, my grandmother will be already anticipating the following week. (18) "I think Uncle Benny is coming," she might say to me, "He always likes it when we have lamb."

11. The writer is considering deleting the first paragraph. Should the writer make this deletion?

(A) Yes, because this paragraph includes irrelevant information that distracts the reader from the author's main point.

(B) Yes, because this paragraph is too informally written in comparison with the rest of the essay.

(C) No, because this paragraph establishes an important difference between traditional Italian and Italian-American culture.

(D) No, because this paragraph sets the tone by establishing the essay as an informal first-person narrative.

12. Which sentence blurs the focus of the second paragraph and should therefore be removed?

(A) Sentence 4

(B) Sentence 5

(C) Sentence 6

(D) Sentence 7

GO ON TO NEXT PAGE

13. Which is the best version of the underlined portion of Sentence 8 (reproduced here)?

Sundays, however were reserved for the traditional Italian trifecta of family, food, and fun.

(A) (as it is now)

(B) Sundays, however,

(C) Sunday's, however,

(D) Sundays', however

14. Which is the best version of the underlined portion of Sentence 15 (reproduced here)?

All of the relatives would leave at the end of the evening with their hearts and bellies equally full.

(A) (as it is now)

(B) with his heart and his belly

(C) with his or her hearts and bellies

(D) with our heart and our belly

15. Which is the best version of the underlined portion of Sentence 17 (reproduced here)?

Yet it seemed that before the last dish was washed and the leftovers placed precariously into the over-stuffed fridge, my grandmother will be already anticipating the following week.

(A) (as it is now)

(B) already anticipated

(C) was already anticipating

(D) would have already been anticipating

(1) Optimists are "glass half-full" people, who tend to view believe that even a difficult predicament will work out well. (2) Likewise, pessimists are "glass half-empty" folks who often believe that a relatively good situation won't last.

(3) Why do some people tend to be optimists and others pessimists? (4) In his book *Learned Optimism*, psychologist, Martin Seligman addresses this question. (5) He finds that in many situations, an optimistic perspective can help propel a person toward a favorable outcome in an otherwise neutral situation. (6) He then devotes the greatest share of the book to teaching people how to *choose* optimism when appropriate.

(7) Seligman cites three dimensions of optimism and pessimism: permanence, personalization, and pervasiveness. (8) When faced with a troubling turn of events, pessimists tend to *maximize* the problem along these three dimensions. (9) They tend to tell themselves, "This situation can't be fixed (permanence), it's my fault (personalization), and it's typical of my whole life (pervasiveness)." (10) Optimists, on the other hand, tend to *minimize*: "This problem is fixable, I'm not solely to blame for it, and it's not typical of the rest of my life."

(11) Seligman identifies these *attributions* — explanations that people give about why things went wrong — as a key factor in optimism. (12) The first step is noticing the difference between an event (for example, the car won't start) and an attribution about it: "Things never work out for me! I never should have bought this car. That salesman probably sold it to me because he knew I was stupid." (13) Noticing this difference opens up the possibility of creating a new and more empowering attribution: "Cars break down. It's nothing personal. It's probably because of the cold weather. And anyway, it can be fixed."

(14) In a variety of experimental settings, Seligman and others have shown that optimism can, in fact, be taught. (15) Moreover, teaching people how to reinterpret events in an even more optimistic sort of way can have widespread benefits. (16) For example, people who learn this skill are measurably healthier: Their anxiety and depression tend to be reduced, and they report having fewer physical health problems.

(17) In summary, *Learned Optimism* is an important book with information that the reader can apply immediately to his or her own life.

16. Which is the best version of the underlined portion of sentence 2 (reproduced below)?

    Likewise, pessimists are "glass half-empty" folks who often believe that a relatively good situation won't last.

    (A) (as it is now)
    (B) For instance
    (C) In contrast
    (D) Therefore

17. Which is the best version of the underlined portion of sentence 4 (reproduced below)?

    In his book Learned Optimism, *psychologist, Martin Seligman addresses this question.*

    (A) (as it is now)
    (B) In his book *Learned Optimism*, psychologist Martin Seligman
    (C) In his book *Learned Optimism*, psychologist Martin Seligman,
    (D) In his book, *Learned Optimism* psychologist Martin Seligman,

18. The writer is considering adding the following sentence after sentence 11:

    In 1998, Seligman was elected the president of the American Psychological Association in recognition of his ongoing contribution to the field.

    Assuming that this information is true, should the writer add this sentence?
    (A) Yes, because it provides an interesting piece of information biographical information.
    (B) Yes, because it helps to convince the reader that Seligman is an important psychologist.
    (C) No, because it interrupts the discussion of attributions.
    (D) No, because it makes a statement about Seligman that is contradicted elsewhere in the essay.

19. Which is the best version of the underlined portion of sentence 15 (reproduced below)?

    Moreover, teaching people how to reinterpret events in an even more optimistic sort of way can have widespread benefits.

    (A) (as it is now)
    (B) in a more optimistical way
    (C) more optimistically
    (D) more optimistic

GO ON TO NEXT PAGE

20. Which is the best version of the underlined portion of sentence 16 (reproduced below)?

For example, people who learn this skill are measurably healthier: Their anxiety and depression tend *to be reduced, and they report having fewer physical health problems.*

(A) (as it is now)

(B) tends

(C) tending

(D) is tending

(1) For years, though her face was rarely seen, she could be heard in dozens of major Hollywood movies as the voice behind such talented actresses as Marilyn Monroe, Audrey Hepburn, and Sophia Loren. (2) In some cases, her name wasn't even mentioned in the credits. (3) And in one case, her voice was recorded in secret to replace vocals by Natalie Wood.

(4) Her stage name was Marni Nixon, and her career in movies, television, and the stage spanned more than 60 years.

(5) Born Margaret Nixon McEathron in 1930, the operatically-trained singer found an unusual niche as "ghost singer": the singing voice behind the actress seen on the screen, who lip-syncs the words to songs instead of singing them herself. (6) Many of Marni Nixon's roles that are now considered legendary were at the time kept secret. (7) This choice helped to preserve the illusion that the actress appearing on screen was performing the role herself. (8) One of Nixon's early roles was in *The King and I*, singing the lead part of Anna, played by Deborah Kerr. (9) Other prominent roles included singing for Audrey Hepburn in *My Fair Lady* and Natalie Wood in *West Side Story*.

(10) In other cases, Nixon provided an assist to actresses whose singing talents were limited in range. (11) Additionally, in *Gentlemen Prefer Blondes*, she gave a boost to Marilyn Monroe's breathy rendition of Diamonds Are a Girl's Best Friend by filling in the difficult high notes.

(12) Nixon contributed unseen and often uncredited in almost a dozen film roles before being cast, in 1965, as Sister Sophia in *The Sound of Music*, her first onscreen performance. (13) As the popularity of movie musicals waned in the late 1960s, however, her career direction shifted to solo concerts at Carnegie Hall and other prominent venues. (14) She also performed in numerous operas, on the Broadway stage, and on television, where she received four Emmy awards.

(15) Her final uncredited ghost singing performance was in 1998, as Grandmother Fa in Disney's *Mulan*. (16) Her stage singing career continued until only a few years before her death in 2016, at the age of 86, in New York City.

21. The writer is considering adding the following sentence after sentence 4:

Surprisingly, however, she was neither seen on screen nor credited for her work in many major Hollywood movies.

Should the writer make this addition, and why?

(A) Yes, because this sentence provides information that is important for the reader to understand.

(B) Yes, because this sentence explains why Marni Nixon is not better known as a singer.

(C) No, because the information presented here is provided elsewhere in the essay.

(D) No, because this sentence contradicts information presented elsewhere in the essay.

22. The writer wishes to combine sentences 6 and 7 (reproduced below).

Many of Marni Nixon's roles that are now considered legendary <u>were at the time kept secret. This choice</u> helped to preserve the illusion that the actress appearing on screen was performing the role herself.

Which of the following best achieves this purpose?

(A) were at the time kept secret, and

(B) were at the time kept secret, being what

(C) were at the time kept secret, so this

(D) were at the time kept secret, which

23. Which is the best version of the underlined portion of sentence 8 (reproduced below)?

<u>One of Nixon's early roles was in</u> The King and I, <u>singing the lead part of Anna, played by Deborah Kerr.</u>

(A) (as it is now)

(B) An early role of Nixon's was in *The King and I*, and she sang the lead part of Anna.

(C) In *The King and I*, an early role for Nixon, singing the lead part of Anna.

(D) *The King and I*, singing the lead part of Anna, was one of Nixon's early roles.

24. Which is the best version of the underlined portion of sentence 11 (reproduced below)?

<u>Additionally</u>, in Gentlemen Prefer Blondes, she gave a boost to Marilyn Monroe's breathy rendition of "Diamonds Are a Girl's Best Friend" by filling in the difficult high notes.

(A) (as it is now)

(B) For example

(C) Notwithstanding

(D) Therefore

25. Which of the following would NOT be a suitable replacement for the word *performed* sentence 14 (reproduced below)?

She also <u>performed</u> in numerous operas, on the Broadway stage, and on television, where she received four Emmy awards.

(A) accomplished

(B) acted

(C) appeared

(D) played

STOP

Arithmetic Test

DIRECTIONS: This test contains items that measure your ability to solve basic arithmetic problems. Choose the best answer to each question. Feel free to use scrap paper. You may not use a calculator.

1. Which of the following decimals is equivalent to $\frac{3}{16}$?

(A) 0.125

(B) 0.1875

(C) 0.275

(D) 0.3125

2. On a flight from Milan to New York, a total of 254 passengers each ordered a dinner whose main dish was beef, chicken, or salmon. The number of passengers who ordered beef was 79, while 88 ordered salmon. How many passengers ordered chicken?

(A) 76

(B) 77

(C) 78

(D) 87

3. $2.89 \times 4.7 =$

(A) 12.5830

(B) 13.583

(C) 125.83

(D) 135.83

4. Which of the following is the nearest estimate of the result when you multiply 29,873 by 714?

(A) 210,000

(B) 2,100,000

(C) 21,000,000

(D) 210,000,000

5. Which of the following inequalities is correct?

(A) $\frac{5}{7} < \frac{7}{9} < \frac{8}{11}$

(B) $\frac{5}{7} < \frac{8}{11} < \frac{7}{9}$

(C) $\frac{7}{9} < \frac{8}{11} < \frac{5}{7}$

(D) $\frac{7}{9} < \frac{5}{7} < \frac{8}{11}$

6. Jeanine used an exercise app to chart her progress and found that she had exercised 64 of the last 80 days. What percent of these days did she exercise?

(A) 62.5%

(B) 64%

(C) 72.5%

(D) 80%

7. What is the value of $4.32 + 0.071 + 0.58$?

(A) 1.083

(B) 4.449

(C) 4.971

(D) 5.61

8. In April, which has 30 days, Matthew worked 4 out of every 5 days at one of two different jobs. He worked at a supermarket for 15 days, and at a mobile phone kiosk for each of the remaining days that he worked. How many days did he work at the mobile phone kiosk?

(A) 9 days

(B) 10 days

(C) 12 days

(D) 15 days

9. What is 2.7654 rounded to the nearest hundredth?

(A) 2.76

(B) 2.77

(C) 2.765

(D) 2.766

10. A two-day hike to Bear Lake was measured at 21.6 kilometers. Gene hiked the first 13.85 kilometers on the first day and then set up his tent. How many kilometers did he walk on the second day in order to reach the lake?

(A) 7.21

(B) 7.75

(C) 11.69

(D) .35.45

11. Devin has been assigned to watch a series of 12-minute videos. He looks at his watch and sees that he has 1 hour and 20 minutes before he has to go to his next class. What is the maximum number of *complete* videos he can watch before he has to leave? (Assume that there is no time lost between videos.)

(A) 5

(B) 6

(C) 7

(D) 8

12. Which of the following inequalities is correct?

(A) $45\% < \dfrac{45}{1,000} < 0.45\%$

(B) $45\% < 0.45\% < \dfrac{45}{1,000}$

(C) $0.45\% < \dfrac{45}{1,000} < 45\%$

(D) $0.45\% < 45\% < \dfrac{45}{1,000}$

GO ON TO NEXT PAGE

13. $\frac{5}{6} \div \frac{2}{3} =$

    (A) $\frac{5}{2}$

    (B) $\frac{5}{4}$

    (C) $\frac{5}{9}$

    (D) $\frac{9}{5}$

14. What is $\frac{7}{100} + \frac{6}{1,000}$?

    (A) $\frac{13}{100}$

    (B) $\frac{76}{100}$

    (C) $\frac{13}{1,000}$

    (D) $\frac{76}{1,000}$

15. What percent of 400 is 50?

    (A) 8%

    (B) 8.25%

    (C) 12.5%

    (D) 80%

16. When you divide 555 by 7, what is the remainder?

    (A) 2

    (B) 3

    (C) 4

    (D) 6

17. $-4 + 8 \div 2 - (-3) \times 7 =$

    (A) 21

    (B) 35

    (C) −21

    (D) −35

18. On the first day of May, Marco measured the height of a tulip at $10\frac{7}{8}$ inches. One week later, it had grown to $14\frac{3}{4}$ inches. By how much did the tulip grow during that week?

    (A) $3\frac{1}{8}$ inches

    (B) $3\frac{1}{4}$ inches

    (C) $3\frac{7}{8}$ inches

    (D) $4\frac{1}{8}$ inches

19. Beth ate lunch in a restaurant, and when the check arrived she found that she owed $12.40. If she leaves an 18% tip for the server, what will be the total amount that she pays for lunch, rounded to the nearest penny?

(A) $14.20

(B) $14.35

(C) $14.63

(D) $14.88

20. Which of the following numbers has the greatest value?

(A) 0.249

(B) $\frac{2}{5}$

(C) 0.31

(D) $\frac{37}{100}$

STOP

Quantitative Reasoning, Algebra, and Statistics Test

DIRECTIONS: This test contains items that measure your ability to do intermediate math, including basic algebra, geometry, and statistics. Choose the best answer to each question. Feel free to use scrap paper. You may not use a calculator. However, for some problems on the computer-based ACCUPLACER, you may be given access to an onscreen calculator with limited functionality.

1. $(x^2 - 7)(x + 5) =$

(A) $x^3 + 5x^2 - 7x - 35$

(B) $x^3 - 7x^2 + 5x - 35$

(C) $x^2 + 2x - 35$

(D) $x^2 - 2x - 35$

2. The number line shows the solution set for which of the following inequalities?

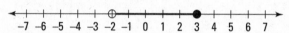

(A) $-2 \le x < 3$

(B) $-2 < x \le 3$

(C) $x \le -2$ or $x > 3$

(D) $x \le -2$ or $x \ge 3$

3. Simplify $\dfrac{x^2}{y^2} \cdot \dfrac{y^3}{x^4 y^2}$.

(A) $x^2 y$

(B) $\dfrac{x^2}{y}$

(C) $\dfrac{y}{x^2}$

(D) $\dfrac{1}{x^2 y}$

4. What is $|x - 8| - |x|$ when $x = 5$?

(A) -8

(B) -2

(C) 2

(D) 8

5. $5x - 2y = 36$

$\quad\ x = y + 3$

This system has a solution in the xy-plane. Which of the following statements is true about this solution?

(A) Both x and y are positive.

(B) Both x and y are negative.

(C) x is positive and y is negative.

(D) x is negative and y is positive.

6. An entrance to a movie theater is just wide enough to allow 42 people per minute to file into the theater. At this rate, what is the minimum number of *whole minutes* required to allow the theater to fill to its 725-person capacity?

(A) 17

(B) 18

(C) 19

(D) 20

7. Mr. Syed asked each of his 24 fifth-grade students to write down how many siblings each of them has. According to the data, which of the following numbers is the median number of siblings?

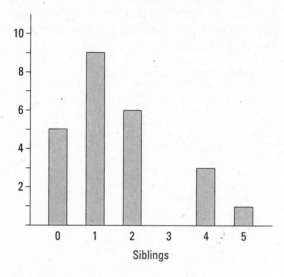

Siblings

(A) 1 sibling

(B) 1.5 siblings

(C) 2 siblings

(D) 2.5 siblings

8. The number of dollars in Geoffrey's retirement account can be modeled by the equation $D(x) = 6000 + 750x$, where x is the number of months after the account was opened. Which of the following is the best interpretation of what the number 750 stands for in this equation?

(A) The number of dollars that Geoffrey initially placed in the account.

(B) The number of dollars that Geoffrey initially placed in the account plus the amount that he places in it each month.

(C) The number of dollars that Geoffrey places in the account each month.

(D) The number of dollars that Geoffrey places in the account every x months.

9. To raise money for his soccer team, Kamal sold boxes of cookies in two varieties: Peanut Butter Chocolate for $3.50 per box, and Vanilla Strawberry for $2.50 per box. He sold a total of 109 boxes and earned $317.50 in revenue. How many boxes of Vanilla Strawberry did he sell?

(A) Fewer than 40

(B) More than 40 and fewer than 50

(C) More than 50 and fewer than 60

GO ON TO NEXT PAGE

(D) More than 60

10. $A = \{1,2,3,4,5,6,7\}$
$B = \{2,4,6,8,10\}$
$C = \{5,6,7,8,9,10\}$

Sets A, B, and C are shown here. Which of the following represents $A \cap (B \cup C)$ (the intersection of set A with the union of sets B and C)?

(A) $\{1,2,3,4,5,6,7,8,10\}$

(B) $\{2,4,5,6,7,8,9,10\}$

(C) $\{2,4,5,6,7\}$

(D) $\{5,6,7,8,10\}$

11. The following table provides information about which of the 144 seniors currently enrolled at Mike Nesmith High School are on the honor roll and which play sports on one of the school's teams. If you pick a student at random from among all those who are on a sports team, what is the probability that they will also be on the honor roll?

	Sports Team	No Sports Team	Total
Honor Roll	17	14	31
Not Honor Roll	66	47	113
Total	83	61	144

(A) $\frac{17}{31}$

(B) $\frac{17}{83}$

(C) $\frac{31}{144}$

(D) $\frac{83}{144}$

12. If 48 divided by the quantity of a number minus 9 equals –6, what is the number?

(A) –17

(B) –3

(C) 1

(D) 17

13. Which of the following is equivalent to $\frac{1}{64}$?

(A) 8^2

(B) 6^{-2}

(C) 4^{-3}

(D) $\frac{1}{2^{-6}}$

14. The following table shows data on the five most widely spoken languages in the world. Which of the following statements is true?

Language	First Language Speakers	Second Language Speakers	Total
English	380	740	1120
Chinese	910	200	1110
Hindustani	330	370	700

Language	First Language Speakers	Second Language Speakers	Total
Spanish	440	70	510
Arabic	290	130	420

Data is shown in millions of speakers.

(A) More than twice as many people speak English as a second language than as a first language.

(B) Chinese is the only language spoken more widely as a first language than as a second language.

(C) More people speak Hindustani as a second language than speak Spanish as a first language.

(D) More than twice as many people speak Arabic as a first language than as a second language.

15. If $\frac{3}{4}x - \frac{1}{3} = \frac{1}{2}x + 1$, what is the value of x?

(A) $\frac{7}{3}$

(B) $\frac{11}{3}$

(C) $\frac{14}{3}$

(D) $\frac{16}{3}$

16. The rectangular game room at a local YMCA has a length of 13 meters and an area of 91 square meters. If p equals the perimeter of the room in meters, which of the following is true?

Area = 91 m²

13 m

(A) $0 < p < 25$

(B) $25 < p < 50$

(C) $50 < p < 75$

(D) $75 < p < 100$

17. Below are the 20 midterm grades for Ms. Callahan's Algebra I class. Which of the following statements about this data is true?

87, 73, 91, 79, 95, 84, 83, 32, 68, 72, 94, 88, 76, 79, 85, 84, 93, 75, 81, 73

(A) There is an outlier that skews the data left, so the median is a more reliable indicator of center than the mean.

(B) There is an outlier that skews the data right, so the mean is a more reliable indicator of center than the median.

(C) There is no outlier, so the mean is a more reliable indicator of center than the median.

(D) There is no outlier, so the mean and median are equally reliable indicators of center.

GO ON TO NEXT PAGE

18. If $c = \dfrac{2a-3b}{5b}$, what is the value of a in terms of b and c?

(A) $\dfrac{5bc+3b}{2}$

(B) $\dfrac{5bc-3b}{2}$

(C) $\dfrac{5bc+2b}{3}$

(D) $\dfrac{5bc-2b}{3}$

19. What is the length of the line segment PQ?

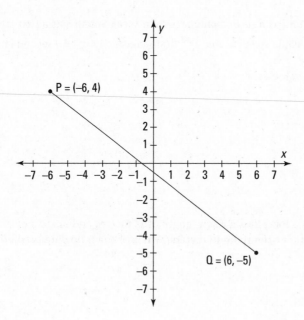

(A) 12

(B) 13

(C) 14

(D) 15

20. What is the y-intercept of a line on the xy-plane that has a slope of -4 and passes through the point $(-3, 5)$?

(A) 17

(B) 12

(C) -2

(D) -7

STOP

Advanced Algebra and Functions Test

DIRECTIONS: This test contains items that measure your ability to do math at an introductory college level. Choose the best answer to each question. Feel free to use scrap paper. You may not use a calculator. However, for some problems on the computer-based ACCUPLACER, you may be given access to an onscreen calculator with limited functionality.

1. If $\dfrac{1}{a} + \dfrac{1}{2a} = b$, what is the value of a in terms of b?

(A) $\dfrac{2}{3b}$

(B) $\dfrac{3}{2b}$

(C) $\dfrac{2b}{3}$

(D) $\dfrac{3b}{2}$

2. $(2x - 3)(x^2 + 4x - 5)$ is equivalent to which of the following expressions?

(A) $2x^3 + 5x^2 + 2x + 15$

(B) $2x^3 - 5x^2 - 2x + 15$

(C) $2x^3 + 5x^2 - 22x + 15$

(D) $2x^3 - 5x^2 + 22x + 15$

3. In the following figure, lines AB and CD are parallel, with AD and BC intersecting at E. If the lengths of AE, BE, and EC are 2, 4, and 5 respectively, what is the length of ED?

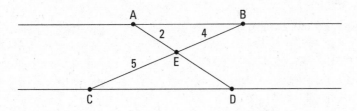

(A) 2.4

(B) 2.5

(C) 2.6

(D) 2.8

4. If $f(x) = x^2 - 4x$ and $g(x) = 3x + 6$, what is the value of $\dfrac{f(6)}{g(-4)}$?

(A) 2

(B) 1

(C) -1

(D) -2

GO ON TO NEXT PAGE

5. What is the equation of the line that is parallel to the following line and passes through the origin?

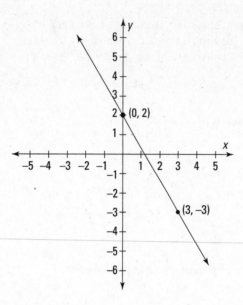

(A) $y = \frac{3}{5}x$

(B) $y = \frac{5}{3}x + 2$

(C) $y = -\frac{3}{5}x + 2$

(D) $y = -\frac{5}{3}x$

6. All of the following are functions EXCEPT

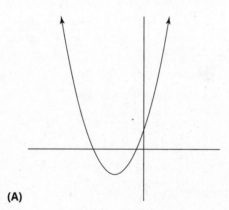

(A)

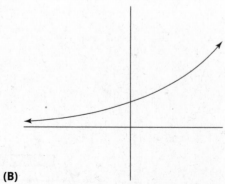

(B)

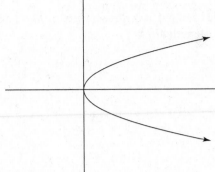

(C)

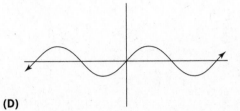

(D)

7. If a and b are the two solutions to the equation $2x^2 - 14x = -20$, which of the following is a possible value of $a - b$?

(A) 3

(B) 5

(C) 7

(D) 10

8. Which of the following equations is the function shown in this xy-graph?

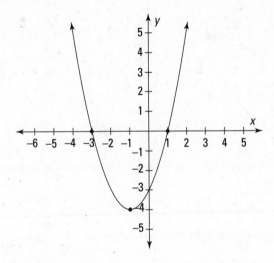

(A) $y = (x+1)(x+3)$

(B) $y = (x+1)(x-3)$

(C) $y = (x+1)^2 - 4$

(D) $y = (x-1)^2 - 4$

GO ON TO NEXT PAGE

9. Points *A* and *B* are two points on the circle centered at *O*, shown here. If $OB = 4$ and the measure of angle *AOB* is 40 degrees, what is the arc length *AB*?

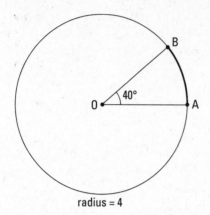

radius = 4

(A) $\dfrac{2\pi}{9}$

(B) $\dfrac{4\pi}{9}$

(C) $\dfrac{8\pi}{9}$

(D) $\dfrac{16\pi}{9}$

10. Jonathan invested $2000 in a CD that paid 4% interest compounded annually. He used the equation $f(t) = 2000(n)^t$ to calculate the amount his CD will be worth in *t* years. Which of the following values should he use for *n*?

(A) 0.04

(B) 0.4

(C) 1.04

(D) 1.4

11. What is the value of *x* if $\dfrac{x+2}{x-3} = \dfrac{x-5}{x-1}$?

(A) $\dfrac{13}{7}$

(B) $\dfrac{13}{9}$

(C) $\dfrac{17}{7}$

(D) $\dfrac{17}{9}$

12. Which value(s) are excluded from the domain of the function $f(x) = \dfrac{1}{x^2 - 10x + 25}$?

(A) 5

(B) –5

(C) 5 and –5

(D) No values are excluded from this function.

13. The equation $\dfrac{x-4}{4} = \dfrac{2}{x-6}$ has two possible values for x. What is the sum of these two values?

    (A) 8

    (B) 10

    (C) 12

    (D) 16

14. If $\log_6 11x = 2$, then x equals

    (A) $\dfrac{11}{36}$

    (B) $\dfrac{11}{64}$

    (C) $\dfrac{36}{11}$

    (D) $\dfrac{64}{11}$

15. In triangle XYZ, angle Y is a right angle. If $\sin X = \dfrac{7}{25}$, then $\sin Z =$

    (A) $\dfrac{7}{24}$

    (B) $\dfrac{24}{25}$

    (C) $\dfrac{24}{7}$

    (D) $\dfrac{25}{7}$

16. Which of the following is the equation for this circle?

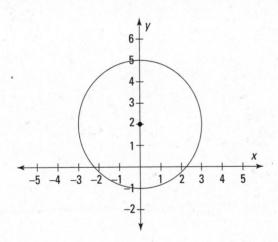

    (A) $(x-2)^2 + y^2 = 3$

    (B) $x^2 + (y+2)^2 = 3$

    (C) $(x+2)^2 + y^2 = 9$

    (D) $x^2 + (y-2)^2 = 9$

17. If $f(x) = x^2 - 4x$, what is the value of $2f(a-1)$?

    (A) $2a^2 - 8a - 6$

    (B) $2a^2 - 12a - 6$

    (C) $2a^2 - 8a + 10$

    (D) $2a^2 - 12a + 10$

GO ON TO NEXT PAGE

18. $\sqrt{x^2-1}-2=x$. Solve for x.

(A) $\dfrac{4}{5}$

(B) $\dfrac{5}{4}$

(C) $-\dfrac{4}{5}$

(D) $-\dfrac{5}{4}$

19. If $125^{x-1}=\sqrt{5}^{\,x+1}$, what is the value of x?

(A) $\dfrac{5}{7}$

(B) $\dfrac{7}{5}$

(C) $-\dfrac{5}{7}$

(D) $-\dfrac{7}{5}$

20. If $x-y=3$ and $x^2-y^2=15$, then $x^3-y^3=$

(A) 27

(B) 63

(C) 75

(D) 255

STOP

Chapter 22

Answers to Practice Test 1

Use this answer key to score the practice test in Chapter 21. To help you make sense of the test, each answer includes a detailed explanation of why it's correct.

Reading Test Answers

I hope you did well on this Reading Test. (I was crossing my fingers the whole time!) If not, flip back to Chapters 4 through 6 to refresh your knowledge of reading passages and questions.

1. **D.** Hazel Morse is described as not "given to recollections." Her past is termed as "a blurred and flickering sequence dealing with the actions of strangers." Even her mother is depicted as "hazy." And the passage continues with her portrayal as a woman who "never pondered if she might not be better occupied doing something else." These all point to Hazel's character as essentially unreflective — that is, not looking inward for insight or answers.

2. **B.** By Paragraph 3, the author has already established Hazel as relatively unimaginative and not motivated by her own inner promptings, ruling out Answer C. Her primary motive in her actions with men seems to be to flatter and charm them so as to get along, which rules out Answer A. There is no evidence in the passage that she is in any particular danger, so Answer D is also ruled out. Throughout the passage, the author is presenting both Hazel herself and her relationships as superficial, so Answer B is correct.

3. **C.** In Paragraph 7, the author tells us: "Other offers of marriage she had had, and not a few of them," which directly contradicts the statement that Hazel had had no other offers of marriage. In contrast, the statements made in Answers A, B, and D are all directly confirmed in the last three paragraphs of the passage. Thus, Answer C is correct.

4. B. The words "spontaneous" and "independent" are not synonymous, which rules out Answer A. Similarly, the words "spontaneous" and "rehearsed" are opposites, so this rules out Answer C. Although being conscientious can potentially be seen as a sign of maturity, this paraphrasing doesn't convey what the author is saying. Her implication is that after years of playing the part she thinks she must play, behavior that was once natural has now become forced. Therefore, Answer B is correct.

5. A. The passage gives general information about how fashion merchandising utilizes TV advertising. The passage is largely positive toward its subject, which rules out Answer B. It's informational rather than persuasive, which rules out Answer C. But this information focuses primarily on TV advertising, which rules out Answer D.

6. C. Passage 2 expresses great positive sentiment toward advertising, calling it "anything but superficial" and "one of the main gears in the machinery of a huge economy," which rules out Answer A. It reveals no suspicion toward the people who work in advertising, applauding them as "intensely involved in improving the client's day-to-day business," ruling out Answer B. But there is no recommended action spoken of in the passage, which rules out Answer D. The authors want to discredit the broad-strokes argument against advertising by casting light on the negative epithets typically thrown at it, so Answer C is correct.

7. D. Passage 1 mentions "Marks and Spencer (M&S), Gap, Nike, and Levi's," specifically citing "Levi's iconic advert" for "the 501 style of jeans." Passage 2 mentions the Nike+ GPS watch and Diet Coke; therefore, Answer D is correct.

8. B. Passage 1 provides information, introducing and defining a number of commonly used advertising terms (such as *reach*, *demographics*, and *adverts*), and discussing a variety of issues involved with advertising. But nowhere is an argument put forth or a counterargument refuted. In contrast, Passage 2 pushes back on the notion that advertising is of little importance or that its practitioners are cranks, arguing strongly to persuade the reader that nothing could be further from the truth. The correct answer is B.

9. C. The author opens the passage by describing a norm in software and hardware markets: "the latest and greatest product is the one that everyone wants." But then it states that "in the world of cryptography, almost the exact opposite is true." Thus, cryptography is unlike most other software and hardware products in this respect, and possibly others.

10. A. The author begins the passage with the question, "Where do antibiotic-resistant bacteria come from?" She then immediately explains the biological mechanism that answers this question, which rules out Answers C and D. This answer is not just a supporting idea, but rather the most important idea in the passage, which rules out Answer B; therefore, the Answer is A.

11. C. The passage specifically cites "the entry of voice and music into people's homes" as the decisive feature of the radio boom, in contrast to "the exchange of Morse code between wireless operators." Thus, despite the impact on the lives of those on a sinking ship, a distress call in Morse code would not be considered part of the radio boom because it did not involve voice or music and did not enter into people's homes.

12. A. The author states that "we might think" the *Morte* "is untouched by the dogma that the Church in 15th-century England expected people like Malory to swallow." This implies that the author finds this dogma implausible and imagines that his readers will as well. But then he goes on to say that the *Morte* does, in fact, "complement that dogma," implying that the tale accepts Church dogma. This acceptance would have tended to evoke approval from 15th-century Church elders, who themselves upheld this dogma.

13. **A.** The passage states the importance of "reading well out loud," and explains how to improve, which rules out Answer B. Although the passage mentions that some "book-worms have trouble" reading aloud, this isn't the main focus of the passage, which rules out Answer C. And it recommends a variety of types of texts as appropriate for practice, including "books and newspaper and magazine articles" and even "a cereal box," so Answer D is ruled out. Answer A is correct.

14. **A.** The passage opens discussing "processes that formed the infant Earth." It then states that "probing these early events is problematic" because "over four billion years of subsequent evolution have obscured most of Earth's primary features." This statement clearly indicates that Answer A is correct.

15. **D.** The words "ruminants and other herbivores" imply that ruminants are herbivores, but that herbivores are not necessarily ruminants, which rules out Answer A. Ruminants are described as producing "milk and meat products" and "non-food products," ruling out Answer B. And ruminants are herbivores, which in turn are animals, so ruminants are animals; therefore, people who don't eat animals don't eat ruminants, which rules out Answer C. Ruminants are herbivores, so they can digest the fibrous tissue of plants, making Answer D a false statement; so the correct answer is D.

16. **C.** Sentences 4 and 5 provide information about the spread of bed bug infestations, but offer no explanation, which rules out Answers A and B. Sentence 8 explains the actions described in Sentence 7, but not why the bed bug outbreak worsened, so Answer D is ruled out. Sentence 6 explains that bed bugs "had not been a problem for almost 40 years" so "most pest management companies had no experience controlling infestations." The correct answer is C.

17. **C.** The quoted words "plethora of long names" highlight that the author has heard these words stated many times with essentially the same phrasing. This is underscored by his statement that this is "the most common accusation" leveled at paleontology. He goes on to say that paleontology is "more than a catalogue of fusty sounding names," acknowledging that there is some truth to this accusation, which rules out Answer A. But he then goes on to call it "a living subject of fundamental importance to both geology and evolutionary biology," which rules out Answers B and D.

18. **C.** The words "theorized" and "according to this thesis" make clear *The Megaphone Effect* that the author is describing is a theory rather than an objective fact. The author doesn't mention whether she personally agrees with it, which rules out Answers A and B. But it would be too strong to suggest that calling this a theory denounces the idea entirely, which rules out Answer D. Answer C is correct.

19. **B.** *Accustomed* means *adapted*, but is used only in reference to an agent of change, not an abstract noun such as *plans*, so Answer A is ruled out. *Misrepresented* means *spoke falsely about*, which doesn't fit the context, so that rules out Answer C. *Regulated* means *controlled*, which also doesn't fit the context, so Answer D is ruled out. *Amended* means *changed*, so Answer B is correct.

20. **C.** *Capitulated* means *gave in*, which is the opposite of what the sentence implies, so that rules out Answer A. *Concurred* means *agreed*, which doesn't fit the context, so Answer B is ruled out. *Tolerated* means *put up with*, but is applied to a situation outside of one's own control, ruling out Answer D. *Persevered* means *pushed through hardship*, so Answer C is correct.

Writing Test Answers

How did you do on the Writing Test? Have a look at these answers, each of which includes an explanation. If you need more help, flip back to Chapters 7 through 9.

1. **D.** The word "Japan's" is possessive and, thus, requires an apostrophe, which rules out Answers A and B. The parenthetical expression, "not to mention its virtually endless supply of opportunities for shopping, dining, and entertainment," begins with a dash (—), so it must also end with one; therefore, Answer D is correct.

2. **C.** Sentence 3 tells why traveling to Tokyo might be problematic, in contrast to the opening, which discusses the positive features. Thus, a contrasting conjunction is required, and "however" is the only option given; therefore, Answer C is correct.

3. **D.** The word "deterrents" is followed by a list of three examples. This construction requires a colon (:), so Answer D is correct.

4. **A.** The passage is written in the third person and with journalistic formality, so the intro-duction of a first-person pronoun, such as "me" or "I," would be out of place; thus, Answers B and D are ruled out. And the words "This is a downer" are too informal for the tone of the passage, which rules out Answer C. The words "This is unfortunate" provide a tone of appropriate formality, so Answer A is correct.

5. **D.** The author's viewpoint is that a trip to Tokyo is so positive that it's worth the small difficulties that an American traveler may encounter along the way. Answer A misinterprets this point, and Answers B and C go astray by introducing potentially interesting but irrelevant new topics, so Answer D is correct.

6. **A.** The parenthetical expression, "especially liquid foods, such as milk, fruit juices, and beer," opens with a parenthesis, so it must close with one as well; therefore, Answer A is correct.

7. **A.** Sentence 2 begins with a dangling modifier, so what follows must begin with the subject of the sentence that is being modified: "pasteurization." Additionally, Answers B, C, and D also include a variety of awkward and non-standard phrasings, so Answer A is correct.

8. **D.** Answer A includes an incorrectly placed apostrophe; the correction would be "*humans'* bodies need," but even this is awkward. Answer B includes a mistake in subject-verb agreement; the correction would be "a human's body *needs*," but this, too, is awkward. Answer C also includes a mistake in subject-verb agreement: the correction would be "humans' bodies *need*," but once again, this is awkward. The words "the human body needs" correct subject-verb agreement and bring the sentence up to the level of standard English, so Answer D is correct.

9. **C.** Answer A ("resounding with") and Answer B ("forthcoming from") are simply incorrect word choices in the context of the sentence. Answer D is a diction error; possible correc-tions would be "unavoidable *in*" or "unavoidable *with*." "Inherent in" is proper diction with a word choice that's correct in context, so Answer C is correct.

10. **B.** Answers A and C include incorrect verb tenses and awkward phrasing. Answer D is technically acceptable but lacks the conciseness of Answer B, so Answer B is correct.

11. **D.** The first paragraph is very informal and includes the author's use of the word "my" as a cue to the reader that this essay is written in the first person. If it were left out, the essay would open with the second paragraph, which is explanatory and lacks another first-person word until its final sentence. Therefore, the paragraph should not be deleted, ruling out Answers A and B. However, the first paragraph mentions only Italian-American culture, which rules out Answer C; therefore, Answer D is correct.

12. **C.** Sentences 4, 5, and 7 all contribute to the reader's understanding of how many Italians changed their eating patterns after they immigrated to the U.S. In contrast, Sentence 6 goes off topic with an irrelevant discussion of loyalties in the Second World War. Thus, Answer C is correct.

13. **B.** In this context, the noun "Sundays" needs to be plural but not possessive, which rules out Answers C and D. The word "however" is parenthetical, so it requires a comma both before and after it. Therefore, Answer B is correct.

14. **A.** The subject of the sentence, "all of the relatives" is plural, so the possessive determiners that follow must also be plural, which rules out Answers B and C. Answer D awkwardly implies that all the relatives share one heart and one belly, so Answer A is correct.

15. **C.** The narrative takes place in the past, which rules out Answer A. The action described in Sentence 17 is already in progress as the narrator recounts it, which rules out Answer B. And Answer D incorrectly describes an action conditional upon some other unstated circumstance; therefore, Answer C is correct.

16. **C.** Sentences 1 and 2 discuss optimists and pessimists, which are opposite personality types. So sentence 2 should begin with a word or phrase that highlights this contrast; therefore, Answer C is correct.

17. **B.** The sentence begins with the dangling modifier "In his book *Learned Optimism*" that must be set off from the rest of the sentence by a comma, so Answer D is incorrect. The subject of the sentence is "psychologist Martin Seligman," so this phrase should not be broken up by a comma; therefore, Answer A is incorrect. And the verb is "addresses," so no comma is needed between the subject and the verb; therefore, Answer C is incorrect. Thus, Answer B is correct.

18. **C.** Although this sentence may contain true information, placing it just after sentence 11 interrupts the discussion of attributions, which is the focus of the paragraph. Therefore, Answer C is correct.

19. **C.** The underlined words modify (describe) the verbal *reinterpret*, which requires an adverb modifier; therefore, the adjective phrase "more optimistic" is incorrect, which rules out answer D. Answers A and B are unnecessarily wordy, so these answers are both ruled out; therefore, Answer C is correct.

20. **A.** The noun phrase "Their anxiety and depression" is the subject of the sentence, and this subject is plural, so it requires a plural verb; therefore, Answers B and D are ruled out. The present participle "tending" requires an auxiliary verb to be used as the verb in a sentence, so Answer B is incorrect; therefore, Answer A is correct.

21. **C.** The information that Marni Nixon was in many "major Hollywood movies" also appears in sentence 1. And the information that she was neither seen on screen nor credited in many films is also in the first paragraph of the essay. Therefore, Answer C is correct.

22. **D.** Sentence 7 begins by essentially repeating the information that ends sentence 6, so you can use a relative pronoun such as *which* or *who* to bring the sentences together. The repeated information is inanimate (not a person), so the correct relative pronoun is *which;* therefore, Answer D is correct.

23. **A.** The awkward phrasing of "an early role of Nixon's" rules out Answer B. Sentence C is a sentence fragment, which rules out this answer. In Answer D, the *The King and I* is inappropriately referenced as "one of Nixon's early roles" rather than as a movie, which rules out this answer. Thus, Answer A is correct.

24. **B.** Sentence 10 makes a general statement that Marni Nixon sometimes assisted singers by singing the difficult parts. Then in sentence 11 gives a specific example when she did this for Marilyn Monroe, so Answer B is correct.

25. **A.** In the given context, in front of the words "in numerous operas," the words *acted, appeared,* and *played* could all be substituted in for *performed.* However, the word *accomplished* isn't typically used in front of the word *in,* so Answer A is correct.

Arithmetic Test Answers

How did you make out with your first math test? Find out in this section, where I give not only the answers but an explanation of how I arrived at it. If you still need work on this subtest, I recommend reviewing Chapters 10 through 13. Good luck!

1. **B. 0.1875.**

$$
\begin{array}{r}
0.1875 \\
16)\overline{3.0000} \\
\underline{16} \\
140 \\
\underline{128} \\
120 \\
\underline{112} \\
80 \\
\underline{80} \\
0
\end{array}
$$

2. **D. 87.** $254 - 79 - 88 = 87$.

3. **B. 13.583.**

$$
\begin{array}{r}
2.89 \\
\times \quad 4.7 \\
\hline
2023 \\
11560 \\
\hline
13.583
\end{array}
$$

4. **C. 21,000,000.** $29{,}873 \times 714 \approx 30{,}000 \times 700 = 21{,}000{,}000$.

5. B. $\frac{5}{7} < \frac{8}{11} < \frac{7}{9}$. Use cross-multiplication to compare all three pairs of fractions:

$$\frac{5}{7} < \frac{7}{9} \qquad \frac{7}{9} > \frac{8}{11} \qquad \frac{8}{11} > \frac{5}{7}$$
$$5 \times 9 < 7 \times 7 \quad 7 \times 11 > 9 \times 8 \quad 8 \times 7 > 11 \times 5$$
$$45 < 49 \qquad\quad 77 > 72 \qquad\quad 56 > 55$$

So the smallest value is $\frac{5}{7}$, the next-smallest is $\frac{8}{11}$, and the greatest value is $\frac{7}{9}$.

6. D. 80%. Divide 64 by 80: $64 \div 80 = 0.8$. Now, multiply by 100 to convert this decimal to percent: $0.8 \times 100 = 80\%$.

7. C. 4.971. To add decimals, line up the decimal points:

$$
\begin{array}{r}
4.32 \\
0.071 \\
+\quad 0.58 \\
\hline
4.971
\end{array}
$$

8. A. 9 days. Matthew worked 4 out of every 5 days for 30 days, so he worked 24 days $(30 \times \frac{4}{5} = 24)$. He worked 15 days at the supermarket, so he worked the remaining 9 days at the mobile phone kiosk $(24 - 15 = 9)$.

9. B. 2.77. Starting with 2.7654, round 6 up to 7 (because 5 is in the thousandths place), then drop the last two digits.

10. B. 7.75. To subtract decimals, line up the decimal points.

$$
\begin{array}{r}
21.60 \\
-\quad 13.85 \\
\hline
7.75
\end{array}
$$

11. B. 6. Devin has 80 minutes $(60 + 20 = 80)$, and each video takes 12 minutes, so he can watch approximately 6.67 videos $(80 \div 12 \approx 6.67)$. Thus, he can watch 6 *complete* videos.

12. C. $0.45\% < \frac{45}{1,000} < 45\%$. Convert all three values to decimals: $45\% = 0.45$, $\frac{45}{1,000} = 0.045$, and $0.45\% = 0.0045$. Thus, the smallest value is 0.45%, the next-smallest is $\frac{45}{1,000}$, and the largest is 45%.

13. B. $\frac{5}{4}$. Turn fraction division into multiplication with "Keep-Change-Flip," then cancel factors in the numerator and denominator, and solve:

$$\frac{5}{6} \div \frac{2}{3} = \frac{5}{6} \times \frac{3}{2} = \frac{5}{\cancel{6}^2} \times \frac{\cancel{3}^1}{2} = \frac{5}{4}$$

14. D. $\frac{76}{1,000}$. Increase the terms of the first fraction by a factor of 10, then add the numerators and keep the denominator the same:

$$\frac{7}{100} + \frac{6}{1,000} = \frac{70}{1,000} + \frac{6}{1,000} = \frac{76}{1,000}$$

15. **C. 12.5%.** Begin by making a word equation and then solve:

$$n\% \times 400 = 50$$
$$n \times 0.01 \times 400 = 50$$
$$n \times 4 = 50$$
$$n = \frac{50}{4} = 12.5$$

16. **A. 2.** $55 \div 7 = 79r2$ (79 with a remainder of 2).

17. **A. 21.** Evaluate using the order of operations (PEMDAS):

$$-4 + 8 \div 2 - (-3) \times 7$$
$$= -4 + 4 - (-3) \times 7$$
$$= -4 + 4 + 21$$
$$= 0 + 21$$
$$= 21$$

18. **C. $3\frac{7}{8}$ inches.** To find the increase in height, calculate $14\frac{3}{4} - 10\frac{7}{8}$ inches. Begin by increasing the terms of $\frac{3}{4}$ to $\frac{6}{8}$:

$$14\frac{3}{4} - 10\frac{7}{8} = 14\frac{6}{8} - 10\frac{7}{8}$$

Next, subtract $14 - 10 = 4$ and $\frac{6}{8} - \frac{7}{8} = -\frac{1}{8}$, and complete the subtraction:

$$14\frac{6}{8} - 10\frac{7}{8} = 4 - \frac{1}{8} = 3\frac{7}{8}$$

19. **C. $14.63.** To calculate a percent increase of 18%, multiply as follows:

```
        12.40
  ×      1.18
        9920
       12400
      124000
      14.6320
```

To complete the problem, round to two decimal places: $14.6320 \approx 14.63$.

20. **B. $\frac{2}{5}$.** Convert all fractions to decimals: $\frac{2}{5} = 0.4$ and $\frac{37}{100} = 0.37$. Thus, $\frac{2}{5}$ has the greatest value.

Quantitative Reasoning, Algebra, and Statistics Test Answers

On this test, the math ramps up a bit from the Arithmetic test that precedes it. How did you do? This section provides you the answers to all the questions, including an explanation of how each answer is reached. If you need more practice, please turn to Chapters 13 through 16.

1. A. $x^3 + 5x^2 - 7x - 35$. Multiply by FOILing (distribution):

$$(x^2 - 7)(x + 5) = x^3 + 5x^2 - 7x - 35$$

2. B. $-2 < x \le 3$. The solution set includes values that are between -2 and 3. The open circle on -2 indicates that this value is excluded from the solution set, so $-2 < x$. The closed circle on 3 indicates that this value is included in the set, so $x \le 3$.

3. D. $\dfrac{1}{x^2 y}$. Multiply, divide, and simplify as follows:

$$\frac{x^2}{y^2} \cdot \frac{y^3}{x^4 y^2} = \frac{x^2 y^3}{x^4 y^4} = x^{-2} y^{-1} = \frac{1}{x^2 y}$$

4. B. -2. Begin by plugging in 5 for x, then simplify as follows:

$$|x - 8| - |x| = |5 - 8| - |5| = |{-3}| - |5| = 3 - 5 = -2$$

5. A. Both x and y are positive. Begin by plugging in $y + 3$ for x, then solve for y.

$$5(y + 3) - 2y = 36$$
$$5y + 15 - 2y = 36$$
$$3y + 15 = 36$$
$$3y = 21$$
$$y = 7$$

Now, $x = y + 3 = 7 + 3 = 10$. Thus, both x and y are positive.

6. B. 18. Divide $725 \div 42 \approx 17.26$. Thus, 17 minutes wouldn't be enough time for 725 people to enter the theater ($17 \times 42 = 714$), so 18 minutes is the minimum whole number of minutes required to allow the theater to fill to capacity.

7. A. 1. To find the median, begin by arranging the number of siblings that each student has in a single row, from least to greatest:

$$0, 0, 0, 0, 0, 1, 1, 1, 1, 1, 1, \mathbf{1, 1,} 1, 2, 2, 2, 2, 2, 2, 4, 4, 4, 5$$

The two middle numbers are both 1, so the median is the mean of these numbers, which is 1.

8. C. The number of dollars that Geoffrey places in the account each month. In the linear equation $D(x) = 6000 + 750x$, 750 is the slope of the line, which is the increment value of the function — that is, the amount that is regularly added to the bank account each month.

9. D. More than 60. Let p represent the number of Peanut Butter Chocolate boxes that Kamal sold, and let v represent the number of Vanilla Strawberry boxes. From the information in the problem, you can set up the following system of equations:

$$p + v = 109$$
$$3.5p + 2.5v = 317.5$$

To solve by substitution, rewrite the first equation as follows:

$$p = 109 - v$$

Next, plug $109 - v$ into the second equation for p:

$$3.5(109 - v) + 2.5v = 317.5$$

Simplify the left side of the equation by distributing and combining like terms, and solve for v:

$$3.5(109 - v) + 2.5v = 317.5$$
$$381.5 - 3.5v + 2.5v = 317.5$$
$$381.5 - v = 317.5$$
$$-v = -64$$
$$v = 64$$

10. **C. $\{2, 4, 5, 6, 7\}$.** First, find the value of $B \cup C$:

$$B \cup C = \{2, 4, 6, 8, 10\} \cup \{5, 6, 7, 8, 9, 10\} = \{2, 4, 5, 6, 7, 8, 9, 10\}$$

Next, find $A \cap (B \cup C)$:

$$A \cap (B \cup C) = \{2, 4, 5, 6, 7, 8, 9, 10\} \cap \{1, 2, 3, 4, 5, 6, 7\} = \{2, 4, 5, 6, 7\}$$

11. **B. $\frac{17}{83}$.** According to the table, 83 students are on sports teams, so this is the total outcome number. Of these students, 17 are on the honor roll, so this is the target outcome number.

$$\text{Probability} = \frac{\text{Target outcomes}}{\text{Total outcomes}} = \frac{17}{83}$$

12. **C. 1.** Let x stand for the number you're looking for, then set up and solve the following equation:

$$\frac{48}{x - 9} = -6$$
$$48 = -6(x - 9)$$
$$48 = -6x + 54$$
$$-6 = -6x$$
$$1 = x$$

13. **C.** $4^{-3} = \dfrac{1}{4 \times 4 \times 4} = \dfrac{1}{64}$.

14. **D. More than twice as many people speak Arabic as a first language than as a second language.** Two hundred and ninety million people speak Arabic as a first language, which is more than twice the number (130 million) who speak it as a second language.

15. **D. $\frac{16}{3}$.** Multiply both sides of the equation by a common denominator of 12 to eliminate the fractions, then solve for x:

$$\frac{3}{4}x - \frac{1}{3} = \frac{1}{2}x + 1$$
$$12\left(\frac{3}{4}x - \frac{1}{3}\right) = 12\left(\frac{1}{2}x + 1\right)$$
$$9x - 4 = 6x + 12$$
$$3x = 16$$
$$x = \frac{16}{3}$$

16. **B. $25 < p < 50$.** The room has a length of 13 meters and an area of 91 square meters, so plug these values into the area formula for a rectangle to find the width:

$$A = l \times w$$
$$91 = 13w$$
$$7 = w$$

Now plug the values for the length and width into the formula for the perimeter of a rectangle:

$$P = 2(l + w) = 2(13 + 7) = 2(20) = 40$$

The perimeter is 40 meters, which is between 25 and 50 meters.

17. **A. There is an outlier that skews the data left, so the median is a more reliable indicator of center than the mean.** The 20 grades include 19 values clustered from 68 to 95, plus a low outlier of 32. This low outlier skews the data left, so the median is a more reliable indicator of center than the mean.

18. **A. $\dfrac{5bc + 3b}{2}$.** Find a in terms of b and c by isolating a on one side of the equation:

$$c = \frac{2a - 3b}{5b}$$
$$5bc = 2a - 3b$$
$$5bc + 3b = 2a$$
$$\frac{5bc + 3b}{2} = a$$

19. **D. 15.** Plug the values $P = (-6,\ 4)$ and $Q = (6, -5)$ into the distance formula and simplify:

$$d = \sqrt{(-6 - 6)^2 + (4 - (-5))^2} = \sqrt{(-12)^2 + 9^2} = \sqrt{144 + 81} = \sqrt{225} = 15$$

20. **D. −7.** A line on the xy-plane has an equation of $y = mx + b$, where b is the y-intercept. The slope $m = -4$, so $y = -4x + b$. Using the point $(-3, 5)$, plug −3 for x and 5 for y into this equation and solve for b:

$$5 = -4(-3) + b$$
$$5 = 12 + b$$
$$-7 = b$$

Advanced Algebra and Functions Test Answers

How did you do with this tough math? Remember, some community colleges don't require you to take this section of the ACCUPLACER, so be sure to check with your school administration. In this section, you find not only the answers but also a detailed explanation of each to help you make sense of them. If you need more help, Chapters 17 through 20 focus exclusively on these types of problems.

1. **B. $\dfrac{3}{2b}$.** Begin by converting $\dfrac{1}{a}$ to an equivalent fraction with $2a$ in the denominator, then add the two fractions on the left side of the equation:

$$\frac{1}{a} + \frac{1}{2a} = b$$
$$\frac{2}{2a} + \frac{1}{2a} = b$$
$$\frac{3}{2a} = b$$

Now, multiply both sides of the equation by $2a$ to remove the fraction, then isolate a:

$$3 = 2ab$$
$$\frac{3}{2b} = a$$

2. C. $2x^3 + 5x^2 - 22x + 15$. Multiply by distributing:

$$(2x - 3)(x^2 + 4x - 5)$$
$$= 2x^3 + 8x^2 - 10x - 3x^2 - 12x + 15$$
$$= 2x^3 + 5x^2 - 22x + 15$$

3. B. 2.5. Lines AB and CD are parallel, so angles A and D are congruent, as are angles B and C. Therefore, triangles ABE and DCE are similar, so the ratio of BE to EC is equal to the ratio of AE to ED. $BE = 4$ and $EC = 5$, so the ratio of BC to EC is 4:5. $AE = 2$, so $ED = 2.5$.

4. D. –2. Substitute the given values into the functions and simplify:

$$\frac{f(6)}{g(-4)} = \frac{6^2 - 4(6)}{3(-4) + 6} = \frac{36 - 24}{-12 + 6} = \frac{12}{-6} = -2$$

5. D. $y = -\frac{5}{3}x$. The line shown has a slope of $-\frac{5}{3}$, so any parallel line also has this slope. And a line passing through the origin has a y-intercept of 0. Plug these values into the slope-intercept form for a linear equation and simplify:

$$y = mx + b$$
$$y = -\frac{5}{3}x + 0$$
$$y = -\frac{5}{3}x$$

6. C. A function is defined as having no more than one output value y for every input value x. That is, it passes the vertical line test: No vertical line can pass through the function in more than one place. Therefore, the correct answer is C.

7. A. 3. Solve the quadratic equation by factoring:

$$2x^2 - 14x = -20$$
$$2x^2 - 14x + 20 = 0$$
$$x^2 - 7x + 10 = 0$$
$$(x - 2)(x - 5) = 0$$
$$x = 2 \text{ and } 5$$

Therefore, if a is 5 and b is 2, $a - b = 3$.

8. C. $y = (x + 1)^2 - 4$. The graph has x-intercepts at 1 and –3, so its factored form is $y = (x - 1)(x + 3)$, which rules out Answers A and B. The graph also has a vertex at $(-1, -4)$, so its vertex form is $y = (x + 1)^2 - 4$, so Answer C is correct.

9. C. $\frac{8\pi}{9}$. To begin, convert 40° to radians:

$$40° \times \frac{\pi}{180°} = \frac{2\pi}{9}$$

Now, plug this value and the radius of 4 into the formula for arc length:

$$\text{Arc length} = \text{Radius} \times \text{Radians} = 4 \times \frac{2\pi}{9} = \frac{8\pi}{9}$$

10. **C. 1.04.** For exponential growth, the value n should be 100% + the percent increase of 4%, which equals 104%. This value expressed as a decimal is 1.04.

11. **D. $\frac{17}{9}$.** Cross-multiply to remove the fractions, then simplify both sides by distributing:

$$\frac{x+2}{x-3} = \frac{x-5}{x-1}$$
$$(x+2)(x-1) = (x-5)(x-3)$$
$$x^2 + x - 2 = x^2 - 8x + 15$$

Now, subtract x^2 from both sides of the equation, and solve for x:

$$x - 2 = -8x + 15$$
$$9x = 17$$
$$x = \frac{17}{9}$$

12. **A. 5.** The domain of the function $f(x) = \frac{1}{x^2 - 10x + 25}$ doesn't include the value of x where the denominator of the fraction equals 0. To find this value, set $x^2 - 10x + 25$ to 0 and solve:

$$x^2 - 10x + 25 = 0$$
$$(x-5)(x-5) = 0$$
$$x = 5$$

13. **B. 10.** Begin solving for x by cross-multiplying and simplifying:

$$\frac{x-4}{4} = \frac{2}{x-6}$$
$$(x-4)(x-6) = 8$$
$$x^2 - 10x + 24 = 8$$
$$x^2 - 10x + 16 = 0$$

Now, solve this quadratic equation by factoring:

$$(x-2)(x-8) = 0$$
$$x = 2 \text{ and } x = 8$$

Therefore, the sum of the two values of x is $2 + 8 = 10$.

14. **C. $\frac{36}{11}$.** You can rewrite $\log_6 11x = 2$ as $6^2 = 11x$, so:

$$6^2 = 11x$$
$$36 = 11x$$
$$\frac{36}{11} = x$$

15. **B. $\frac{24}{25}$.** You know that $\sin X = \frac{7}{25}$, so one leg of the triangle and the hypotenuse are in a 7:25 ratio. You can use the Pythagorean theorem to find the length of the remaining side:

$$a^2 + b^2 = c^2$$
$$7^2 + b^2 = 25^2$$
$$49 + b^2 = 625$$
$$b^2 = 484$$
$$b = 24$$

Thus, the three legs of the triangle are in a 7:24:25 ratio.

Thus, $\sin Z = \dfrac{24}{25}$.

16. **D.** $x^2 + (y-2)^2 = 9$. The circle shown has a radius of 3 and is centered at (0, 2). Plug this information into the formula for a circle with a radius of r centered at (h, k) and simplify as needed:

$$(x-h)^2 + (y-k)^2 = r^2$$
$$(x-0)^2 + (y-2)^2 = 3^2$$
$$x^2 + (y-2)^2 = 9$$

17. **D.** $2a^2 - 12a + 10$. To find the value of $2f(a-1)$, substitute $a-1$ for x into the function, and multiply the entire result by 2, then simplify:

$$2f(a-1)$$
$$= 2[(a-1)^2 - 4(a-1)]$$
$$= 2[(a-1)(a-1) - 4(a-1)]$$
$$= 2[a^2 - 2a + 1 - 4a + 4]$$
$$= 2[a^2 - 6a + 5]$$
$$= 2a^2 - 12a + 10$$

18. **D.** $-\dfrac{5}{4}$. To solve, isolate the radical, then square both sides of the equation, and solve for x:

$$\sqrt{x^2 - 1} - 2 = x$$
$$\sqrt{x^2 - 1} = x + 2$$
$$(\sqrt{x^2 - 1})^2 = (x+2)^2$$
$$x^2 - 1 = x^2 + 4x + 4$$
$$-1 = 4x + 4$$
$$-5 = 4x$$
$$-\frac{5}{4} = x$$

19. **B.** $\dfrac{7}{5}$. Begin by expressing 125 and $\sqrt{5}$ in terms of a common base ($125 = 5^3$ and $\sqrt{5} = 5^{\frac{1}{2}}$), then use the multiplication rule for finding exponents of exponents:

$$125^{x-1} = \sqrt{5}^{\,x+1}$$
$$(5^3)^{x-1} = (5^{\frac{1}{2}})^{x+1}$$
$$5^{3(x-1)} = 5^{\frac{1}{2}(x+1)}$$

Now, the bases are both 5, so you can set the exponents equal to each other and solve for x:

$$3(x-1) = \frac{1}{2}(x+1)$$
$$3x - 3 = \frac{1}{2}x + \frac{1}{2}$$
$$6x - 6 = x + 1$$
$$5x = 7$$
$$x = \frac{7}{5}$$

20. B. 63. Begin by factoring the left side of the equation $x^2 - y^2 = 15$:

$$x^2 - y^2 = 15$$
$$(x - y)(x + y) = 15$$
$$3(x + y) = 15$$
$$x + y = 5$$

Now, solve the following system of equations:

$$x + y = 5$$
$$x - y = 3$$

Adding these two equations results in the following:

$$2x = 8$$
$$x = 4$$

Now, solve for y:

$$4 + y = 5$$
$$y = 1$$

So $x^3 - y^3 = 4^3 - 1^3 = 64 - 1 = 63$

Chapter **23**

Practice Test 2

'’ve designed the second practice test so you can see how much you’ve improved. This exam is exactly like the first one from Chapter 21, except (of course) the questions are different. I hope you used the results from the first practice exam to pinpoint your weak areas, and then spent some time hitting the books and recharging your thinking cap.

TIP

To get the most out of this practice exam, take it like you’d take the real ACCUPLACER under the same conditions:

>> Find a quiet place where nothing will interrupt you.

>> Feel free to take the five separate sections of the test in any order you choose. There is no time limit, so take as long as you need. (But bear in mind that when you really take the test, you’ll need to start and finish each section of the test in one sitting.)

After you complete the entire sample test, check your answers against the answer explanations and key in Chapter 24.

Answer Sheet for Practice Exam 2

Reading Test

1. Ⓐ Ⓑ Ⓒ Ⓓ
2. Ⓐ Ⓑ Ⓒ Ⓓ
3. Ⓐ Ⓑ Ⓒ Ⓓ
4. Ⓐ Ⓑ Ⓒ Ⓓ

5. Ⓐ Ⓑ Ⓒ Ⓓ
6. Ⓐ Ⓑ Ⓒ Ⓓ
7. Ⓐ Ⓑ Ⓒ Ⓓ
8. Ⓐ Ⓑ Ⓒ Ⓓ

9. Ⓐ Ⓑ Ⓒ Ⓓ
10. Ⓐ Ⓑ Ⓒ Ⓓ
11. Ⓐ Ⓑ Ⓒ Ⓓ
12. Ⓐ Ⓑ Ⓒ Ⓓ

13. Ⓐ Ⓑ Ⓒ Ⓓ
14. Ⓐ Ⓑ Ⓒ Ⓓ
15. Ⓐ Ⓑ Ⓒ Ⓓ
16. Ⓐ Ⓑ Ⓒ Ⓓ

17. Ⓐ Ⓑ Ⓒ Ⓓ
18. Ⓐ Ⓑ Ⓒ Ⓓ
19. Ⓐ Ⓑ Ⓒ Ⓓ
20. Ⓐ Ⓑ Ⓒ Ⓓ

Writing Test

1. Ⓐ Ⓑ Ⓒ Ⓓ
2. Ⓐ Ⓑ Ⓒ Ⓓ
3. Ⓐ Ⓑ Ⓒ Ⓓ
4. Ⓐ Ⓑ Ⓒ Ⓓ
5. Ⓐ Ⓑ Ⓒ Ⓓ

6. Ⓐ Ⓑ Ⓒ Ⓓ
7. Ⓐ Ⓑ Ⓒ Ⓓ
8. Ⓐ Ⓑ Ⓒ Ⓓ
9. Ⓐ Ⓑ Ⓒ Ⓓ
10. Ⓐ Ⓑ Ⓒ Ⓓ

11. Ⓐ Ⓑ Ⓒ Ⓓ
12. Ⓐ Ⓑ Ⓒ Ⓓ
13. Ⓐ Ⓑ Ⓒ Ⓓ
14. Ⓐ Ⓑ Ⓒ Ⓓ
15. Ⓐ Ⓑ Ⓒ Ⓓ

16. Ⓐ Ⓑ Ⓒ Ⓓ
17. Ⓐ Ⓑ Ⓒ Ⓓ
18. Ⓐ Ⓑ Ⓒ Ⓓ
19. Ⓐ Ⓑ Ⓒ Ⓓ
20. Ⓐ Ⓑ Ⓒ Ⓓ

21. Ⓐ Ⓑ Ⓒ Ⓓ
22. Ⓐ Ⓑ Ⓒ Ⓓ
23. Ⓐ Ⓑ Ⓒ Ⓓ
24. Ⓐ Ⓑ Ⓒ Ⓓ
25. Ⓐ Ⓑ Ⓒ Ⓓ

Arithmetic Test

1. Ⓐ Ⓑ Ⓒ Ⓓ
2. Ⓐ Ⓑ Ⓒ Ⓓ
3. Ⓐ Ⓑ Ⓒ Ⓓ
4. Ⓐ Ⓑ Ⓒ Ⓓ

5. Ⓐ Ⓑ Ⓒ Ⓓ
6. Ⓐ Ⓑ Ⓒ Ⓓ
7. Ⓐ Ⓑ Ⓒ Ⓓ
8. Ⓐ Ⓑ Ⓒ Ⓓ

9. Ⓐ Ⓑ Ⓒ Ⓓ
10. Ⓐ Ⓑ Ⓒ Ⓓ
11. Ⓐ Ⓑ Ⓒ Ⓓ
12. Ⓐ Ⓑ Ⓒ Ⓓ

13. Ⓐ Ⓑ Ⓒ Ⓓ
14. Ⓐ Ⓑ Ⓒ Ⓓ
15. Ⓐ Ⓑ Ⓒ Ⓓ
16. Ⓐ Ⓑ Ⓒ Ⓓ

17. Ⓐ Ⓑ Ⓒ Ⓓ
18. Ⓐ Ⓑ Ⓒ Ⓓ
19. Ⓐ Ⓑ Ⓒ Ⓓ
20. Ⓐ Ⓑ Ⓒ Ⓓ

Quantitative Reasoning, Algebra, and Statistics Test

1. Ⓐ Ⓑ Ⓒ Ⓓ
2. Ⓐ Ⓑ Ⓒ Ⓓ
3. Ⓐ Ⓑ Ⓒ Ⓓ
4. Ⓐ Ⓑ Ⓒ Ⓓ

5. Ⓐ Ⓑ Ⓒ Ⓓ
6. Ⓐ Ⓑ Ⓒ Ⓓ
7. Ⓐ Ⓑ Ⓒ Ⓓ
8. Ⓐ Ⓑ Ⓒ Ⓓ

9. Ⓐ Ⓑ Ⓒ Ⓓ
10. Ⓐ Ⓑ Ⓒ Ⓓ
11. Ⓐ Ⓑ Ⓒ Ⓓ
12. Ⓐ Ⓑ Ⓒ Ⓓ

13. Ⓐ Ⓑ Ⓒ Ⓓ
14. Ⓐ Ⓑ Ⓒ Ⓓ
15. Ⓐ Ⓑ Ⓒ Ⓓ
16. Ⓐ Ⓑ Ⓒ Ⓓ

17. Ⓐ Ⓑ Ⓒ Ⓓ
18. Ⓐ Ⓑ Ⓒ Ⓓ
19. Ⓐ Ⓑ Ⓒ Ⓓ
20. Ⓐ Ⓑ Ⓒ Ⓓ

Advanced Algebra and Functions Test

1. Ⓐ Ⓑ Ⓒ Ⓓ
2. Ⓐ Ⓑ Ⓒ Ⓓ
3. Ⓐ Ⓑ Ⓒ Ⓓ
4. Ⓐ Ⓑ Ⓒ Ⓓ

5. Ⓐ Ⓑ Ⓒ Ⓓ
6. Ⓐ Ⓑ Ⓒ Ⓓ
7. Ⓐ Ⓑ Ⓒ Ⓓ
8. Ⓐ Ⓑ Ⓒ Ⓓ

9. Ⓐ Ⓑ Ⓒ Ⓓ
10. Ⓐ Ⓑ Ⓒ Ⓓ
11. Ⓐ Ⓑ Ⓒ Ⓓ
12. Ⓐ Ⓑ Ⓒ Ⓓ

13. Ⓐ Ⓑ Ⓒ Ⓓ
14. Ⓐ Ⓑ Ⓒ Ⓓ
15. Ⓐ Ⓑ Ⓒ Ⓓ
16. Ⓐ Ⓑ Ⓒ Ⓓ

17. Ⓐ Ⓑ Ⓒ Ⓓ
18. Ⓐ Ⓑ Ⓒ Ⓓ
19. Ⓐ Ⓑ Ⓒ Ⓓ
20. Ⓐ Ⓑ Ⓒ Ⓓ

Reading Test

DIRECTIONS: This test contains items that measure your ability to understand what you read. Read each passage and select the choice that best completes the statement or answers the question. Mark your choice on your answer sheet, using the correct letter with each question number.

"A hundred years hence!" he murmured, as in a trance.

"We shall not be here," I briskly, but fatuously, added.

"We shall not be here. No," he droned, "but the museum will still be just where it is. And the reading-room just where it is. And people will be able to go and read there." He inhaled sharply, and a spasm as of actual pain contorted his features.

I wondered what train of thought poor Soames had been following. He did not enlighten me when he said, after a long pause, "You think I haven't minded."

"Minded what, Soames?"

"Neglect. Failure."

"*Failure*?" I said heartily. "Failure?" I repeated vaguely. "Neglect — yes, perhaps; but that is quite another matter. Of course you haven't been — appreciated. But what then? Any artist who — who gives —" What I wanted to say was, "Any artist who gives truly new and great things to the world has always to wait long for recognition"; but the flattery would not out: in the face of his misery — a misery so genuine and so unmasked — my lips would not say the words.

And then he said them for me. I flushed. "That's what you were going to say, isn't it?" he asked.

"How did you know?"

"It's what you said to me three years ago." I flushed the more. I need not have flushed at all. "It's the only important thing I ever heard you say," he continued. "But — d'you remember what I answered? I said, 'I don't care a sou for recognition.' And you believed me. You're shallow. What should *you* know of the feelings of a man like me? You imagine that a great artist's faith in himself and in the verdict of posterity is enough to keep him happy. You've never guessed at the bitterness and loneliness, the" — his voice broke; but presently he resumed, speaking with a force that I had never known in him. "Posterity! What use is it to *me*? A dead man doesn't know that people are visiting his grave, visiting his birthplace, putting up tablets to him, unveiling statues of him. A hundred years hence! Think of it! If I could come back to life *then* — just for a few hours — and go to the reading-room and *read*! Think of the pages and pages in the catalogue: 'Soames, Enoch' endlessly — endless editions, commentaries, prolegomena, biographies" ("Enoch Soames" by Max Beerbohm — Pages 20–21)

1. In **the line reproduced here**, which of the following is most likely the reason why the narrator does not complete what he is saying to Soames?

"What I wanted to say was, "Any artist who gives truly new and great things to the world has always to wait long for recognition"; but the flattery would not out: in the face of his misery — a misery so genuine and so unmasked — my lips would not say the words."

(A) He believes that Soames deserves his failure and, therefore, deserves to suffer for it.

(B) He pities Soames for his distress and can't bring himself to heap dishonest praise on him.

(C) He doesn't want to risk flattering a man whose work he has not read but, in fact, only pretended to read.

(D) He hopes that Soames will realize for himself how great his talent is, without another person having to confirm it.

2. Which of the following terms best describes Soames?

(A) Self-effacing

(B) Self-destructive

(C) Self-involved

(D) Self-possessed

3. In Soames's statement, "I don't care a *sou* for recognition," the word *sou* refers to something

(A) cheap and worthless.

(B) rare and precious.

(C) gaudy and tasteless.

(D) sad and misunderstood.

4. In the last paragraph, the most plausible reason why the author writes the words *you, me, then,* and *read* in italics is to

(A) make sure that the reader can understand what Soames is saying.

(B) underscore that Soames is screaming in anger.

(C) capture the overly dramatic nature of Soames's speech.

(D) convince the reader that Soames is, in fact, speaking the truth.

Passage 1

As a habitat, water has one major disadvantage for photosynthetic organisms (plants): the deeper the water, the less light there is. Light may be reflected off the water surface, it may be scattered by particles in the water and it is absorbed by the water. The speed at which the latter happens depends on the wavelength of the light; light at the red end of the spectrum is absorbed before light at the blue end of the spectrum. Thus, in clear water, red light penetrates only to about 15 meters, whereas blue light may reach 100 meters. There is therefore a zone — the *euphotic zone* — in which light penetration is adequate to support photosynthesis. In general, shallow water occurs on the margins of land masses and, in this primal history of photosynthetic eukaryotes, the land represented a major niche endowed with a much better light environment.

GO ON TO NEXT PAGE

Although better access to light was an obvious advantage that land habitats offered to plants, there were also obvious disadvantages. The need for water in order to maintain life meant that the possibility of desiccation (shriveling due to water loss) was a serious problem. Water is also the medium into which algae release their gametes, so sexual reproduction on land would be more difficult. Furthermore, immersion in water made for easy uptake of nutrients and also provided support for the larger organisms. (*Functional Biology of Plants*, by Hodson and Bryant — Pages 4–5)

Passage 2

Because water is essential to a plant's functioning on land, it has built-in mechanisms that help prevent it from losing too much water: a cuticle and guard cells.

The cuticle is a layer of cells found on the top surfaces of a plant's leaves. It lets light pass into the leaf but protects the leaf from losing water. Many plants have cuticles that contain waxes that resist the movement of water into and out of a leaf, much like wax on your car keeps water off the paint.

Guard cells are found on the bottom of a plant's leaf, near a stomate, a tiny opening that you can't see with your naked eye. Plants need to keep their stomates open in order to obtain carbon dioxide for photosynthesis and to release oxygen. To prevent water loss from happening, each stomata has two guard cells surrounding it.

Guard cells can swell and contract in order to open and close the stomates. When the sun is shining and photosynthesis is occurring, guard cells swell up with water like full balloons, which stretches them outward and opens the stomates. At night, when photosynthesis isn't occurring, the guard cells release some water and collapse together, closing the stomates. (*Biology For Dummies*, by Rene Fester Kratz — Page 357)

5. All of the following are mentioned in Passage 1 EXCEPT

(A) Blue light penetrates deeper into the ocean than red light.

(B) Some plants use water as a medium for sexual reproduction.

(C) The land offers the potential for plant organisms to grow larger.

(D) Water helps to facilitate the absorption of nutrients in plants.

6. According to Passage 2, guard cells help plants to

(A) maximize water outtake while minimizing carbon dioxide intake.

(B) maximize water intake while minimizing carbon dioxide outtake.

(C) maximize carbon dioxide outtake while minimizing water intake.

(D) maximize carbon dioxide intake while minimizing water outtake.

7. Which of the following provides the best summary of the relationship between Passage 1 and Passage 2?

(A) Passage 1 presents a theory, and Passage 2 presents an opposing theory with a larger body of evidence to support it.

(B) Passage 1 describes an opportunity with some downsides, and Passage 2 describes how one of these downsides was overcome.

(C) Passage 1 discusses a problematic limitation, and Passage 2 discusses an additional and even more problematic limitation.

(D) Passage 1 unveils a new scientific paradigm, and Passage 2 endorses that paradigm and uses it to draw a set of important conclusions.

8. Which of the following is a problem mentioned in Passage 1 that is directly addressed in Passage 2?

(A) Desiccation

(B) The euphotic zone

(C) Photosynthesis

(D) Photosynthetic eukaryotes

Of all the new social and professional networks, LinkedIn is quickly becoming a way of connecting with future employees for an increasing number of recruiters and hiring managers in the graphic design field. In a segment on National Public Radio (NPR), Yuki Noguchi wrote, "Not having a profile on the social networking site LinkedIn is, for some employers, not only a major liability but also a sign that the candidate is horribly out of touch." Monica Bloom, a design industry recruiter for Aquent in Los Angeles, says that it is essential for graphic designers seeking employment to have a LinkedIn profile — more so than Facebook, although that is debatable. (*Becoming a Graphic and Digital Designer: A Guide to Careers in Design*, by Steven Heller and Veronique Vienne — Page xii)

9. The main purpose of this passage is to offer

(A) advice.

(B) comparison.

(C) exemplification.

(D) justification.

Over the coming decades we will witness both increasing numbers and increasing proportions of older adults in the United States and across the globe. At the global level, in 2012, one out of every nine persons was 60 years old or older. By 2050, one out of every five persons is projected to be over 60 years old. Moreover, by 2050 the global population of people 60 years old and older is expected to surpass, for the first time in history, the number of children under 15 years of age. Like much of the rest of the world, the United States is in the midst of a longevity revolution, primarily due to the massive numbers of Baby Boomers marching into later life. Beginning in 2012, nearly 10,000 Americans turn 65 every day, essentially becoming an "older adult." (*Aging and Mental Health*, by Daniel L. Segal, Sara Honn Qualls, and Michael A. Smyer — Pages 23–24)

10. Which of the following best paraphrases the passage?

(A) A trend found in the U.S. is shown to be unusual in the rest of the world.

(B) A trend found outside the U.S. is shown to be unusual in this country.

(C) A trend found both inside and outside the U.S. is expected to continue for the foreseeable future.

(D) A trend found both inside and outside the U.S. is expected to decrease in the near future.

Rubber, in the form of latex, was discovered in the early 16th century in South America, but it gained little acceptance because it became sticky and lost its shape in the heat. Charles Goodyear was trying to find a way to make the rubber stable when he accidentally spilled a batch of rubber mixed with sulfur on a hot stove. He noticed that the resulting compound didn't lose its shape in the heat. Goodyear went on to patent the vulcanization process, which is the chemical process used to treat crude or synthetic rubber or plastics to give them useful properties such as elasticity, strength, and stability. (*Chemistry Essentials For Dummies*, by John T. Moore — Page 316)

GO ON TO NEXT PAGE

11. This passage illustrates which of the following general principles?

    (A) Science is not so much about designing a single grand experiment, but rather about testing a large number of small hypotheses.

    (B) Scientific genius consists of making something useful from components that others see as not useful.

    (C) A small mishap can, if viewed from the proper perspective, lead a scientist to an important discovery.

    (D) A scientist who sets to work without a specific goal in mind is likely to be more successful than one who thinks more rigidly.

> (1) One type of concern regarding the harmful effects of advertising relates to indirect types of effects on parent-child relationships. (2) For example, the role of advertising as a stimulant of conflict, and the effects on the general sense of parents' and children's wellbeing and happiness. (3) The cycle that leads to such unpleasant feelings includes the fact that commercials can stimulate young children's requests of their parents to purchase a particular item. (4) According to parents' and children's self-reports, frequent "buy me" demands (coined *nagging effect* or *pester power*) often lead to arguments, quarrels, and even temper tantrums. (5) Researchers' observations of children's behaviors in supermarkets documented such exchanges. (6) Thus, children exert direct influence over their parents' purchasing habits by requesting products, as well as indirect influence, as parents internalize their children's tastes and make purchases that will please them, even without children making explicit requests or even being present when purchases are made. (*Children and Media: A Global Perspective*, by Dafna Lemish — Page 118)

12. Which of the following would be the best example of the type of "internalizing" described in Sentence 6?

    (A) A son asks for a Nintendo product repeatedly until his mother, exhausted from saying no, allows him to buy it with his own money.

    (B) A mother buys her daughter an expensive sewing machine so that the two of them can pursue an interest in common.

    (C) A father surprises his daughter by buying a LEGO product, which he knows that she likes, and bringing it home to her.

    (D) A son insists that if his father refuses to buy him the bicycle that he wants for his birthday, then he would prefer not to receive any presents at all.

> When people talk about a place (including its language), they often mean a collection of social characteristics, not just a physical location. Look at the term *Cockney*, used for hundreds of years to describe the inhabitants of a region and the language they speak. It technically refers only to central London, but it has strong social connotations — urban, working class, low prestige, street-smart. And these connotations have been around for a while. Even back in 1803, when Joshua Pegge wrote the amazing but little known *Anecdotes of the English Language: Dialect of London* about Cockney, some reviewers assumed that the book was a joke. Why would anyone write about the historical pedigree and internal logic of Cockney? (*What Is Sociolinguistics?* by Gerard Van Herk — Page 39)

13. Which of the following best paraphrases the question that ends this passage?

   (A) It's ludicrous to imagine that anyone would seriously want to read a book about Cockney.

   (B) Although some aspects of Cockney may be interesting, its historical pedigree and internal logic are certainly not.

   (C) Let's explore the reasons why some people might be interested in Cockney, while others clearly would not.

   (D) Please understand that Cockney is a language in its own right that deserves to be explored further.

(1) The food you eat could very likely contain *genetically modified organisms* (GMOs) — living things whose genes have been altered by scientists in order to give them useful traits. (2) For example, crop plants may be engineered to better resist pests, and animals may be treated with hormones to increase their growth or milk production. (3) Some people object to the idea of GMOs in their diets, but genetic modification of organisms has enabled some amazing health breakthroughs. (4) If you know someone who takes insulin to treat diabetes, that insulin is made by bacteria that scientists engineered to contain the human gene for insulin. (*Biology For Dummies,* by Rene Fester Kratz — Page 370)

14. In which sentence does the writer anticipate a possible counterargument to her thesis?

   (A) Sentence 1

   (B) Sentence 2

   (C) Sentence 3

   (D) Sentence 4

One of the main characteristics of predation is that predators have a detrimental effect on the fitness of their individual prey. However, starving to death is not good news from an evolutionary (or personal) perspective. We are thus confronted with the absolute inevitability of an evolutionary arms race between predators and their prey. The pressure of the race is not equally balanced between competitors, since predators will usually live to fight another day if they fail in any particular attack, while failure of a prey's defense system is likely to be fatal to that individual. Nor does predation always seriously damage individual prey fitness; predators will often target old or diseased prey that may have little chance of further reproduction anyway. And, of course, in the long term, predation is a major drive for prey evolution by weeding out less fit individuals. (*The Neuroethology of Predation and Escape,* by Keith T. Sillar, Laurence D. Picton, and William J. Heitler — Page xiii)

15. According to the passage, the effect predators have on prey can be

   (A) positive for both the individual and for the evolution of the species.

   (B) negative for both the individual and for the evolution of the species.

   (C) positive for the individual but negative for the evolution of the species.

   (D) negative for the individual but positive for the evolution of the species.

The Fairness Doctrine is the most significant regulatory statute related to diversity of television content. The doctrine essentially had two basic elements: it required broadcasters to devote some of their airtime to discussing controversial matters of public interest, and to air contrasting views regarding those matters. Stations were given wide latitude as to how to provide contrasting views, and how much time was required. This could be achieved through news segments, public affairs shows, or editorials. However, this responsibility was not supposed to be simply a passive one; in fact, the concept of "ascertainment" was a part of it as well — stations were directed to actively seek out diverse views to broadcast instead of ignoring the hot button issues. The doctrine did not require that each program be internally balanced, nor did it mandate equal time for opposing points of view. It simply prohibited stations from broadcasting from a single perspective, day after day, without presenting opposing views. (*A Companion to the History of American Broadcasting*, edited by Aniko Bodroghkozy)

16. The concept of "ascertainment" in this context most nearly refers to

   (A) the single-minded devotion to one point of view if that outlook is most likely correct.

   (B) the responsibility to seek out a variety of perspectives on controversial issues.

   (C) the freedom to seek out opposing views and present them without undue restriction.

   (D) the loosening of the restriction that each and every broadcasted segment must itself contain multiple viewpoints.

The Icelandic term *jokulhlaup* (sometimes pronounced *yer-kul-hyolp*) refers to a flood from the sudden melting of a glacier. In Iceland, where glaciers cover a volcanic landscape, small jokulhlaups often occur when the ice is heated by volcanic activity, melts, and rushes to the sea. These "glacial outburst floods" cause all the damage associated with massive flooding and occur with the unpredictability of a volcano.

Jokulhlaups, while first named in Iceland, are not restricted to Iceland. Any region where glaciers form is in danger of experiencing a jokulhlaup. A jokulhlaup can occur when a glacier that dams a lake shrinks enough that the lake water spills out suddenly. The eruption of glacier-covered volcanoes, such as Mount Rainier, Washington, would result in jokulhlaup-type flooding, as well as debris flows from the combination of meltwater and volcanic debris produced during the eruption. (*Geology For Dummies*, by Alecia M. Spooner)

17. The reader can reasonably conclude that a jokulhlaup

   (A) could not occur in the United States.

   (B) can now be predicted with reasonable certainty.

   (C) tends to be less dangerous than a flood caused by heavy rainfall.

   (D) can only happen where both glaciers and volcanoes are present.

Researching the ways in which women have been routinely portrayed by mainstream media has been a continuing preoccupation for many media scholars over the past decades. Sadly, most contemporary studies suggest that despite the incursions which women have made into the media industry and the success of the women's movement in challenging some of the wider gender inequalities in society, the media seem stuck in a very traditional and stereotypical groove. Women *have* made and *continue* to make a difference to what we see, read, and hear in the media, but there still remain too many examples of sexist reporting in the news media, and of stereotypical characterizations in entertainment media, for the claim that we are "post-equality" to be seen as anything other than decidedly hollow. (*Women and Media: International Perspectives,* edited by Karen Ross and Carolyn M. Byerly — Page 9)

18. The writer uses the word "sadly" in connection with which of the following statements?

(A) Many scholars over the past decades have been preoccupied with how women have been routinely portrayed in the media.

(B) Women have made incursions into the media industry, and had success in challenging some of the wider gender inequalities in society.

(C) The media seem stuck in a very traditional and stereotypical groove.

(D) To claim that we are "post-equality" cannot be seen as anything other than decidedly hollow.

Directions for Questions 19–20: Each of the following sentences has a blank indicating that something has been left out. Beneath each sentence are four words or phrases. Choose that response that best fits the meaning of the sentence.

19. Although Bianca seemed at first rather _____, as I got to know her I found her to be surprisingly warm and personable, so we became good friends.

(A) abrasive

(B) captivating

(C) precarious

(D) superfluous

20. Because the Axis powers failed to capture Moscow during the Second World War, the Soviet Union was able to _____ a defense that eventually repelled the invaders.

(A) capitalize

(B) envision

(C) mobilize

(D) retreat

STOP

Writing Test

DIRECTIONS: This test contains items that measure your ability to write and improve English sentences in a given context. It includes five early drafts of essays. Read each essay and then choose the best answer to the questions that follow.

(1) Carl Rogers (1902–1987) was an American psychologist and founder of humanistic psychology. (2) Although Rogers is now considered one of the most eminent psychologists of 20th century, some of his most provocative ideas may now seem quite tame, even self-evident.

(3) While working with troubled children and their families in the 1930s and 1940s, Rogers found that he was often more beneficial when he just listened to his clients rather than trying to diagnose them or give them advice. (4) Paradoxically, the less he tried to help, the more helpful he actually was. (5) This experience clashed with other therapeutic methods of the time, especially psychoanalysis. (6) Strongly influenced by the work of Sigmund Freud and his disciples, psychoanalysts usually saw themselves as experts whose job was to "cure" their patients. (7) To do this, they were encouraged to remain as distant, objective, and aloof as possible as their patients struggled with neuroses and other pathologies.

(8) Rogers rebelled directly against this methodology. (9) In his 1951 work *Client-Centered Therapy* — he advocates for a more humane and caring approach. (10) He relieves therapists of their authority as experts, transferring this authority to the client, whom he believes to be eminently capable of handling it. (11) For example, Rogers recasts the therapist not so much as the agent of a cure, but instead as the facilitator of the client's inborn potential for positive change.

(12) Throughout his long and varied career — as a clinician, an author, and the president of the American Psychological Association — Rogers worked tirelessly to further transform psychology. (13) His pioneering influence upon this relatively new area of human study places Carl Rogers at the very forefront of great contributors to the field.

1. Which is the best version of the underlined portion of Sentence 3 (reproduced here)?

While working with troubled children and their families in the 1930s and 1940s, Rogers found that he was often more beneficial when he just listened to his clients rather than trying to diagnose them or give them advice.

(A) (as it is now)

(B) he found that Rogers was often more beneficial when he just listened to them instead of

(C) listening to his clients often proved more beneficial to Rogers than

(D) his clients often found Rogers to be more beneficial when he just listened to them rather than

2. The writer is considering placing the following sentence after Sentence 8:

Although he lived exclusively in the United States, by the end of his life, Rogers had traveled to such varied and often troubled parts of the globe as Brazil, Northern Ireland, South Africa, and the Soviet Union.

Should the writer make this addition there?

(A) Yes, because it provides a perspective on Rogers not shown in the rest of the essay.

(B) Yes, because it supports the argument that the writer is making here.

(C) No, because it interrupts the flow and blurs the focus of the paragraph.

(D) No, because it restates information that is found elsewhere in the essay.

3. Which is the best version of the underlined portion of Sentence 9 (reproduced here)?

In his 1951 work Client-Centered Therapy — *he advocates for a more humane and caring approach.*

(A) (as it is now)

(B) In his 1951 work *Client-Centered Therapy*; he

(C) In his 1951 work, *Client-Centered Therapy*: he

(D) In his 1951 work, *Client-Centered Therapy*, he

4. Which is the best version of the underlined portion of Sentence 11 (reproduced here)?

For example, Rogers recasts the therapist not so much as the agent of a cure, but instead as the facilitator of the client's inborn potential for positive change.

(A) (as it is now)

(B) In contrast,

(C) In this way,

(D) Likewise,

5. Which of the following versions of Sentence 13 (reproduced here) provides the most effective concluding sentence?

His pioneering influence upon this relatively new area of human study places Carl Rogers at the very forefront of great contributors to the field.

(A) (as it is now)

(B) In my opinion, Carl Rogers stands as a giant in the field of psychology, and one whom I would personally emulate.

(C) A true rebel by nature, Carl Rogers was never one to back away from conflict when an important point could be made.

(D) Notwithstanding his inconsistencies, Carl Rogers still deserves a place at the table when psychology is being discussed.

GO ON TO NEXT PAGE

(1) If somebody asks you to recall one of the 27 Amendments to the U.S. Constitution, you'll probably think of one of the first ten Amendments, which comprise the Bill of Rights. (2) Or maybe you'll think of the 14th Amendment, which freed the slaves, or the 19th Amendment, which gave women the right to vote.

(3) You might not even know about the most recently ratified Amendment — the 27th Amendment — which prohibits the current Congress from increasing or decreasing their own salary until the next Congress has been sworn into office. (4) When placed in a side-by side comparison with the monumental achievements of some of the earlier Amendments, the 27th seems pretty tame. (5) So you may be surprised to learn that while most Amendments take about one to three years to be ratified (made law), the 27th took more than 200 years!

(6) The story goes all the way back to 1789, the first year that George Washington was president. (7) The First U.S. Congress passed 12 articles of amendment and sent them to the states for ratification. (8) Ten of these were ratified, and collectively became known as the Bill of Rights. (9) The other two articles failed to be ratified and was all but forgotten.

(10) That is, until 1982.

(11) In that year, a 19-year-old college student named Gregory Watson wrote a paper for a political science course in which he claimed that the nearly 200-year-old Amendment could still be ratified. (12) His professor, Sharon Waite, disagreed — and gave him a grade of C. (13) Curious to see what the result might be, Watson pursued the matter with a series of letters to state legislatures. (14) In state after state, Watson's argument found support, becoming a popular cause that almost everybody — even the lawmakers themselves — could agree upon.

(15) 1n 1992, the ratification process was complete, and the 27th Amendment finally took its place in the U.S. Constitution. (16) However, this model for belated ratification is now under consideration by proponents of other unratified Amendments, such as the women's Equal Rights Amendment and the Washington D.C. Voting Rights Amendment.

(17) And, by the way, in 2017, Professor Waite retroactively changed Watson's grade on that paper to an A-plus!

6. Which is the best version of the underlined portion of Sentence 4 (reproduced here)?

When placed in a side-by-side comparison with the monumental achievements of some of the earlier Amendments, the 27th seems pretty tame.

(A) (as it is now)

(B) placed in comparison

(C) made to be compared

(D) compared

7. Which is the best version of the underlined portion of Sentence 9 (reproduced here)?

The other two articles failed to be ratified and was all but forgotten.

(A) (as it is now)

(B) failed to be ratified and were all but forgotten.

(C) failed to be ratified being all but forgotten.

(D) failed to be ratified; it was all but forgotten.

8. Which is the best version of the underlined portion of Sentence 13 (reproduced here)?

Curious to see what the result might be, <u>Watson pursued the matter with a series of letters to state legislatures.</u>

(A) (as it is now)

(B) Watson's next move was to pursue the matter with a series of letters to state legislatures.

(C) a series of letters to state legislatures followed as written by Watson.

(D) state legislatures began to receive a series of Watson's letters to them.

9. Which is the best version of the underlined portion of Sentence 16 (reproduced here)?

<u>However,</u> this model for belated ratification is now under consideration by proponents of other unratified Amendments, such as the women's Equal Rights Amendment and the Washington D.C. Voting Rights Amendment.

(A) (as it is now)

(B) Moreover

(C) For example

(D) Nevertheless

10. The writer is considering deleting Sentence 17 (reproduced here).

And, by the way, in 2017, Professor Waite retroactively changed Watson's grade on that paper to an A-plus!

Should the writer make this deletion?

(A) Yes, because this sentence is inconsistent with the formal style and tone of the essay.

(B) Yes, because the information provided here blurs the focus of the essay.

(C) No, because this sentence concludes the essay with an interesting fact that connects with information discussed earlier.

(D) No, because the most important information in the essay is contained in this sentence.

---

(1) If you saw a group of masked men entering a bank, what would you do? (2) Call the police? (3) Try to warn people? (4) Run the other way?

(5) If this occurred in Asia during the winter months, the best response might well be to do nothing. (6) In many Asian countries, cold-and-flu season has now become the time to wear a fashion accessory not normally seen in the West, the surgical mask.

(7) Wearing a surgical mask in public has now become commonly accepted in many of the larger cities in Japan, Korea, and China. (8) The practice began in Tokyo about a century ago in reaction to the 1918 influenza pandemic, which killed from 50 to 100 million people worldwide. (9) As with more benign versions of the flu, the airborne virus responsible for the disease spread quickly in cities. (10) In cities, close contact with people who may be infected is virtually unavoidable. (11) Although wearing a face mask perhaps provided minimal protection from the virus, it may have offered the illusion of safety, was helping people to cope better with an otherwise alarming situation.

(12) From a strictly medical perspective, however, the most valid to wear a surgical mask is the reason that surgeons wear them: to prevent the wearer from spreading disease to others. (13) This seems to be one of the most common reasons why many Asian people wear them. (14) In Japan, and especially at work, there is considerable social pressure among coworkers to wear a mask when suffering from a cold. (15) As a result, some people report wearing a mask to protect themselves from air pollution, pollen, and even unpleasant odors.

 GO ON TO NEXT PAGE

(16) Finally, another reason some people wear a mask has nothing to do with physical well-being. (17) Instead, they simply feel more comfortable being a little less visible and approachable in public. (18) Some women report that wearing a mask makes them feel a little less vulnerable to male advances. (19) But even men say that a mask can help to deflect attention, making them feel less self-conscious.

(20) Whatever the reason, the mask-wearing trend is so prevalent in Asia, especially among young people, that it's not likely to disappear anytime soon. (21) Will the surgical mask follow karate, sushi, and anime as the next Japanese export to the United States? (22) Only time will tell.

11. Which is the best version of the underlined portion of Sentence 6 (reproduced here)?

    In many Asian countries, cold-and-flu season has now become the time to wear a fashion accessory *not normally seen in the West, the surgical mask.*

    (A) (as it is now)
    (B) not normally seen in the West; the surgical mask.
    (C) not normally seen in the West: the surgical mask.
    (D) not normally seen in the West.

12. The writer wants to combine Sentences 9 and 10 (reproduced here) into a single sentence.

    As with more benign versions of the flu, the airborne virus responsible for the disease spread quickly *in cities. In cities,* *close contact with people who may be infected is virtually unavoidable.*

    Which is the best version of the underlined portions of these sentences?
    (A) in cities, there being
    (B) in cities, where
    (C) in cities; because
    (D) in cities; where,

13. Which is the best version of the underlined portion of Sentence 11 (reproduced here)?

    Although wearing a face mask perhaps provided minimal protection from the virus, it may have offered the illusion of safety, *was helping* *people to cope better with an otherwise alarming situation.*
    (A) (as it is now)
    (B) were helping
    (C) helping
    (D) helped

14. Which is the best version of the underlined portion of Sentence 15 (reproduced here)?

    As a result, *some people report wearing a mask to protect themselves from air pollution, pollen, and even unpleasant odors.*
    (A) (as it is now)
    (B) Additionally,
    (C) Even so,
    (D) Therefore,

15. Which is the best version of the underlined portion of Sentence 20 (reproduced here)?

Whatever the reason, the mask-wearing trend is so prevalent in Asia, especially among young people, that it's not likely to disappear anytime soon.

(A) (as it is now)

(B) people, thus its not likely

(C) people, where it's not likely

(D) people — that it's not likely

---

(1) The movie *Groundhog Day*, starring comedian Bill Murray, was only a moderate success when it was first released in 1993. (2) It received reasonably good reviews, but broke no records at the box office. (3) In comparison with some of Murray's earlier films, such as the incredibly successful *Ghostbusters* movies, *Groundhog Day* may well have been a letdown to many of those involved in its production.

(4) Yet surprisingly, over time the film has steadily grown in prominence as a land-mark American comedy. (6) Renowned film critic Roger Ebert originally gave it only three stars. (7) He later revised his opinion in his "Great Movies" series of retrospective reviews. (8) In 2006, the National Film Registry added *Groundhog Day* to its list of "culturally, historically, or aesthetically" significant films. (9) Furthermore, the words "Groundhog Day" themselves have come to symbolize any recurrent and usually unpleasant sequence of events: "This job is beginning to look like Groundhog Day – nothing ever changes!"

(10) The plot of the movie is simple and now well-known. (11) Phil Connors, a self-centered and ceaselessly complaining weatherman for a local television station, enters a time loop that forces him to relive Groundhog Day over and over again. (12) To the people around him, nothing is out of the ordinary, to Phil, the experience is sheer misery, as he wakes each morning to the same dreary day in a small dull town he longs to escape.

(13) Unable to escape, Phil attempts to divert himself with a variety of pleasures that eventually fail to satisfy him. (14) Next, he sets his sites on winning the love of coworker, Rita, the only woman in town who seems immune to his false charm – but she doesn't fall for him. (15) Over time, his suffering becomes so intense that he even attempts suicide – only to wake up again in the same place and time, with nothing changed.

(16) Only when Phil begins to accept his predicament and engage with the people around him does his life shift from miserable to happy. (17) However, as he becomes a more caring person, Rita begins to take a genuine interest in him. (18) Ultimately, having discovered joy in life, Phil is released from the time loop – and, ironically, chooses to stay in the town he once hated, with the woman who now loves him.

(19) Beyond the comedy and the romance, *Groundhog Day* is a genuine allegory: a modern-day *Pilgrim's Progress* with a grumpy weatherman cast as the unlikely pilgrim.

---

16. Which is the best version of the underlined portion of sentence 3 (reproduced below)?

In comparison with some of Murrays earlier films, such as the incredibly successful Ghostbusters movies, Groundhog Day may well have been a letdown to many of those involved in its production.

(A) (as it is now)

(B) Murrays earlier films – such as

(C) Murray's earlier films, such as

(D) Murray's earlier films – such as

GO ON TO NEXT PAGE →

17. The writer wants to combine sentences 6 and 7 (reproduced below).

Renowned film critic Roger Ebert originally gave it only three stars. He later revised his opinion in his "Great Movies" series of retrospective reviews.

Which of the following best achieves this purpose?

(A) Renowned film critic Roger Ebert originally gave it only three stars, who later revised his opinion in his "Great Movies" series of retrospective reviews.

(B) Renowned film critic Roger Ebert had originally given it only three stars, having later revised his opinion in his "Great Movies" series of retrospective reviews.

(C) Renowned film critic Roger Ebert, who originally gave it only three stars, later revised his opinion in his "Great Movies" series of retrospective reviews.

(D) Renowned film critic Roger Ebert, who originally gave it only three stars, but later revised his opinion in his "Great Movies" series of retrospective reviews.

18. Which is the best version of the underlined portion of sentence 12 (reproduced below)?

To the people around him, nothing is out of the ordinary, to Phil, the experience is sheer misery, as he wakes each morning to the same dreary day in a small dull town he longs to escape.

(A) (as it is now)

(B) nothing is out of the ordinary, however to Phil, the experience is sheer misery,

(C) nothing is out of the ordinary; to Phil, the experience is sheer misery,

(D) nothing is out of the ordinary: however to Phil, the experience is sheer misery,

19. Which is the best version of the underlined portion of sentence 13 (reproduced below)?

Unable to escape, Phil attempts to divert himself with a variety of pleasures that eventually fail to satisfy him.

(A) (as it is now)

(B) a variety of pleasures that eventually fail to satisfy him become Phil's next attempt to divert himself

(C) the next thing Phil attempts to divert himself with is a variety of pleasures that eventually fail to satisfy him.

(D) Phil's attempt to divert himself becomes a variety of pleasures that eventually fail to satisfy him.

20. Which is the best version of the underlined portion of sentence 17 (reproduced below)?

However, as he becomes a more caring person, Rita begins to take a genuine interest in him.

(A) (as it is now)

(B) Moreover

(C) Nevertheless

(D) Otherwise

(1) Language is a natural phenomenon. (2) And like other natural things – for example, forests, mountains, and oceans), language changes and evolves over time, often in dramatic and unexpected ways.

(3) Why do so many of my students who can easily read and enjoy contemporary fiction, such as the *Harry Potter* books, struggle with *Romeo and Juliet*? (4) Written over 400 years ago, this popular Shakespearean play is now full of English words and phrases that seem almost like a foreign language. (5) Even Juliet's famous question "Wherefore art thou Romeo?", which means "Why are you Romeo?", includes three words that are no longer used by English speakers in the 21st century.

(6) Even 50 years can make a big difference with language. In particular, slang changes very quickly from one decade to the next. (7) In 1969, high school and college students described things they liked as "groovy." (8) Acting cowardly or unfairly was "copping out."

(9) Even grammar changes from one generation to the next. (10) For example, when I was in school, English teacher tested me and my classmates on the correct use of the words *who* and *whom*. (11) Today, the word *whom* has virtually fallen out of the language, with forms like "to whom are you referring?" commonly replaced by "who are you referring to?"

(12) One form that's still in the process of changing is the use of the plural pronouns *they* and *them* to refer to a singular noun of non-specific gender, as in "When your best friend moves away, you really miss *them*." (13) In this context, the word *them* is currently considered ungrammatical in formal written language. (14) But using *them* to mean *him or her* is now so common in spoken language – especially among younger people – that it seems destined to become standard within the next few years.

(15) Speaking personally as an English teacher, I enjoy watching the language change as in response to social and cultural changes. (16) Some purists believe that any grammatical form that was considered non-standard in 1960 is an example of objectively poor grammar. (17) But language is a living thing, and no living thing can flourish for long in an airtight container.

21. Which is the best version of the underlined portion of sentence 2 (reproduced below)?

And like other natural things – for example, forests, mountains, and oceans), language changes and evolves over time, often in dramatic and unexpected ways.

(A) (as it is now)

(B) And like other natural things; for example

(C) And like other natural things: for example

(D) And like other natural things (for example

22. Which is the best version of the underlined portion of sentence 4 (reproduced below)?

Written over 400 years ago, this popular Shakespearean play is now full of English words and phrases that seem almost like a foreign language.

(A) (as it is now)

(B) this popular Shakespearean play is full of English words and phrases that now seem almost like a foreign language.

(C) words and phrases that this popular Shakespearean play is full of now almost seem like a foreign language.

(D) words and phrases that now seem almost like a foreign language fill this now popular Shakespearean play.

23. The writer wishes to add a third example of outdated 1960s slang after sentence 8 in a style that matches the two previous examples. Which of the following best accomplishes this purpose?

(A) Examples like these once sounded cool, but now seem outdated and a little ridiculous.

(B) The 1960s was, of course, a time of significant social upheaval, especially among young people.

(C) And a bad time was "a total bummer."

(D) And as for the word "bread," young people used it to mean *money*.

24. The writer is considering omitting the underlined words in sentence 14 (reproduced here):

But using them to mean him or her is now so common in spoken language – especially among younger people – that it seems destined to become standard <u>within the next few years</u>.

Should the writer make this change?

(A) Yes, because these words blur the focus of the paragraph.

(B) Yes, because the time frame mentioned is merely a speculation.

(C) No, because these words clarify the time frame that the writer is discussing.

(D) No, because these words correct a misperception introduced earlier in the essay.

25. Which is the best version of the underlined portion of sentence 16 (reproduced below)?

Some purists believe that any grammatical form that was considered non-standard in 1960 <u>is an example</u> of objectively poor grammar.

(A) (as it is now)

(B) are examples of

(C) is being an example of

(D) are being examples of

STOP

Arithmetic Test

DIRECTIONS: This test contains items that measure your ability to solve basic arithmetic problems. Choose the best answer to each question. Feel free to use scrap paper. You may not use a calculator.

1. What is $0.801 - 0.22$?

(A) 0.581

(B) 0.823

(C) 1.021

(D) 3.001

2. What is 0.8485 rounded to the nearest hundredth?

(A) 0.84

(B) 0.848

(C) 0.849

(D) 0.85

3. Reggie baked a batch of brownies. While they were cooling, his sister Lorna ate $\frac{1}{4}$ of them, and then his brother Jonas ate $\frac{2}{5}$ of them. What fraction of the brownies was left after that?

(A) $\frac{7}{20}$

(B) $\frac{9}{20}$

(C) $\frac{11}{20}$

(D) $\frac{13}{20}$

4. $5738 + 496 =$

(A) 6234

(B) 6244

(C) 6334

(D) 6344

5. When you divide 453 by 7, what is the remainder?

(A) 2

(B) 3

(C) 4

(D) 5

6. What is 73% of 51?

(A) 37.23

(B) 37.33

(C) 38.23

(D) 38.33

GO ON TO NEXT PAGE

7. A banquet room can hold no more than 45 people, according to the fire safety code. What is the maximum number of groups of 6 people that can be seated in the room?

(A) 6

(B) 7

(C) 8

(D) 9

8. If you multiply 5076×889, which of the following will be the closest estimate to the answer?

(A) 45,000

(B) 450,000

(C) 4,500,000

(D) 45,000,000

9. $\frac{5}{8} \div \frac{3}{4} =$

(A) $\frac{5}{6}$

(B) $\frac{6}{5}$

(C) $\frac{7}{8}$

(D) $\frac{8}{7}$

10. $-9 + 12 \div 3 \times 7 - (-8) =$

(A) 13

(B) 15

(C) 27

(D) 29

11. Last year on his birthday, Sean was exactly $47\frac{5}{8}$ inches in height. This year, he's $5\frac{3}{4}$ inches taller. How tall is he now?

(A) $53\frac{1}{8}$ inches

(B) $53\frac{1}{4}$ inches

(C) $53\frac{3}{8}$ inches

(D) $53\frac{1}{2}$ inches

12. Which of the following correctly orders the values $\frac{4}{1000}$, 0.4, and 4%?

(A) $\frac{4}{1000} < 0.4 < 4\%$

(B) $\frac{4}{1000} < 4\% < 0.4$

(C) $4\% < \frac{4}{1000} < 0.4$

(D) $4\% < 0.4 < \frac{4}{1000}$

13. Erika was assigned a novel to read for English class. She wants to read at least 40% of it over the weekend. If the novel is 370 pages, how many pages should Erika read this weekend?

   (A) 140
   (B) 148
   (C) 152
   (D) 164

14. What percent of 300 equals 120?

   (A) 40%
   (B) 88%
   (C) 150%
   (D) 250%

15. Which of the following inequalities is true?

   (A) $\frac{2}{3} < \frac{5}{8}$
   (B) $\frac{3}{4} < \frac{4}{7}$
   (C) $\frac{4}{9} < \frac{3}{10}$
   (D) $\frac{6}{11} < \frac{5}{8}$

16. Cameron walked 27.2 kilometers on the first day of a three-day hike and 15.9 kilometers on the third day. If he walked a total of 66.5 kilometers, how many kilometers did Cameron walk on the second day?

   (A) 22.4
   (B) 23.2
   (C) 23.4
   (D) 23.6

17. Which of the following numbers has the least value?

   (A) $\frac{4}{7}$
   (B) 47%
   (C) $\frac{4}{70}$
   (D) 0.047

18. Jonathan only got 2 questions out of 40 questions wrong on an exam. What percent of the questions did he get right?

   (A) 94%
   (B) 95%
   (C) 96%
   (D) 98%

GO ON TO NEXT PAGE

19. Evelyn earned a 16% profit on a $12,000 investment on a stock. How much money did she sell the stock for?

(A) $13,420

(B) $13,520

(C) $13,820

(D) $13,920

20. What is 27.1×0.83?

(A) 22.493

(B) 22.503

(C) 23.493

(D) 23.593

STOP

Quantitative Reasoning, Algebra, and Statistics Test

DIRECTIONS: This test contains items that measure your ability to do intermediate math, including basic algebra, geometry, and statistics. Choose the best answer to each question. Feel free to use scrap paper. You may not use a calculator. However, for some problems on the computer-based ACCUPLACER, you may be given access to an onscreen calculator with limited functionality.

1. What is the greatest common factor of $24a^2b^2c^2$ and $30bc^5$?

    (A) $3bc^5$

    (B) $3abc^2$

    (C) $6bc^2$

    (D) $6a^2b^2c^5$

2. A marathon is 26.2 miles in length. Approximately how many kilometers does this equal? (1 km = 0.62 miles)

    (A) Between 0 and 20 kilometers

    (B) Between 20 and 40 kilometers

    (C) Between 40 and 60 kilometers

    (D) More than 60 kilometers

3. If $x = 3$ and $y = -2$, then $8x + 4y^3 - 3x^2y^2 =$

    (A) 116

    (B) 100

    (C) −100

    (D) −116

4. The following table lists data about the 29 children in Ms. Frankel's fifth-grade class, dividing the group into girls and boys, and then dividing each of these two groups into only children and children with siblings. To the nearest whole percent, what percentage of the class are boys who have at least one sibling?

	Only Child	Has at Least 1 Sibling	Total
Girls	5	8	13
Boys	4	12	16
Total	9	20	29

    (A) 14%

    (B) 41%

    (C) 55%

    (D) 69%

GO ON TO NEXT PAGE

5. Which of the following is equivalent to $12x^6y^3 + 8x^7y - 20x^2y^5$?

(A) $2x^2y(6x^4y^2 + 8x^5 - 10y^4)$

(B) $2x^2y^2(3x^4y + 2x^5 - 5y^3)$

(C) $4x^2y(3x^4y^2 + 2x^5 - 5y^4)$

(D) $4x^2y^2(3x^4y + 2x^5 - 5y^3)$

6. The planet Jupiter is approximately 483,000,000 miles from the sun. How do you express this number in scientific notation?

(A) 4.83×10^8

(B) 4.83×10^{-8}

(C) 4.83×10^9

(D) 4.83×10^{-9}

7. Matthew started a tutoring business in 2014. The following graph shows the number of new students his business worked with in each of the years from 2014 to 2018. In which year did his business see the greatest increase in new students as compared with the previous year?

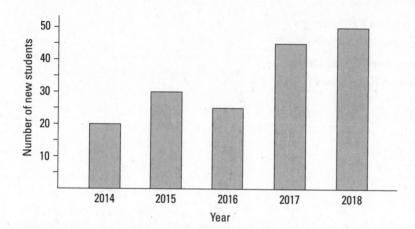

(A) 2015

(B) 2016

(C) 2017

(D) 2018

8. The following figure shows a map of a rectangular field. What is the distance from point *A* to point *C*?

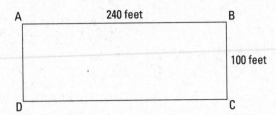

(A) 250 feet

(B) 260 feet

(C) 270 feet

(D) 280 feet

9. When you double a number, then add 7, and then divide by 5, the result is two more than the number you started with. What is the number?

(A) −1

(B) −3

(C) −5

(D) −7

10. $(2x^2)^3(3x^6)^2 =$

(A) $64x^{12}$

(B) $64x^{13}$

(C) $72x^{13}$

(D) $72x^{18}$

GO ON TO NEXT PAGE

11. In the following graph, the line intersects the two points $P = (-5, 1)$ and $Q = (1, -2)$. What is the equation of this line?

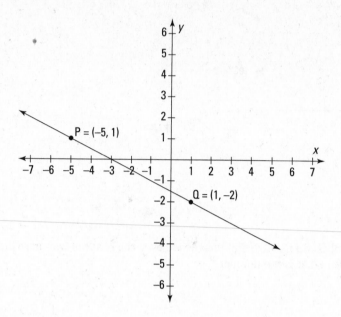

(A) $y = \frac{1}{2}x + \frac{3}{2}$

(B) $y = \frac{1}{2}x - \frac{3}{2}$

(C) $y = -\frac{1}{2}x + \frac{3}{2}$

(D) $y = -\frac{1}{2}x - \frac{3}{2}$

12. If a circle has a circumference of 9π, what is its area?

(A) 6π

(B) 12.25π

(C) 20.25π

(D) 81π

13. Twenty students took a 10-point pop quiz in their history class. Three students received scores of 10, five got scores of 9, two got scores of 8, six got scores of 7, and four got scores of 6. What was the difference between the mean score and the median score?

(A) 0

(B) 0.2

(C) 0.25

(D) 0.35

14. Given that $\frac{5x + 4}{8} = \frac{3x - 8}{10}$, which of the following is true of x?

(A) x is a positive odd number.

(B) x is a positive even number.

(C) x is a negative odd number.

(D) x is a negative even number.

15. Rectangle *ABCD* lies in the *xy*-plane, and the coordinates of point *C* are $(-6, 1)$. This rectangle is then reflected across the *y*-axis and rotated 90° clockwise, to produce rectangle *A'B'C'D'*. What are the coordinates of point *C'*?

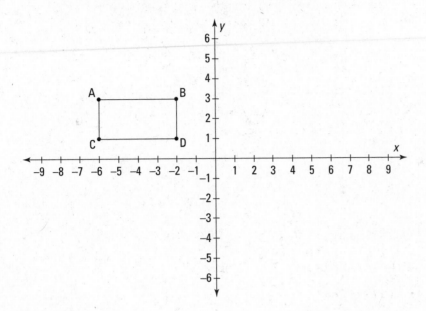

   (A) $(1, -6)$

   (B) $(-1, -6)$

   (C) $(6, -1)$

   (D) $(-6, -1)$

16. If you simplify $(2x + 4)(5x - 3) - 6(x - 1)$, what's the result?

   (A) $10x^2 + 8x$

   (B) $10x^2 - 8x$

   (C) $10x^2 + 8x - 6$

   (D) $10x^2 - 8x - 6$

17. Which of the following sets is NOT equivalent to $\{3, 4, 5, 7\}$?

   (A) $\{3, 7\} \cup \{4, 5\}$

   (B) $\{3, 7\} \cup \{4, 5, 7\}$

   (C) $\{1, 2, 3, 4, 5, 7, 9\} \cap \{3, 4, 5, 6, 7, 8\}$

   (D) $\{4, 5, 7\} \cap \{1, 2, 3, 4, 5, 6, 7, 8, 9\}$

18. If $4.6x - 0.21 = 2.3x + 1.63$, then *x* equals

   (A) 0.6

   (B) 0.7

   (C) 0.8

   (D) 0.9

GO ON TO NEXT PAGE

19. Which of the following systems of inequalities is depicted by the shaded area in this figure?

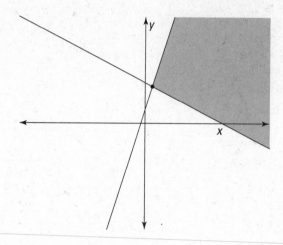

(A) $y \geq -\frac{1}{2}x + 3$ and $y \leq 3x + 1$

(B) $y \leq -\frac{1}{2}x + 3$ and $y \geq 3x + 1$

(C) $y \leq -\frac{1}{2}x + 3$ and $y \leq 3x + 1$

(D) $y \geq -\frac{1}{2}x + 3$ and $y \geq 3x + 1$

20. If $y = mx + b$ is a linear function with a slope of 4 that intersects the point $(-3, -11)$, then $m + b =$

(A) 3

(B) 5

(C) −10

(D) −11

STOP

Advanced Algebra and Functions Test

DIRECTIONS: This test contains items that measure your ability to do math at an introductory college level. Choose the best answer to each question. Feel free to use scrap paper. You may not use a calculator. However, for some problems on the computer-based ACCUPLACER, you may be given access to an onscreen calculator with limited functionality.

1. If $2ab - b = 3a$, then a equals which of the following?

(A) $\dfrac{b}{2b-3}$

(B) $-\dfrac{b}{2b+3}$

(C) $\dfrac{b}{3b-2}$

(D) $-\dfrac{b}{2+3b}$

2. In the following figure, $\angle DAB = (x+10)°$ and $\angle BCE = (x+25)°$. What is the value of y in terms of x?

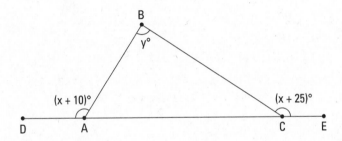

(A) $y = 35 - 2x$

(B) $y = 2x - 35$

(C) $y = 145 - 2x$

(D) $y = 2x - 145$

3. The year she was born, Jeannine's grandfather placed $5000 in a bank account, and then every year after that added $1500 to it. Her uncle also opened a bank account with a lump sum, and after that added a constant amount to it every year, according to the function $y = 750x + 8000$, where x is Jeannine's age and y is the number of dollars in the account. What will be the sum in both accounts when Jeannine is 18 years old?

(A) Less than $30,000

(B) Between $30,000 and $40,000

(C) Between $40,000 and $50,000

(D) More than $50,000

GO ON TO NEXT PAGE

4. What is the equation of a line that passes through the point $(0, -5)$ and is perpendicular to the line shown in the following figure?

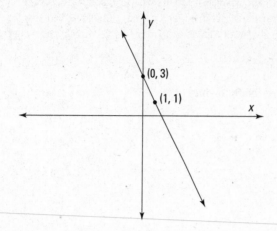

(A) $y = 2x - 5$

(B) $y = 5x + 2$

(C) $y = \frac{1}{2}x - 5$

(D) $y = -5x - \frac{1}{2}$

5. $32a^3c - 50ab^2c =$

(A) $2ac(2a + 5b)(8a - 5b)$

(B) $2ac(4a + 5b)(4a - 5b)$

(C) $-2ac(4a + b)(-4a + 25b)$

(D) $-2ac(2a - 5b)(8a - 5b)$

6. In the figure shown here, $g(x)$ is a transformation of $f(x) = x^2$ such that $f(x)$ is reflected across the x-axis and its vertex translated to $(3, -2)$. Which of the following does $g(x)$ equal?

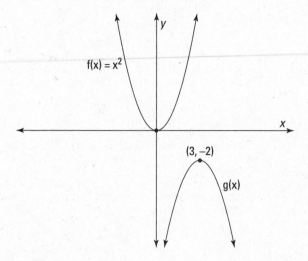

(A) $g(x) = (-x + 3)^2 + 2$

(B) $g(x) = -(x - 3)^2 + 2$

(C) $g(x) = -(x - 3)^2 - 2$

(D) $g(x) = (-x + 3)^2 - 2$

7. Which of the following is equivalent to $\cos 70° \tan 70°$?

(A) $\sin 20°$

(B) $\cos 20°$

(C) $\tan 20°$

(D) $\cot 20°$

8. A scientist observes that a population of 50 fruit flies doubles every 36 hours. Which of the following expressions models the number of fruit flies after 6 days?

(A) $4(50)^2$

(B) $0.25(50)^2$

(C) $50(2)^4$

(D) $50(2)^{0.25}$

GO ON TO NEXT PAGE

9. In the following figure, $\angle JKL$ is a right angle, where $KM = 3$, $KL = 5$, and $JM = n$. What is the value of n?

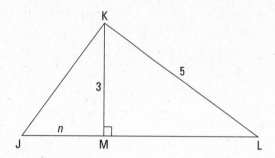

(A) $\dfrac{9}{2}$

(B) $\dfrac{9}{4}$

(C) $\dfrac{9}{5}$

(D) $\dfrac{9}{8}$

10. If a, b, and c are the zeros of the polynomial $y = (x-2)(x+5)(4x-1)$, then $a+b+c =$

(A) $3\dfrac{1}{4}$

(B) $2\dfrac{3}{4}$

(C) $-2\dfrac{3}{4}$

(D) $-3\dfrac{1}{4}$

11. What is $2\sqrt{20} + \sqrt{45}$?

(A) $4\sqrt{5}$

(B) $5\sqrt{6}$

(C) $5\sqrt{7}$

(D) $7\sqrt{5}$

12. In the figure shown here, $f(x) = x^2 - 2x - 3$ and $g(x) = -5x + 1$ intersect at points P and Q. Which of the following are the coordinates of P and Q?

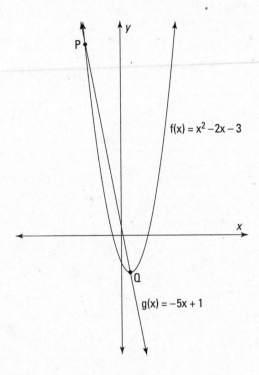

(A) $P = (-3, 21)$ and $Q = (2, -4)$

(B) $P = (-4, 20)$ and $Q = (1, -3)$

(C) $P = (-3, 20)$ and $Q = (2, -3)$

(D) $P = (-4, 21)$ and $Q = (1, -4)$

13. What are the solutions of the quadratic function $y = 5x^2 - 3x - 1$?

(A) $\dfrac{3 \pm \sqrt{11}}{10}$

(B) $\dfrac{3 \pm \sqrt{29}}{10}$

(C) $\dfrac{-3 \pm \sqrt{11}}{10}$

(D) $\dfrac{-3 \pm \sqrt{29}}{10}$

14. If $x + y = m$ and $x^2 - y^2 = n$, then $y - x =$

(A) $\dfrac{m}{n}$

(B) $\dfrac{n}{m}$

(C) $-\dfrac{m}{n}$

(D) $-\dfrac{n}{m}$

GO ON TO NEXT PAGE

15. The following figure shows a circle with a radius of 3 centered at the point (1,–5). What is the equation of this circle?

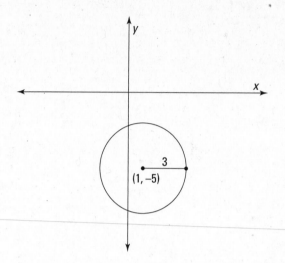

(A) $(x-1)^2 - (y+5)^2 = 3$

(B) $(x+1)^2 + (y-5)^2 = 3$

(C) $(x-1)^2 + (y+5)^2 = 9$

(D) $(x+1)^2 - (y-5)^2 = 9$

16. If $f(x) = 2x - 11$ and $g(x) = 3x^2 + 1$, what is the value of $g(f(4))$?

(A) 26

(B) 28

(C) –26

(D) –28

17. If $\log_3 81 = 2x$, then which of the following is true?

(A) $x < 0$

(B) $0 < x < 1$

(C) $1 < x < 5$

(D) $x > 5$

18. What is the value of 25° in radian measure?

(A) $\frac{\pi}{8}$

(B) $\frac{2\pi}{15}$

(C) $\frac{3\pi}{25}$

(D) $\frac{5\pi}{36}$

19. The figure shown here displays the function $h(x) = ax^4 + bx^3 + cx^2 + dx + e$, where a, b, c, d, and e are all integer values. Which of the following gives possible values for both a and e?

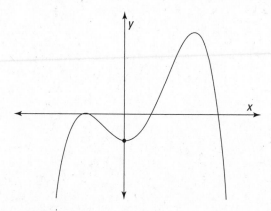

(A) $a = 2$ and $e = 3$

(B) $a = 2$ and $e = -3$

(C) $a = -2$ and $e = 3$

(D) $a = -2$ and $e = -3$

20. What is the domain of the function $f(x) = \sqrt{14 - 2x}$?

(A) $x \leq 7$

(B) $x \geq 7$

(C) $x \leq -7$

(D) $x \geq -7$

STOP

Chapter **24**

Answers to Practice Test 2

Use this answer key to score the practice test in Chapter 23. To help you make sense of the test, each answer includes a detailed explanation why it's correct.

Reading Test Answers

How did you do with your second Reading Test? Have a look at the following answers, each of which includes an explanation. If you need more help, flip back to Chapters 4 through 6.

1. **B.** The narrator initially describes his friend as "poor Soames." He then tells us that "the flattery would not out" — that is, he can't complete what he's saying — because he observes Soames's misery. It's reasonable to infer that the narrator's reaction to this observation is that he pities Soames his distress; thus, Answer B is correct.

2. **C.** Soames considers himself to be a great artist who has been overlooked in his time. He further imagines that in the future, the world will have finally recognized his talent and heaped well-deserved praise upon his work. These are traits of a person who is self-involved, so Answer C is correct.

3. **A.** In context, when Soames says, "I don't care a *sou* for recognition," he's saying that recognition is worthless to him, so Answer A is correct. (In fact, a *sou* is an old French coin that would no longer have much monetary value.)

4. **C.** In the last paragraph, Soames is describing a fantasy of what the world will be like when it finally recognizes his talent. This fantasy isn't credible, and the author signals the reader of this fact by writing the speech in an overly dramatic way that includes extravagant language, exclamation points, and words emphasized with italics. Answer C is correct.

5. **C.** Passage 1 states that "immersion in water provided support for the larger organisms." This directly contradicts the assertion that the land offers the potential for plant organisms to grow larger, so Answer C is correct.

6. D. According to Passage 2, "Plants need to keep their stomates open in order to obtain carbon dioxide for photosynthesis." It further states that "to prevent water loss from happening, each stomate has two guard cells surrounding it." Thus, guard cells around stomates help plants to maximize carbon dioxide intake while minimizing water outtake, so Answer D is correct.

7. B. Passage 1 explains why plants tended to thrive in "shallow water on the margins of land masses." From here, the possibility of growth onto land was an opportunity for access to more light for photosynthesis, but with some downsides such as a lack of water. Passage 2 discusses some ways in which plants on land have adapted mechanisms for holding onto water. Thus, Answer B is correct.

8. A. Passage 1 states that "desiccation [shriveling due to water loss] was a serious problem" for plants. And Passage 2 discusses how cuticles and guard cells "help prevent [a plant] from losing too much water." Answer A is correct.

9. A. The passage offers advice, including two quotes from professionals, on the importance of having a LinkedIn profile. Answer A is correct.

10. C. The passage opens by discussing the trend of "both increasing numbers and increasing proportions of older adults in the United States and across the globe." This trend is shown to be pervasive both inside and outside the U.S., with no expectation that it will decrease anytime soon, so Answer C is correct.

11. C. The turning point of the story recounted in the passage occurs after Goodyear "accidentally spilled a batch of rubber mixed with sulfur on a hot stove." The passage goes on to say, "He noticed that the resulting compound didn't lose its shape in the heat." Thus, Goodyear noticed the result of his small mishap, and this led him to his important discovery, so Answer C is correct.

12. C. The passage states that "parents *internalize* their children's tastes and make purchases that will please them, even without children making explicit requests or even being present when purchases are made." This concept is exemplified by the father who surprises his daughter, buying her a present based on his knowledge of what she likes, so Answer C is correct.

13. A. The passage states of Joshua Pegge's book that "some reviewers assumed the book was a joke." The question that follows this statement underscores the statement, implying the opinion that nobody would seriously want to read a book about Cockney. Therefore, Answer A is correct.

14. C. The writer's thesis is that GMOs have enabled some amazing health breakthroughs. Thus, the counterargument is that GMOs are harmful and shouldn't be eaten. The writer anticipates this counterargument in Sentence 3 — "Some people object to the idea of GMOs in their diets" — so Answer C is correct.

15. D. The passage states that "failure of a prey's defense system is likely to be fatal to that individual." And although "predation [does not] always seriously damage individual prey fitness," the passage indicates no positive effect of predation on individual prey, which rules out Answers A and C. However, the passage says that "predation is a major drive for prey evolution by weeding out less fit individuals," which can be a positive effect on the evolution of the species, so Answer B is ruled out; therefore, Answer D is correct.

16. B. The passage describes "ascertainment" in this way: "Stations were directed to actively seek out diverse views to broadcast instead of ignoring the hot button issues." This is best paraphrased as the responsibility to seek out a variety of perspectives on controversial issues, so Answer B is correct.

17. **D.** The passage states that the eruption of Mount Rainier, Washington, would result in a jokulhlaup, which rules out Answer A. It also says that jokulhlaups "cause all the damage associated with massive flooding and occur with the unpredictability of a volcano," which rules out Answers B and C. All examples of jokulhlaups in the passage mention both glaciers and volcanoes, so Answer D is correct.

18. **C.** The word "sadly" begins a long sentence that takes a detour with the word "despite," but which eventually returns to its main point that "the media seem stuck in a very traditional and stereotypical groove"; therefore, Answer C is correct.

19. **A.** The word *although* at the beginning of the sentence indicates a contrast between the first and second parts. In the second part, Bianca is described as "warm and personable," so the correct contrast is a word that means "cold" and "disagreeable." The word *abrasive* is close in meaning, so Answer A is correct.

20. **C.** The sentence emphasizes the ability of the Soviet Union to *activate* or *marshal* a defense. The nearest synonym to these words is *mobilize*, so Answer C is correct.

Writing Test Answers

How did you do on your second Writing Test? Each of the following answers includes an explanation to help you understand it. If you need more help, flip back to Chapters 7 through 9.

1. **A.** The sentence begins with the dangling modifier, "While working with troubled children and their families in the 1930s and 1940s." This modifier describes Rogers, so the words following the comma should clarify that the subject of the sentence is Rogers, ruling out Answers C and D. Answer B contains awkward phrasing, so this answer is incorrect; therefore, Answer A is correct.

2. **C.** The information about Rogers having traveled to many countries may be interesting, but it doesn't fit in with the essay at this point, which rules out Answers A and B. This information isn't found elsewhere in the essay, however, so Answer D is incorrect. In this context, the proposed sentence interrupts the flow and blurs the focus of the paragraph, so Answer C is correct.

3. **D.** The title of the book, *Client-Centered Therapy,* is a parenthetical element, so it requires a matched pair of punctuation marks both before and after it. Only Answer D offers this option, so Answer D is correct.

4. **C.** The transitional words need to weave the information in Sentence 10 together with the information in Sentence 11. Sentence 10 explains Carl Rogers's approach, and Sentence 11 further enhances this idea. Thus, Sentence 11 doesn't provide an example or a contrast, which rules out Answers A and B. Nor does Sentence 11 provide additional information, so Answer D is ruled out. The words "in this way" work best to connect the two sentences, so Answer C is correct.

5. **A.** The essay is relatively formal and includes no first-person (*I*) references, so Answer B is ruled out. Although Carl Rogers is said to have rebelled in Sentence 8, this trait isn't the focus of the essay, so Answer C is incorrect. And the essay mentions few if any inconsistencies in his work, which rules out Answer D. Answer A wraps up the essay in much the same manner as it begins, so Answer A is correct.

6. **D.** Answers A, B, and C all use several words where one word will serve the purpose, and in the case of Answer C, introduce awkward phrasing into the essay. The most concise answer here uses one word, so Answer D is correct.

7. **B.** Parallel construction requires that the verb *failed* and the verb that follows it both need to be past-tense verbs, so Answer C is incorrect. Additionally, the subject of the sentence is plural, so subject-verb agreement requires both verbs to accept a plural subject, so Answer A is incorrect. Answer D uses a semicolon to split the sentence into two clauses; however, the subject of the second clause (*it*) should be plural, not singular, which rules out Answer D. Therefore, Answer B is correct.

8. **A.** The sentence starts with the dangling modifier, "Curious to see what the results might be." The subject of the sentence must immediately follow this modifier, and must be what is being described — that is, *Watson*. Therefore, Answer A is correct.

9. **B.** Sentence 16 provides an additional effect of Watson's work rather than a contrast, so Answers A and D are incorrect. Additionally, Sentence 16 doesn't provide an example, so Answer C is ruled out; therefore, Answer B is correct.

10. **C.** Sentence 17 is consistent with the relatively informal style and tone of the essay, which rules out Answer A. Nor does this sentence blur the focus of the essay, so Answer B is incorrect. However, the sentence contains an interesting fact rather than the most important information in the essay, so Answer D is incorrect; thus, Answer C is correct.

11. **C.** The words "the surgical mask" are important to the sentence, so Answer D is ruled out. Neither a comma nor a semicolon is correct for setting off this type of important reveal, which rules out Answers A and B. The punctuation required here is the colon, so Answer C is correct.

12. **B.** All the possible choices offer a replacement for the repeated words "in cities." These words refer to a location, so the proper choice is a comma followed by the relative pronoun "where." Therefore, Answer B is correct.

13. **C.** The part of the sentence that follows the last comma is intended to be a phrase that modifies the word *illusion*. Among the four answer choices, the only way to achieve this goal is to use the word *helping*, so Answer C is correct.

14. **B.** Sentence 15 provides another in a sequence of reasons why Japanese people wear masks. It doesn't reach a conclusion, so the words *as a result* and *therefore* are both inappropriate, which rules out Answers A and D. It also doesn't provide a contrast with previous information, so Answer C is ruled out; thus, Answer B is correct.

15. **A.** The contraction *it's* (which means "it is") rather than the possessive *its* is required in this context, so Answer B is ruled out. The parenthetical element "especially among young people" begins with a comma, so it must also end with a comma, which rules out Answer D. But the use of the word *where* in this context is inappropriate, because the words before the comma don't refer to a location, so Answer C is incorrect. Therefore, Answer A is correct.

16. **C.** The word "Murray's" is used as a possessive, so it requires an apostrophe; therefore, Answers A and B are incorrect. The parenthetical element "such as the incredibly successful *Ghostbusters* movies" ends with a comma, so it must begin with a comma; therefore, Answer D is incorrect, so Answer C is correct.

17. C. In Answer A, the relative pronoun *who* appears too late in the sentence to refer to the subject, ruling out this answer. Answer B incorrectly uses the tenses of the verb "had originally given" and "having later revised." Answer D is sentence fragment, so it's incorrect. Answer C correctly subjoins the clause "who originally gave it only gave it only three stars" so that the subject "Renowned film critic Roger Ebert" can connect with the verb "revised"; therefore, Answer C is correct.

18. C. Answer A is a run-on sentence — a pair of clauses with a comma splice — so this is incorrect. In Answer B, the conjunction *however* appears incorrectly punctuated (it should read "; however,"), so it's incorrect. Answer D incorrectly uses a colon to connect the two clauses. The correct Answer is C, which correctly uses a semicolon to connect the two clauses.

19. A. The sentence begins with the dangling modifier "Unable to escape," which modifies the noun *Phil*. Thus, this noun must appear directly after the comma, so Answers B and C are incorrect. Answer D incorrectly uses the possessive "Phil's," so this is wrong as well. Thus, Answer A is correct.

20. B. Sentence 16 introduces a paragraph describing Phil's character transformation. Sentence 17 continues this portion of the narrative, so contrasting conjunctions *however* and *nevertheless* are inappropriate, ruling out Answers A and C. The word *otherwise* is also contrasting, and additionally cannot properly begin a sentence, so Answer B is ruled out. Answer B correctly uses the conjunction *moreover* to move the narrative forward without changing direction.

21. D. The parenthetical element "for example, forests, mountains, and oceans" ends with a parenthesis, so it must begin with one as well; therefore, Answer D is correct.

22. B. The sentence begins with the dangling modifier "Written over 400 years ago," which modifies the noun phrase "this popular Shakespearean play." Thus, this noun phrase must appear directly after the comma, so Answers C and D are both incorrect. In Answer A, the placement of the word *now* implies that the words and phrases in the play have changed, which is misleading, so Answer A is incorrect. In Answer B, the word *now* is correctly placed before the word *seem,* implying that the words haven't changed but now seem unusual; therefore, Answer B is correct.

23. C. Answers A and B both fail to provide a third example of 1960s slang, so both are ruled out. Answer D provides a third example, but in a form that doesn't match the two previous examples because the slang word is introduced before rather than after its description; therefore, Answer D is incorrect. Answer C matches both the purpose and the tone of the two previous examples, so this answer is correct.

24. C. The essay discusses a variety of changes to the English language, each within a context of time, such as "over 400 years ago" and "even 50 years." Thus, also presenting this prediction within a time frame is appropriate and clarifying, so Answer C is correct.

25. A. The clause beginning "that any grammatical form " and running to the end of the sentence functions as the direct object of the sentence. Within this clause, the phrase "any grammatical form that was considered non-standard in 1960" functions as the subject. This subject is singular, so the verb must also be singular; therefore, Answers B and D are ruled out. Answer C includes the words "is being," which implies that the subject of the sentence is acting by choice, so this answer is ruled out. Thus, Answer A is correct.

Arithmetic Test Answers

How did you do on your second time around with arithmetic? In this section, I provide all the answers, in each case with explanations of how to find it. If you still need work on this subtest, I recommend reviewing Chapters 10 through 13. Good luck!

1. **A. 0.581.** Line up the decimal points, place the decimal point in the answer in the same position, and then subtract as usual:

$$\begin{array}{r} 0.801 \\ -\ \ 0.220 \\ \hline 0.581 \end{array}$$

2. **D. 0.85.** Focus on the hundredths digit (4) and the digit to the right of it. Because the latter digit is 8, round 4 up to 5 and change the remaining digits to 0s:

$$0.8485 \rightarrow 0.8500 \rightarrow 0.85$$

3. **A.** $\frac{7}{20}$. To solve, let r stand for the fraction of the brownies that was left for Reggie, then set up the following equation:

$$\frac{1}{4} + \frac{2}{5} + r = 1$$

To begin, add $\frac{1}{4} + \frac{2}{5} = \frac{5+8}{20} = \frac{13}{20}$. So:

$$\frac{13}{20} + r = 1$$

Now, subtract $\frac{13}{20}$ from each side of the equation and simplify the result:

$$r = 1 - \frac{13}{20} = \frac{7}{20}$$

Therefore, $\frac{7}{20}$ of the brownies were left for Reggie.

4. **A. 6234.** Line up the columns correctly and add down:

$$\begin{array}{r} 5738 \\ +\ 496 \\ \hline 6234 \end{array}$$

5. **D. 5.** Divide 453 by 7:

$$\begin{array}{r} 64 \\ 7{\overline{\smash{)}453}} \\ \underline{42} \\ 33 \\ \underline{28} \\ 5 \end{array}$$

Therefore, the remainder is 5.

6. **A. 37.23.** Change 73% to the equivalent decimal 0.73, then multiply by 51:

$$\begin{array}{r} 0.73 \\ \times\ 51 \\ \hline 73 \\ 3650 \\ \hline 37.23 \end{array}$$

7. B. 7. Divide 45 by 6:

$$45 \div 6 = 7.5$$

Therefore, a maximum of 7 tables of 6 can fit in the room.

8. C. 4,500,000. Round 5076 down to 5000, and round 889 up to 900, then multiply:

$$5000 \times 900 = 4,500,000$$

9. A. $\frac{5}{6}$. To divide fractions, use Keep–Change–Flip, then cancel common factors and multiply across:

$$\frac{5}{8} \div \frac{3}{4} = \frac{5}{8} \times \frac{4}{3} = \frac{5}{\cancel{8}^2} \times \frac{\cancel{4}^1}{3} = \frac{5}{6}$$

10. C. 27. Begin by evaluating parentheses, then multiplication and division from left to right, then move on to addition and subtraction:

$$-9 + 12 \div 3 \times 7 - (-8)$$
$$= -9 + 12 \div 3 \times 7 + 8$$
$$= -9 + 4 \times 7 + 8$$
$$= -9 + 28 + 8$$
$$= 19 + 8$$
$$= 27$$

11. C. $53\frac{3}{8}$ inches. To add $47\frac{5}{8} + 5\frac{3}{4}$, begin by adding the whole-number parts of the mixed numbers: $47 + 5 = 52$. Next, add the two fractional parts and convert to a mixed number:

$$\frac{5}{8} + \frac{3}{4} = \frac{5}{8} + \frac{6}{8} = \frac{11}{8} = 1\frac{3}{8}$$

To complete the problem, add the results from the first two steps:

$$52 + 1\frac{3}{8} = 53\frac{3}{8}$$

12. B. $\frac{4}{1000} < 4\% < 0.4$. To solve, change all values to decimals and compare:

$$\frac{4}{1000} = 0.004 \quad 0.4 = 0.4 \quad 4\% = 0.04$$

Thus, $0.004 < 0.04 < 0.4$, so $\frac{4}{1000} < 4\% < 0.4$.

13. B. 148. To solve, find 40% of 370 by multiplying:

$$370 \times 0.4 = 148$$

14. A. 40%. To solve this problem, change the words in this question into an equation as follows:

What percent of 300 equals 120?

$$n(0.01) \times 300 = 120$$

The parentheses around (0.01) remind you to multiply, and then solve the problem:

$$3n = 120$$
$$\frac{3n}{3} = \frac{120}{3}$$
$$n = 40$$

So the answer is 40%.

15. **D.** $\frac{6}{11} < \frac{5}{8}$. Cross-multiply to compare values:

$$\frac{2}{3} \qquad\qquad \frac{5}{8}$$
$$2 \times 8 = 16 \qquad 5 \times 3 = 15$$

$$\frac{3}{4} \qquad\qquad \frac{4}{7}$$
$$3 \times 7 = 21 \qquad 4 \times 4 = 16$$

$$\frac{4}{9} \qquad\qquad \frac{3}{10}$$
$$4 \times 10 = 40 \qquad 3 \times 9 = 27$$

$$\frac{6}{11} \qquad\qquad \frac{5}{8}$$
$$6 \times 8 = 48 \qquad 5 \times 11 = 55$$

The only case in which the first value is less than the second is $48 < 55$, so $\frac{6}{11} < \frac{5}{8}$.

16. **C. 23.4.** Cameron walked $27.2 + 15.9 = 43.1$ kilometers on the first and third days of the hike, and walked a total of 66.5 kilometers, so on the second day he walked $66.5 - 43.1 = 23.4$ kilometers.

17. **D. 0.047.** Convert all four values to decimals:

$$\frac{4}{7} = 0.571$$
$$47\% = 0.47$$
$$\frac{4}{70} \approx 0.057$$
$$0.047 = 0.047$$

Therefore, the least value is 0.047.

18. **B. 95%.** Jonathan got 2 questions wrong out of 40, so he got 38 right out of 40. Write this information as a fraction and convert it to first a decimal and then a percent:

$$\frac{38}{40} = 0.95 = 95\%$$

19. **D. \$13,920.** Calculate 16% profit as a percent increase of 116%, which converts to 1.16, and multiply 12,000 by this value:

$$12,000 \times 1.16 = 13,920$$

20. **A. 22.493.** Multiply just like whole numbers, but remember to include 3 decimal places:

$$
\begin{array}{r}
27.1 \\
\times \quad 0.83 \\
\hline
813 \\
2168 \\
\hline
22.493
\end{array}
$$

Quantitative Reasoning, Algebra, and Statistics Test Answers

How did you do on this intermediate-level math? This section provides you the answers to all the questions, including an explanation of how each answer is reached. If you need more practice, please turn to Chapters 13 through 16.

1. **C. $6bc^2$.** The greatest common factor of 24 and 30 is 6, so the answer has a coefficient of 6. The second expression has no a factor, so this factor must be excluded from the greatest common factor. Therefore, the answer is C.

2. **C. Between 40 and 60 kilometers.** Set up a proportion as follows:

   $$\frac{1 \text{ km}}{0.62 \text{ mi.}} = \frac{x \text{ km}}{26.2 \text{ mi.}}$$

   Next, cross-multiply and solve for x:

   $$26.2 = 0.62x$$
   $$\frac{26.2}{0.62} = \frac{0.62x}{0.62}$$
   $$42.26 \approx x$$

3. **D. –116.** To begin, plug in 3 for x and –2 for y:

   $$8x + 4y^3 - 3x^2y^2 = 8(3) + 4(-2)^3 - 3(3)^2(-2)^2$$

   Now, use the order of operations (PEMDAS), beginning with exponents, then multiplication, and finally addition and subtraction:

   $$= 8(3) + 4(-2)(-2)(-2) - 3(3)(3)(-2)(-2)$$
   $$= 24 - 32 - 108$$
   $$= -116$$

4. **B. 41%.** The class has 29 children. Out of these, 12 are boys who have at least one sibling. Make a fraction out of these two values and change to a percent:

   $$\frac{12}{29} \approx 0.41 = 41\%$$

5. **C. $4x^2y(3x^4y^2 + 2x^5 - 5y^4)$.** The polynomial has coefficients with a greatest common factor of 4. Each term includes the variable x with an exponent that is at least 2, so you can also factor out an x^2. And each term includes the variable y with an exponent that is at least 2, so you can also factor out a y:

   $$12x^6y^3 + 8x^7y - 20x^2y^5 = 4x^2y(3x^4y^2 + 2x^5 - 5y^4)$$

6. **A. 4.83×10^8.** Starting with 483,000,000, move the decimal point 8 places to the left and multiply the result by 10^8.

7. **C. 2017.** The graph shows approximately 20 new students in 2014, 31 in 2015, 27 in 2016, 47 in 2017, and 51 in 2018. Thus, in 2017, the increase was 20 new students as compared with the previous year, which is the greatest increase.

8. **B. 260 feet.** Triangle *ABC* is a right triangle, so apply the Pythagorean theorem to find the length of *AC*:

$$a^2 + b^2 = c^2$$
$$100^2 + 240^2 = c^2$$
$$10{,}000 + 57{,}600 = c^2$$
$$67{,}600 = c^2$$
$$\sqrt{67{,}600} = \sqrt{c^2}$$
$$260 = c$$

9. **A. −1.** Let *n* equal the number that you started with. So, you can write the following equation:

$$\frac{2n + 7}{5} = n + 2$$

Remove the fraction by multiplying both sides of this equation by 5, then simplify and isolate *n*:

$$2n + 7 = 5(n + 2)$$
$$2n + 7 = 5n + 10$$
$$7 = 3n + 10$$
$$-3 = 3n$$
$$-1 = n$$

10. **D. $72x^{18}$.** Begin by evaluating $(2x^2)^3$ and $(3x^6)^2$ separately, using the rules for evaluating exponents of exponents:

$$(2x^2)^3(3x^6)^2 = (8x^6)(9x^{12})$$

Next, multiply $(8x^6)$ and $(9x^{12})$:

$$= 72x^{18}$$

11. **D. $y = -\frac{1}{2}x - \frac{3}{2}$.** The line connecting these two points is of the form $y = mx + b$. To find the slope *m*, use the slope formula:

$$m = \frac{y_2 - y_1}{x_2 - x_1} = \frac{-2 - 1}{1 - (-5)} = \frac{-3}{6} = -\frac{1}{2}$$

So the equation of the line is $y = -\frac{1}{2}x + b$ for some value *b*. To find this value, plug in the *x*-value and *y*-value for either point and solve for *b*:

$$-2 = -\frac{1}{2}(1) + b$$
$$-2 = -\frac{1}{2} + b$$
$$-2 + \frac{1}{2} = b$$
$$-\frac{3}{2} = b$$

So the equation is $y = -\frac{1}{2}x - \frac{3}{2}$

12. **C. 20.25π.** Begin by plugging 9π into the circumference formula $C = 2\pi r$ and solving for the radius:

$$9\pi = 2\pi r$$
$$\frac{9\pi}{2\pi} = \frac{2\pi r}{2\pi}$$
$$4.5 = r$$

Now, plug 4.5 for r into the area formula $A = \pi r^2$:

$$A = \pi(4.5)^2 = 20.25\pi$$

13. **D. 0.35.** To find the median, place all the scores in order and find the mean of the two middle numbers:

10, 10, 10, 9, 9, 9, 9, 9, 8, **8, 7,** 7, 7, 7, 7, 7, 6, **6, 6, 6**

Thus, the median is 7.5.

To calculate the mean score, multiply each score by the number of students who received it and divide by 20:

$$\frac{30 + 45 + 16 + 42 + 24}{20} = \frac{157}{20} = 7.85$$

Therefore, the difference between the mean and the median is 7.85 – 7.5 = 0.35.

14. **D. x is a negative even number.** Begin by cross-multiplying, then distribute and isolate x:

$$\frac{5x + 4}{8} = \frac{3x - 8}{10}$$
$$10(5x + 4) = 8(3x - 8)$$
$$50x + 40 = 24x - 64$$
$$26x + 40 = -64$$
$$26x = -104$$
$$x = -4$$

15. **A. (1, –6).** Reflection across the y-axis equals the transformation $(x, y) \rightarrow (-x, y)$:

$$(-6, 1) \rightarrow (6, 1)$$

This 90° clockwise rotation equals the transformation $(x, y) \rightarrow (y, -x)$:

$$(6, 1) \rightarrow (1, -6)$$

16. **C. $10x^2 + 8x - 6$.** Begin by removing the parentheses, FOILing the first part of the expression, and distributing –6 to the second part:

$$(2x + 4)(5x - 3) - 6(x - 1)$$
$$= 10x^2 - 6x + 20x - 12 - 6x + 6$$

Next, combine like terms:

$$= 10x^2 + 8x - 6$$

17. **D. {4,5,7} ∩ {1,2,3,4,5,6,7,8,9}.** The union of sets (∪) includes all members of *either* set, and the intersection of sets (∩) includes members of *both* sets. However, the first set in {4,5,7} ∩ {1,2,3,4,5,6,7,8,9} is missing a 3, so its intersection is {4,5,7} rather than {3,4,5,7}.

18. **C. 0.8.** To begin, multiply every term by 100 to remove the decimals:

$$4.6x - 0.21 = 2.3x + 1.63$$
$$460x - 21 = 230x + 163$$

Now, isolate x:

$$230x - 21 = 163$$
$$230x = 184$$
$$\frac{230x}{230} = \frac{184}{230}$$
$$x = 0.8$$

19. **A.** $y \geq -\frac{1}{2}x + 3$ **and** $y \leq 3x + 1$. The shaded area is above $y = -\frac{1}{2}x + 3$ (the line with the negative slope), so this area is $y \geq -\frac{1}{2}x + 3$. The shaded area is below $y = 3x + 1$ (the line with the positive slope), so this area is $y \leq 3x + 1$.

20. **B. 5.** The slope is 4, so $m = 4$. Plug 4 for m, –3 for x, and –11 for y into the equation and solve for b:

$$y = mx + b$$
$$-11 = 4(-3) + b$$
$$-11 = -12 + b$$
$$1 = b$$

Thus, $b = 1$, so $m + b = 5$.

Advanced Algebra and Functions Test Answers

How did you do? Please recall that some community colleges don't make you take this section of the ACCUPLACER, so check with your school administration if you're in any doubt. In this section, you find not only the answers but also a detailed explanation of each one to help you make sense of them. If you need more help, Chapters 17 through 20 focus exclusively on these types of problems.

1. **A.** $\frac{b}{2b-3}$. Begin isolating a by grouping all a-terms on the left side of the equation:

$$2ab - b = 3a$$
$$2ab = 3a + b$$
$$2ab - 3a = b$$

Factor a on the left side and then divide out the non-a factor:

$$a(2b - 3) = b$$
$$a = \frac{b}{2b - 3}$$

2. **D.** $y = 2x - 145$. $\angle DAB = (x + 10)°$, and $\angle BAC$ is supplementary with this angle, so:

$$\angle BAC = 180° - (x + 10)° = 180° - x° - 10° = 170° - x°$$

Similarly, $\angle BCE = (x + 25)°$, and $\angle BCA$ is supplementary with this angle, so:

$$\angle BCA = 180° - (x + 25)° = 180° - x° - 25° = 155° - x°$$

These two angles plus $\angle B$ form the three angles of a triangle, so their sum equals $180°$, so:

$$170° - x° + 155° - x° + y° = 180°$$
$$325° - 2x° + y° = 180°$$
$$-2x° + y° = -145°$$
$$y° = 2x° - 145°$$

Therefore, $y = 2x - 145$.

3. **D. More than $50,000.** The amount of money in the grandfather's account is $y = 1500x + 5000$ and in the uncle's account is $y = 750x + 8000$. The sum of the two accounts is $y = 2250x + 13,000$. When Jeannine is 18, the amount of money in both accounts will be:

$$y = 2250(18) + 13,000 = 40,500 + 13,000 = 53,500$$

4. **C. $y = \frac{1}{2}x - 5$.** The line in the figure has a slope of -2, so any line perpendicular to it has a slope of $\frac{1}{2}$.

5. **B. $2ac(4a + 5b)(4a - 5b)$.** Begin by factoring out the GCF:

$$32a^3c - 50ab^2c = 2ac(16a^2 - 25b^2)$$

Now, factor $16a^2 - 25b^2$ as the difference of two squares:

$$= 2ac(4a + 5b)(4a - 5b)$$

6. **C. $g(x) = -(x - 3)^2 - 2$.** A vertical reflection across the x-axis changes $f(x)$ to $-f(x)$. Then, translating this function to $(3, -2)$ changes $-f(x)$ to $-f(x - 3) - 2$. Find $g(x)$ by applying this transformation to $f(x) = x^2$, which results in the following:

$$g(x) = -(x - 3)^2 - 2$$

7. **B. $\cos 20°$.** Begin by applying the identity $\tan x° = \frac{\sin x°}{\cos x°}$:

$$\cos 70° \tan 70° = \cos 70° \frac{\sin 70°}{\cos 70°} = \sin 70°$$

Now, apply the identity $\sin x° = \cos(90 - x)°$:

$$\sin 70° = \cos(90 - 20)° = \cos 20°$$

8. **C. $50(2)^4$.** Using the exponential function $y = ab^x$, plug in 50 for a and 2 for b:

$$y = ab^x = 50(2)^x$$

To find x, consider that 6 days equals 144 hours. Thus, a population of fruit flies that doubles every 36 hours will double 4 times in 144 hours, so set x to 4:

$$= 50(2)^4$$

9. **B.** $\frac{9}{4}$. To begin, notice that triangle *LMK* is a right triangle with a leg of 3 and a hypotenuse of 5, so triangle *LMK* is a 3-4-5 right triangle; therefore, *ML* = 4.

Triangle *KMJ* is also a right triangle, so the next step is to show that it's similar to triangle *LMK*. Let $\angle J = x°$. $\angle JKL$ is a right angle, so $\angle L = (90 - x)°$, and $\angle JKM = (90 - x)°$. Thus, triangle *KMJ* and triangle *LMK* are both right triangles that have one angle that equals $(90 - x)°$, so the larger angle of each equals $x°$.

Thus, because triangles *KMJ* and *LMK* are similar, the ratio of *JM* to *KM* equals 3:4. Therefore:

$$\frac{n}{3} = \frac{3}{4}$$
$$n = \frac{9}{4}$$

10. **C.** $-2\frac{3}{4}$. To find the zeros of the polynomial $y = (x - 2)(x + 5)(4x - 1)$, set each of the three factors to 0 and solve each resulting equation:

$$
\begin{array}{ccc}
x - 2 = 0 & x + 5 = 0 & 4x - 1 = 0 \\
x = 2 & x = -5 & 4x = 1 \\
& & x = \frac{1}{4}
\end{array}
$$

These are the three values *a*, *b*, and *c*, so

$$a + b + c = 2 - 5 + \frac{1}{4} = -3 + \frac{1}{4} = -2\frac{3}{4}$$

11. **D.** $7\sqrt{5}$. Begin by simplifying both radicals, then add the whole number parts and keep the radical part:

$$2\sqrt{20} + \sqrt{45} = 2\sqrt{4}\sqrt{5} + \sqrt{9}\sqrt{5} = 4\sqrt{5} + 3\sqrt{5} = 7\sqrt{5}$$

12. **D.** $P = (-4, 21)$ and $Q = (1, -4)$. To find where $f(x)$ and $g(x)$ intersect, begin by setting these two functions equal and solve for *x*:

$$x^2 - 2x - 3 = -5x + 1$$
$$x^2 + 3x - 4 = 0$$
$$(x + 4)(x - 1) = 0$$
$$x = -4, 1$$

Now, plug these two *x*-values into either of the original functions and solve for *y*:

$$y = -5(-4) + 1 = 20 + 1 = 21$$
$$y = -5(1) + 1 = -5 + 1 = -4$$

So $P = (-4, 21)$ and $Q = (1, -4)$.

13. **B.** $\frac{3 \pm \sqrt{29}}{10}$. To solve this non-factorable quadratic equation, use the quadratic formula, plugging in 5 for *a*, –3 for *b*, and –1 for *c*:

$$x = \frac{-b \pm \sqrt{b^2 - 4ac}}{2a} = \frac{3 \pm \sqrt{(-3)^2 - 4(5)(-1)}}{2(5)} = \frac{3 \pm \sqrt{9 + 20}}{10} = \frac{3 \pm \sqrt{29}}{10}$$

14. **D.** $-\frac{n}{m}$. Begin by recalling that $x^2 - y^2 = (x + y)(x - y)$, then substitute as follows:

$$n = m(x - y)$$

Now, divide both sides by m, and multiply both sides by -1:

$$\frac{n}{m} = x - y$$
$$-\frac{n}{m} = -(x - y)$$
$$-\frac{n}{m} = y - x$$

15. **C. $(x - 1)^2 + (y + 5)^2 = 9$.** Use the formula for a circle centered at the point (h, k) with a radius of r:

$$(x - h)^2 + (y - k)^2 = r^2$$
$$(x - 1)^2 + (y - (-5))^2 = 3^2$$
$$(x - 1)^2 + (y + 5)^2 = 9$$

16. **B. 28.** Begin by finding the value of $f(4)$:

$$f(4) = 2(4) - 11 = 8 - 11 = -3$$

Substitute -3 for $f(4)$ into $g(f(4))$ and evaluate:

$$g(f(4)) = g(-3) = 3(-3)^2 + 1 = 3(9) + 1 = 27 + 1 = 28$$

17. **C. $1 < x < 5$.** Begin by writing $\log_3 81 = 2x$ as an equivalent exponential equation:

$$3^{2x} = 81$$

Next, rewrite 81 as a power of 3:

$$3^{2x} = 3^4$$

Now, drop the equivalent bases, setting the exponents equal to each other, and solve for x:

$$2x = 4$$
$$x = 2$$

Therefore, $x = 2$, so $1 < x < 5$.

18. **D. $\frac{5\pi}{36}$.** Using the equivalence $180° = \pi$, write a proportion as follows, and solve for x:

$$\frac{180}{\pi} = \frac{25}{x}$$
$$180x = 25\pi$$
$$x = \frac{25\pi}{180} = \frac{5\pi}{36}$$

19. **D. $a = -2$ and $e = -3$.** The quartic (4th-degree) polynomial shown in the figure displays end behavior that is negative for both very large and very small values of x, so a is a negative number. Additionally, the y-intercept is negative, so e is a negative number.

20. **A. $x \leq 7$.** Because you can't take the square root of a negative number, create an inequality in which the radicand (the value inside the radical) is greater than or equal to 0, and simplify:

$$14 - 2x \geq 0$$
$$-2x \geq -14$$
$$x \leq 7$$

8

The Part of Tens

Understand ten ways in which college is different from high school.

Discover ten things you can do to boost your score on the ACCUPLACER.

Chapter **25**

Ten Ways College Is Different from High School

You already know that things are going to be different when you advance from high school to college. But what, *specifically*, changes when you enter a college classroom for the first time?

And why does nobody ever tell you?

Okay, I'm telling you. Here's my personal take on ten expectations of college students as you make this transition into adulthood. Please don't take this as gospel, but rather as a set of guidelines to consider and explore.

Greater Personal Freedom

In a high school classroom, the school's faculty are responsible for you *in loco parentis* — "in place of a parent" — until you are at least 18. This means that *legally*, they really have to keep an eye on you and know your whereabouts at all times. In day-to-day terms, this usually means that a high school student needs to be present or accounted for in every class, and must ask for permission to leave the room to get a drink of water or use the restroom.

In contrast, in a college classroom, the assumption is that you're old enough to look after yourself. So, you can pretty much enter and leave a college classroom at your own discretion, without raising your hand.

Greater Personal Responsibility

With greater freedom, however, comes greater responsibility. While college professors will not expect you to explain your behavior regarding absences or being late or unprepared for class, they will take note of this behavior and figure it into their assessment of you as a student. (That said, some professors in smaller classes do take attendance, and automatically fail students who miss, say, three classes.)

So, if you're on the fence between an A and a B (or between a D and an F!), the deciding factor will probably be the ever-elusive "class participation" factor. Don't say you weren't warned!

Professional Behavior

A college classroom resembles a business meeting more than it resembles a high school classroom. So, when you're in the room, you need to be fully in the room, present to the conversation, and on task.

Non-professional behavior — extended texting, having a side conversation with another student, putting your head on your desk, slouching in your chair, or wearing a bored or hostile look on your face — drains the room of energy, affecting the professor and other students.

Be smart — don't do that.

A Civil Conversation

Professors tend to like passion as it relates to their subject. However, when passion tips over into self-righteousness — even in the name of what you perceive to be a worthy cause — it can shut down discourse instead of opening it up.

Here's a good rule of thumb: If you think everyone else in the room is stupid, wrong, or ill-informed, you're probably not going to be successful at changing hearts and minds at that moment. See if you can become curious and just listen for a while.

Due Respect for and Engagement with the Subject of the Class

Most professors at four-year universities have a terminal degree in their field — often a PhD but in some cases an MD (medical degree) or MFA (master of fine arts). In any case, that's usually five to seven years of graduate school, capped off with a book-length dissertation. As you can imagine, this usually goes hand in hand with a genuine devotion to what they're teaching.

Whether the subject is world history, sociology, organic chemistry, or Hawaiian dancing, you can assume that your professors care about what they're teaching, and hope to spark your own interest in it.

That said, professors also understand that sometimes students are stuck taking required courses they don't love. For example, if you want to be a physical therapist, you may have to suffer through a statistics class or two to achieve your goal, even if math isn't your strong suit.

So, if you simply do your best to actively engage with the class material, you'll probably enjoy the class and succeed in it more than if you fold your arms and visibly resent every second you're stuck in that seat.

On the plus side, if your default personality trait is baseline respect, you might make a genuine friend who can also be an ally and a mentor.

Fewer — But Longer — Tests and Projects

You probably noticed that as you progressed in school, assignments tended to get longer and more involved, and you were given more time to complete them. This trend continues in college. In some college classes, only two or three tests may determine your final grade. In others, the same number of papers, plus perhaps a final exam, serve the same purpose.

A smaller number of longer assignments shifts the responsibility for discipline and initiative onto you. If you don't keep up with the workload, the second half of the semester will begin to close in on you.

Access to the Professor only during Office Hours

In most high schools, you see your teacher every day, from Monday through Friday. And most teachers have a classroom where you can find them easily.

In contrast, most college classes meet no more than twice a week. Aside from these times, professors tend to have relatively short office hours — say, on Tuesday from 2:30 to 4:30 p.m. Beyond this, they have their own lives and careers to worry about, so tracking them down — especially to beg them for an extension on a paper you haven't started working on — won't be easy.

That said, most professors are happy when good, interested students visit them during office hours. In most cases, you don't have to make an appointment as long as you don't mind waiting outside the office until your turn is called. Plan to spend no more than five to ten minutes with the professor, especially if other students are waiting. And you probably want to bring a question or topic to discuss that's related to the coursework — either something you're struggling with, or something that you find especially interesting.

This goes double with professors who teach for the department in which you plan to major. If you go on to graduate school, or even need a letter of recommendation to keep in your back pocket, a professor who genuinely likes and respects you can be a godsend.

College Costs Money

Although your parents may have paid for your school, just about every young person in the U.S. can get a K-12 education without spending a penny. And while there's talk of extending this benefit to community colleges, for the moment college is where you — or someone who loves you – have to start paying for your education.

The good news is that community college is still a lot less expensive than most four-year schools. And if you can demonstrate any sort of financial hardship, you're very likely to receive partial or even full financial assistance to attend. Furthermore, community colleges tend to coordinate with state schools to keep their curriculum consistent with state standards. So most if not all of the classes you take and pass will give you college credit that's transferrable to your state university system, and quite possibly other universities, both public and private. That's 60 out of 120 degree credits you can attain without breaking the bank!

Moreover, many state schools have a policy of automatically accepting in-state community college graduates who maintain a decent grade point average.

And just a touch more good news, if you're willing to accept it: Please consider that colleges and universities are businesses and, as such, need customers — that is, students. Sure, Princeton, Stanford, and MIT are extremely selective in their choice of students, and demanding of those they do select. But there are a wide variety of other good schools that are perfectly motivated to accept you as a paying student.

So please put out of your mind that as a community college graduate you'll be lucky to receive a single college acceptance letter. More often than not, serious students who are willing to thread the needle to pay for their education find the college that's right for them, and make the most of it.

More Opportunities — and Challenges — and Distractions

If you go on to a four-year college in another state, or even another country, the college experience will offer you opportunities you haven't dreamed of yet. Many of these will be outside the classroom, among new friends and a variety of people from all sorts of backgrounds. There will be an endless supply of experiences waiting for you, both good and bad.

At times, in fact, your classes may seem like one of the less important parts of your life. Whether this is true or false, from the perspective of the college administration, if you let your grades slip too far, you'll be placed on academic probation (as I was after my tumultuous first semester!). And if the trend continues, you'll ultimately be expelled so that another more motivated student can take your place.

So, try to maintain a balance between curricular and extra-curricular activities. Do take advantage of all the new opportunities that college provides. Just don't let them derail you from your one job as a student, which is to maintain reasonable grades until you graduate.

Simple Human Decency

In high school, you see your teachers every day, and usually get to know them pretty well. In contrast, college professors are often a bit more remote — you see them once or twice a week, often in the context of a huge lecture, and have to hunt them down otherwise.

Even so, college professors are people. And, like all people, they prefer to be treated with regard. The relationship between a professor and a student is, first and foremost, a *relationship*, even if only for one semester, and even if it remains more formal than your relationships with your high-school teachers.

So, consider this expectation a catch-all for anything not explicitly stated earlier in this list.

On the negative side, if you treat professors with contempt or try to outsmart, manipulate, use, or falsely flatter them, you probably won't like what you receive from them in return.

On the positive side, if you treat them nicely, they'll most likely do the same for you!

Chapter **26**

Ten Ways to Boost Your Performance (and Your Score) on the ACCUPLACER

This book is all about test prep, and *prep* stands for *preparation*.

But as you know — especially if you play a sport or if you're an actor, musician, or dancer — when your preparation is done, your *performance* is what counts. Even a well-prepared player can have an off day. And on the flipside, a relaxed player can more easily get into *the zone* — that effortless space where you can seemingly do no wrong and perform far beyond your normal capacity.

Following are ten ways to help improve your performance on the ACCUPLACER. Some are specific to the test, and others are just a good idea no matter what type of performance you're trying to enhance. Good luck!

Know Which ACCUPLACER Sections Your School Requires

The ACCUPLACER includes five separate sections: Reading, Writing, and three separate Math tests. But not every community college requires all five tests.

For example, the community college where I work doesn't typically require its students to take the ACCUPLACER Advanced Algebra and Functions Test (AAF). That's the most difficult of the three Math tests!

So before you spend a moment worrying about (or studying for!) a test that you may not have to take, be sure to find out exactly which sections of the ACCUPLACER your school will require you to take.

Take Care of Yourself the Night before and the Day of the ACCUPLACER Exam

This goes for any situation when you want to turn in a good performance: Take care of yourself! This includes getting enough sleep the night before (no big parties!), and perhaps doing something you enjoy that relaxes you. (For me, the movie *Groundhog Day* usually does the trick!)

Additionally, don't make any sudden changes to your usual habits the day of your ACCUPLACER exam. For example, eat the same type of breakfast you typically eat. If you usually start the day with coffee, do so on your test day. Conversely, if you don't usually drink coffee, don't pick your test day to experiment with it!

Pick <u>Your</u> Best Time of Day to Take the Test

If you're familiar with the SAT or ACT, you probably know that these tests are administered only on a preset number of very specific times and days throughout the year — usually, on Saturdays starting at around 8:00 a.m. As a non-morning person, I can tell you that I'm sure I would have gained a few points on my SAT if the test had started at 8:00 p.m.

In contrast, the ACCUPLACER is administered directly by your community college rather than by a third party. This usually makes the process a lot more informal and friendly. For example, many community colleges allow you to take the ACCUPLACER anytime during administrative hours — say, from 9:00 a.m. to 5:00 p.m. You may not even have to make an appointment — just show up ready to sit in front of a computer for a few hours. (Obviously, you'll want to call your school and find out the specific details.)

If your community college offers this level of flexibility, by all means take advantage of it. If you're a morning person, show up early; if not, plan to arrive in the early afternoon. In either case, be sure to arrive early enough that the office isn't getting ready to close!

Consider Taking the ACCUPLACER in Two or More Sessions

Here's an important tidbit: You don't have to take the entire ACCUPLACER in one sitting. Each section of the ACCUPLACER stands discretely on its own. Furthermore, your score on each section is independent of your other scores.

So, consider taking your ACCUPLACER exam on several separate days. In any case, I suggest that you start with what you believe will be the easiest test. Most of my students find this to be the Writing Test. But if your native language isn't English, or if you're confident with basic math, you may well find the Arithmetic Test to be the best place to start.

As You Start the Test, Breathe! And Keep Breathing

It's natural to feel nervous as you start taking a test. You may find, for example, that your heartbeat increases, your hands shake, and your mind races.

These effects are all physical manifestations of the fight-or-flight response, your body's response to anxiety. Adrenalin and other hormones pump into your bloodstream to give you immediate energy to, say, face down a wolf or run away from a bear.

Unfortunately, adrenalin isn't terribly helpful as you take the ACCUPLACER. To counter it, simply focus on your breathing for a moment. (The test will wait.) Notice your breath and then *slightly* deepen it — not so much as to hyperventilate. Do this for two or, at most, three breaths — say, for 20 seconds. Then resume the test.

If you notice your nervousness returning, take another 20 seconds and repeat this exercise. You may want to do this half a dozen times during the course of the first section of the test. After that, I'm betting that the worst of your anxiety will begin to settle down.

Take Your Time

Unlike most other tests, the ACCUPLACER isn't a timed test. Each section includes 20 to 25 questions, which you can answer at your own pace.

So take the time that *you* need to do your best. If you're taking the Reading Test, read slowly and carefully, and allow the information to seep in. On the Writing Test, take the time to read the entire passage, and then for each question focus on the sentences you're being asked about. And on each of the Math sections, take the time to proceed slowly and check your work before entering your final answer.

Check In with Yourself Before You Start Each New Section

As you finish each section of the ACCUPLACER, take a few moments to check in with yourself. How are you feeling? Strong? Comfortable? Discouraged? Tired? Hungry?

As I mention earlier, you don't have to take the entire test in just one sitting. Each section stands on its own. So if you're not in good shape as you finish a section, feel free to tell the nice person behind the desk who's giving you the test that you'd like to stop for the day.

Keep the ACCUPLACER in Perspective

I know this is easy to say, but I'll say it anyway: The ACCUPLACER is just a test, not a firing squad. And by far, the ACCUPLACER is not even the most important test you'll ever take.

Of course, you'd love to ace the ACCUPLACER and kick off community college with courses that give you college credit. But if the specter of the test has you worried enough to affect your performance, I recommend putting the importance of the ACCUPLACER in perspective.

If you have test anxiety, I understand that *any* test can be scary. So remind yourself as you take the ACCUPLACER that no matter what score you get, you're not in any danger of being thrown out of school or having your financial aid package revoked. It's just not that kind of a test.

The very worst outcome is that you'll have to attend a few remedial, no-credit courses to get up to speed so that you're ready to do well in your later college courses.

Sometimes, Two Is a Charm!

More good news: Most community colleges allow you to take the ACCUPLACER twice, and in some cases possibly more. So before you take the test, ask to see what your school's policy is.

Assuming your school permits two tries, you may want to take the first round without too much preparation. If you kept up with your work in high school, you may well pass some or even all of the ACCUPLACER sections without extra work.

When you've received your scores for the first round, find out what your community college considers a passing grade on each section. This number varies from one school to the next, so be sure to ask. Knowing how many more points you need to pass each section of the ACCUPLACER gives you a lot of information about how much studying you need to do, and in which areas.

Repeat after Me: "I'm Doing My Best."

I can say this with certainty: Your life will contain wonders you haven't even dreamed of yet.

When you're standing at the altar on your wedding day, or holding your first child in your arms, or climbing Mount Everest, or completing a marathon, or taking a call from your new boss telling you about your promotion — when any of these things happen for you, I guarantee that you won't find yourself saying, "I could be happy right now, if only I'd done better on my ACCUPLACER exam!"

So here's the thing. Just do your best. And keep telling yourself, "I'm doing my best."

That's all you can do.

Index

relative pronouns, 65

remedial courses, skipping, 8, 10–11

retaking tests, 25

rhetoric

  analyzing text structure, 39–40

  analyzing word choice, 39

  defined, 38

  overview, 14

  questions focusing on, 38–42

right triangles. *See also* triangles

  30-60-90, 301

  45-45-90, 300

roots

  integer, 273

  irrational, 273–274

rotations, 199–200

rounding, 106

S

SAT (Scholastic Assessment Test), 8

science and technology, 32

scientific notation, 163–164

scores, passing, 10

scrap paper, 9

secant, 298

second quartile. *See* median

sections (ACCUPLACER)

  deciding when to take your first section, 23–24

  failing to pass, 11

  overview, 2

  repeating the steps for the other, 24–25

  required by your school, 22

  retaking tests, 25

  scoring, 10

  study schedule for, 24

  taking the section easiest for you first, 23

semicolon

  building compound sentences with, 64

  separating clauses with, 69

sentences

  complex, 64–65

  compound, 63–64

  overview, 14

  parallel construction, 74

  parenthetical elements in, 70–71

  simple, 62–63

  subject-verb agreement in, 72–73

  transitional, 87

sets

  combining union and intersection of, 214

  defined, 213

  elements of, 213

  intersection of, 213

  overview, 17

  practice questions, 220–221

  union of, 213

set theory, 213

signature point, 251

similar statements, 86

similar triangles, 296

simple probability, 210, 211

simple sentences, 62–63

simplification (math)

  combining like terms, 171–172

  distributing to remove parentheses, 171–172

  factoring, 172–173

  FOIL method, 172

sine, 251, 254, 298, 308–310

SOH-CAH-TOA mnemonic, 299–300

special right triangles

  30-60-90, 301

  45-45-90 right triangle, 300

  trig functions for, 301

sphere, volume of, 290

spread of data sets, 206

square root, 162

squares

  defined, 162

  difference of, 242–243

  formula, 200

Standard English conventions, 14

state universities, 34

statistics

  answers to practice questions, 234–235

  boxplot, 298

  calculating, 206–207

  data sets, 206–207

  defined, 205

  descriptive, 206–209

  displaying, 207–208

  maximum, 206

  mean (or mean average), 206

  median, 206

  minimum, 206

  outliers, 208–209

  practice questions, 220–221

  spread (or range), 206

About the Author

Mark Zegarelli is the author of nine other For Dummies books, including *Basic Math and Pre-Algebra For Dummies*. He is a test-prep teacher and tutor for the ACCUPLACER, and is an enthusiastic supporter of American community colleges. He also prepares students to take the SAT, ACT, GRE, GMAT, and LSAT.

Mark lives in Tokyo, Japan, and Long Branch, New Jersey.

Dedication

This is for Zach, with love and admiration.

Author's Acknowledgments

This is my tenth book for Wiley, and as always I would like to extend my many thanks to this wonderful team of professionals who make writing a book not only possible, but joyful!

Many thanks, as always, to acquisitions editor Lindsay Lefevere, who immediately saw the possibility in this book and ran with it. Thank you again to project editor Chrissy Guthrie, whom I've now worked with on three *For Dummies* books, and who never fails to "herd the cats," keeping me motivated and inspired. Thank you to Sheri Gilbert for obtaining the rights to the printed material used for Reading Test practice questions, and for helping to clarify and expedite this process. Thanks to copy editor Marylouise Wiack for making it look like I write well — a clever illusion for sure! And thanks to technical editors Amy Nicklin, Suzanne Langebartels, Caleb Leggett, Katherine Schnell, and Jordan Langebartels for catching a ton of errors and unpremeditated obfuscations before you, the reader, had a chance to see them.

Finally, thank you to production editor Siddique Shaik for making the finished book look as good as we all imagined it would.

I have so many thanks for my friend and assistant, Chris Mark, that you get your own paragraph! You are just so awesome to work with, buddy, and I truly, truly hope you know how much I appreciate you!

Finally, thanks so much to my ACCUPLACER students at Asbury Park High School and Raritan High School. You all helped and influenced me more than you know and, I think, made me a better teacher and person. I deeply appreciate it and hope that I offered something worthwhile in return.

Publisher's Acknowledgments

Executive Editor: Lindsay Sandman Lefevere

Editorial Project Manager and Development Editor: Christina N. Guthrie

Copy Editor: Marylouise Wiack

Technical Editors: Amy Nicklin, Suzanne Langebartels, Caleb Leggett, Katherine Schnell, Jordan Langebartels

Proofreader: Debbye Butler

Production Editor: Siddique Shaik

Cover Photos: Nikada / Getty Images

Take dummies with you everywhere you go!

Whether you are excited about e-books, want more from the web, must have your mobile apps, or are swept up in social media, dummies makes everything easier.

Find us online!

dummies.com

Dummies is the global leader in the reference category and one of the most trusted and highly regarded brands in the world. No longer just focused on books, customers now have access to the dummies content they need in the format they want. Together we'll craft a solution that engages your customers, stands out from the competition, and helps you meet your goals.

Advertising & Sponsorships

Connect with an engaged audience on a powerful multimedia site, and position your message alongside expert how-to content. Dummies.com is a one-stop shop for free, online information and know-how curated by a team of experts.

- Targeted ads
- Video
- Email Marketing

- Microsites
- Sweepstakes sponsorship

20 **MILLION** PAGE VIEWS EVERY SINGLE MONTH

15 MILLION UNIQUE VISITORS PER MONTH

43% OF ALL VISITORS ACCESS THE SITE VIA THEIR MOBILE DEVICES

700,000 NEWSLETTE SUBSCRIPTIO TO THE INBOXES OF 300,000 UNIQUE INDIVIDUALS EVERY WEEK

# of dummies

Custom Publishing

Reach a global audience in any language by creating a solution that will differentiate you from competitors, amplify your message, and encourage customers to make a buying decision.

- Apps
- Books
- eBooks
- Video
- Audio
- Webinars

Brand Licensing & Content

Leverage the strength of the world's most popular reference brand to reach new audiences and channels of distribution.

For more information, visit dummies.com/biz

PERSONAL ENRICHMENT

Staying Sharp	Facebook	Guitar	Investing	Beekeeping	Digital Photography
9781119187790	9781119179030	9781119293354	9781119293347	9781119310068	9781119235606
USA $26.00	USA $21.99	USA $24.99	USA $22.99	USA $22.99	USA $24.99
CAN $31.99	CAN $25.99	CAN $29.99	CAN $27.99	CAN $27.99	CAN $29.99
UK £19.99	UK £16.99	UK £17.99	UK £16.99	UK £16.99	UK £17.99

Meditation	Pregnancy	Samsung Galaxy S7	iPhone	Crocheting	Nutrition
9781119251163	9781119235491	9781119279952	9781119283133	9781119287117	9781119130246
USA $24.99	USA $26.99	USA $24.99	USA $24.99	USA $24.99	USA $22.99
CAN $29.99	CAN $31.99	CAN $29.99	CAN $29.99	CAN $29.99	CAN $27.99
UK £17.99	UK £19.99	UK £17.99	UK £17.99	UK £16.99	UK £16.99

PROFESSIONAL DEVELOPMENT

Windows 10	AutoCAD	Excel 2016	QuickBooks 2017	macOS Sierra	LinkedIn	Windows 10 All-in-One
9781119311041	9781119255796	9781119293439	9781119281467	9781119280651	9781119251132	9781119310563
USA $24.99	USA $39.99	USA $26.99	USA $26.99	USA $29.99	USA $24.99	USA $34.00
CAN $29.99	CAN $47.99	CAN $31.99	CAN $31.99	CAN $35.99	CAN $29.99	CAN $41.99
UK £17.99	UK £27.99	UK £19.99	UK £19.99	UK £21.99	UK £17.99	UK £24.99

SharePoint 2016	Fundamental Analysis	Networking	Office 2016	Office 365	Salesforce.com	Coding
9781119181705	9781119263593	9781119257769	9781119293477	9781119265313	9781119239314	9781119293323
USA $29.99	USA $26.99	USA $29.99	USA $26.99	USA $24.99	USA $29.99	USA $29.99
CAN $35.99	CAN $31.99	CAN $35.99	CAN $31.99	CAN $29.99	CAN $35.99	CAN $35.99
UK £21.99	UK £19.99	UK £21.99	UK £19.99	UK £17.99	UK £21.99	UK £21.99

dummies.com

dummies®
A Wiley Brand

Learning Made Easy

ACADEMIC

Algebra I dummies

Mary Jane Sterling

9781119293576
USA $19.99
CAN $23.99
UK £15.99

Basic Math & Pre-Algebra dummies

Mark Zegarelli

9781119293637
USA $19.99
CAN $23.99
UK £15.99

Calculus dummies

Mark Ryan

9781119293491
USA $19.99
CAN $23.99
UK £15.99

Chemistry dummies

John T. Moore, EdD

9781119293460
USA $19.99
CAN $23.99
UK £15.99

Physics I dummies

Steven Holzner, PhD

9781119293590
USA $19.99
CAN $23.99
UK £15.99

1,001 Practice Questions
SAT dummies

Ron Woldoff

9781119215844
USA $26.99
CAN $31.99
UK £19.99

Organic Chemistry I dummies

Arthur Winter

9781119293378
USA $22.99
CAN $27.99
UK £16.99

Statistics dummies

Deborah J. Rumsey, PhD

9781119293521
USA $19.99
CAN $23.99
UK £15.99

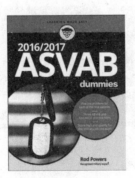

2016/2017
ASVAB dummies

Rod Powers

9781119239178
USA $18.99
CAN $22.99
UK £14.99

Includes Online Practice Tests
1,001 Practice Questions
Praxis Core dummies

Carla Kirkland
Chan Cleveland

9781119263883
USA $26.99
CAN $31.99
UK £19.99

Available Everywhere Books Are Sold

dummies.com

dummies®
A Wiley Brand

Small books for big imaginations

GETTING STARTED WITH Coding
Get Creative with Code!
Camille McCue, PhD

9781119177173
USA $9.99
CAN $9.99
UK £8.99

MODDING Minecraft™
Build Your Own Minecraft Mods!
Sarah Guthals, PhD
Stephen Foster, PhD
Lindsey Handley, PhD

9781119177272
USA $9.99
CAN $9.99
UK £8.99

MAKING YouTube® VIDEOS
Star in Your Own Video!
Nick Willoughby

9781119177241
USA $9.99
CAN $9.99
UK £8.99

DESIGNING Digital Games
Create Games with Scratch™!
Derek Breen

9781119177210
USA $9.99
CAN $9.99
UK £8.99

GETTING STARTED WITH Raspberry Pi®
Program Your Raspberry Pi®!
Richard Wentk

9781119262657
USA $9.99
CAN $9.99
UK £6.99

EXPERIMENTING WITH Science
Think, Test, and Learn!

9781119291336
USA $9.99
CAN $9.99
UK £6.99

CREATING Digital Animations
Animate Stories with Scratch™!
Derek Breen

9781119233527
USA $9.99
CAN $9.99
UK £6.99

GETTING STARTED WITH Engineering
Think Like an Engineer!

9781119291220
USA $9.99
CAN $9.99
UK £6.99

WRITING Computer Code
Learn the Language of Computers!
Chris Minnick and Eva Holland

9781119177302
USA $9.99
CAN $9.99
UK £8.99

Unleash Their Creativity

dummies.com

dummies
A Wiley Brand